Characterisation Methods in Solid State and Materials Science

Characterisation Methods in Solid State and Materials Science

Kelly Morrison
Physics Department, Loughborough University, Leicestershire, UK

IOP Publishing, Bristol, UK

ISBN 978-0-7503-1383-4 (ebook)
ISBN 978-0-7503-1384-1 (print)
ISBN 978-0-7503-1385-8 (mobi)

DOI 10.1088/2053-2563/ab2df5

Version: 20191001

IOP Expanding Physics
ISSN 2053-2563 (online)
ISSN 2054-7315 (print)

British Library Cataloguing-in-Publication Data: A catalogue record for this book is available from the British Library.

Published by IOP Publishing, wholly owned by The Institute of Physics, London

IOP Publishing, Temple Circus, Temple Way, Bristol, BS1 6HG, UK

US Office: IOP Publishing, Inc., 190 North Independence Mall West, Suite 601, Philadelphia, PA 19106, USA

To the students who inspired me to bring all this information together in one place, and to Rob, whose unique combination of patience and impatience helped drag me over the finish line.

Contents

Appendices

Author biography

Kelly Morrison

Dr Kelly Morrison is a senior lecturer and EPSRC Fellow at Loughborough University. She completed her doctorate in Physics at Imperial College, London in 2010, focussing on magnetic materials for solid state refrigeration. Since joining Loughborough University in 2013, she has shifted her research focus towards potential mechanisms for energy harvesting in thin films—in particular the spin Seebeck effect.

Over the course of her career she has worked on superconductors (such as YBCO, or the oxypnictides), magnetocalorics (maganites, intermetallic alloys, rare earth alloys), magnetic frustration (spin glasses), additive manufacturing, energy harvesting (spin thermoelectrics), spintronics (thin films, FMR) and development of thermal characterisation techniques (e.g. AC microcalorimetry and *in situ* spin Seebeck measurements). Along the way this has brought about experience with various forms of neutron and x-ray characterisation techniques, in particular inelastic neutron scattering, neutron reflectivity and resonant x-ray reflectivity.

IOP Publishing

Characterisation Methods in Solid State and Materials Science

Kelly Morrison

Chapter 1

Introduction

In this chapter, we will review the key concepts required to understand the majority of characterisation techniques, namely, diffraction, microscopy and spectroscopy. A discussion of these techniques requires an overview of some general physics concepts such as wave–particle duality, atomic notation, and crystallographic notation. Thus, we will start with a general overview of the key aims of this textbook, followed by a general discussion of characterisation methods, an overview of the required knowledge and finally a discussion of the potential direction(s) of research in materials and solid state science.

1.1 Overview

Solid state science is an interdisciplinary research field that encompasses materials science, solid state chemistry, solid state physics and parts of electrical engineering. It is the study of any solid system for which there are currently numerous applications including: electronics such as computers, mobile phones, TVs, radios, and LEDs; renewable energy technologies such as solar cells, fuel cells and thermoelectrics; and novel phenomena such as superconductivity and giant magnetoresistance. It also underpins fundamental research by providing an experimental framework in which to test new theories such as topological insulators, graphene and magnetic 'monopoles'.

As an indicator of the future of materials research the reader is referred to 'The 2019 materials by design roadmap' [1], which discusses the potential synergy between computational methods (density functional theory and beyond) and experimental data in the search for new materials. Some of the challenges that could be addressed include:

- Perovskite photovoltaics—where efficiencies of the order of 22% have been observed (compared to ~26% for single junction monocrystal silicon). As the perovskite structure—ABX_3—has the potential for flexibility in the choice of

doi:10.1088/2053-2563/ab2df5ch1

elements (A,B,X) there is potential for further increases in efficiency and an overall reduction in cost per watt [2].
- Organic semiconductors—such as organic photovoltaics LEDs, which have the advantage of scalable, low cost production.
- Materials for battery technology—such as Li-ion technology, where there is a need for smaller, higher energy density, faster charge/discharge times in order to accommodate developing renewable and electric vehicle technologies [3].
- Thermoelectrics—where computational methods might be able to predict materials with increased efficiency [1].
- Multifunctional metal alloys and functional ceramics—where the interplay of magnetic, electric, elastic and thermal effects can give rise to phenomena such as the magnetocaloric effect (MCE). The MCE is observed as a change in temperature as a material is magnetised and has been put forward for alternative refrigeration technology [4] as well as harvesting low grade heat [5].
- Transparent conducting materials—which are the backbone of touchscreen technology employed in phones, laptops and tablets. The world leading transparent conducting material (TCM) is indium tin oxide (ITO), which is n-type (electrons as charge carriers). There is significant effort looking at new TCMs to improve efficiency, reduce costs and discover p-type TCMs for increased functionality [6].
- Materials for quantum technologies—such as strongly correlated electron systems [7], where emergent phenomena such as monopoles or skyrmions can be harnessed to produce more efficient data storage or computation.
- Interfacial phenomena—where physical interactions at an interface such as multilayer thin films can yield functional phenomena. For example, magnetic interactions at the interface of bilayer (or multilayer) thin films can result in exchange bias. This manifests as the shift in the coercive field of the thin film device, which has been used to tailor sensors such as the read head in magnet based hard drives [8].
- 2D materials—defined as single atomic layer materials and most widely known for graphene. These crystalline materials have the potential to replace silicon technology with carbon based technology, for example. Implementation in devices such as supercapacitors, solar cells, or p-n junctions, with improved electrical transport properties [9].
- Materials for additive manufacturing—where the capability to assemble complex structures is limited only by the choice and design of suitable materials [10]. For example, additive manufacturing of permanent magnets could be used to produce bigger magnets with lower Nd content, thus reducing cost. It could also, feasibly be used to directly print small magnets in electronic technology such as mobile phones.

The aim of this book will be to familiarise you with the application of various experimental methods in solid state research. This will not be an exhaustive list, but should provide you with the tools to critically appraise an experiment in the context of the capabilities of the equipment involved. It will also give you a taste of the wealth of information available for a given technique. In order to understand the broader concepts discussed in this book it is necessary to develop some grounding in general undergraduate physics, such as atomic theory, general optics, Fourier transforms, magnetism, and crystallographic notation. These will be reviewed on the assumption that the reader has a general scientific or engineering background, to a level appropriate to understand the techniques discussed.

Throughout the book there will be a focus on individual (research) papers, selected to run alongside the concepts discussed in the text; you will be expected to read a selection of these papers to reflect on the context of the work as well as the techniques chosen to study the chosen problem. With a shift in the publication of scientific research towards open access papers and datasets, there is now the opportunity for the student to interrogate published data in order to grasp how the authors analysed raw data to return a specific result.

The rest of this chapter will focus on defining and establishing the key mechanisms employed for the study of materials: microscopy, spectroscopy and diffraction techniques. This will be followed by an overview of general concepts in diffraction and optics, atomic structure, and solid state physics, finishing with a broad discussion of the current trends and future of materials in the 21st century.

1.2 Microscopy, spectroscopy and diffraction

The aim of most characterisation techniques in solid state science is to build a picture of what is happening in a given material or structure. The first step might be to visually inspect it, from which we can determine size, shape, patterns, colour, and transparency; however, we are limited by the available resolution—i.e. the smallest feature that we can reliably see or detect. For example, we can obtain information directly by observation, which is easy enough if we are only interested in details on relatively large lengthscales such as bacteria (~ μm, see figure 1.1). We might be able to resolve features of the order of 100 μm with the naked eye, or with the aid of an optical microscope ($\lambda = 400–650$ nm), such as shown in figure 1.2, we could push this limit to 0.3 μm. Higher resolution microscopy would require further thought.

Other methods for obtaining information include scattering or diffraction of a particle or wave off of the sample to be studied. This can typically provide information on the surface morphology (roughness), uniformity (size, strain, and chemical composition) and crystallinity (whether amorphous, single crystalline, polycrystalline or a textured polycrystal). Diffraction is concerned with the change in direction of incident radiation (or particles) on a material (e.g. x-ray diffraction). It is most commonly discussed in the context of Bragg's law:

$$n\lambda = 2d \sin \theta \tag{1.1}$$

which was originally derived by Bragg to determine the lattice spacing of a crystal structure using x-ray diffraction (see figure 1.3) and holds for elastic (specular) scattering of waves or wave-like particles such as neutrons. The exact nature of this diffraction, however, will depend on a host of factors, most notably the nature of the

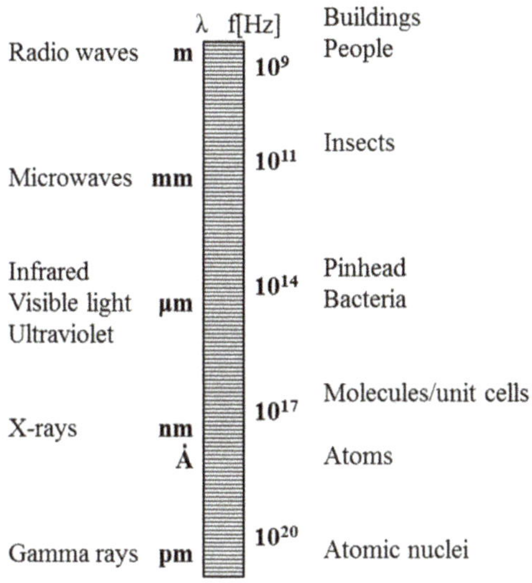

Figure 1.1. Familiar lengthscales in the context of the electromagnetic spectrum.

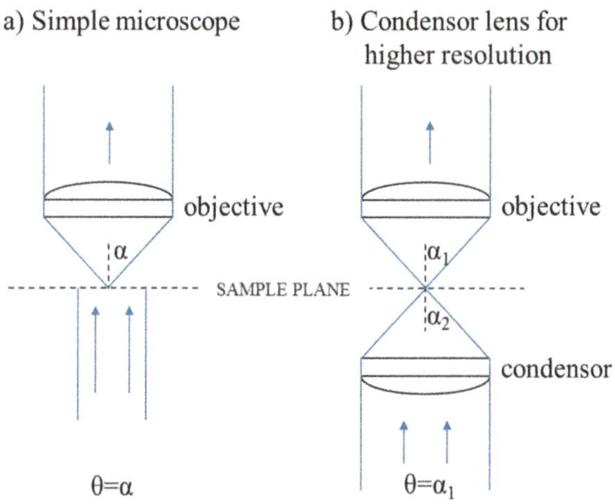

Figure 1.2. Basic microscope set-ups, where a lens is used to magnify the image formed by collimated light incident on a sample. The resolution is determined by the numerical aperture, NA = $n \sin \theta$, where θ describes the maximum angle at which light can be collected.

interaction between the radiation and the material involved. For x-rays the electromagnetic interaction between the passing electromagnetic wave and the orbital electrons typically results in diffraction. This leads to Bragg peaks that characterise the crystal structure of the material being probed. Neutrons, on the other hand, will be diffracted by the nucleus (nuclear interaction) at a rate similar to the dipole–dipole interaction that provides information on the magnetic order of the lattice (summarised in table 1.1) resulting in a combined picture of crystalline and magnetic order.

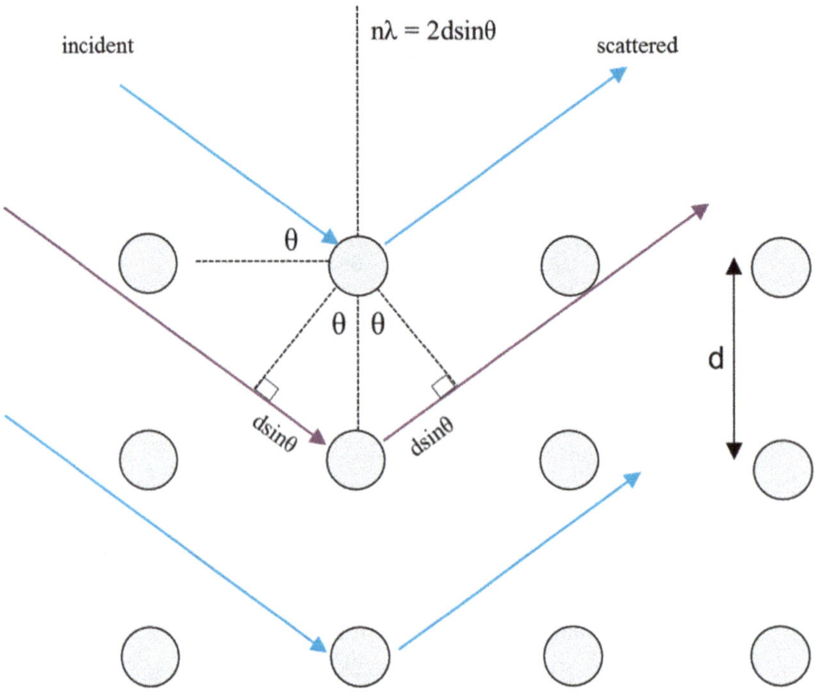

Figure 1.3. Diffraction of x-rays by a regular lattice—Bragg's law.

Table 1.1. Summary of the dominating particle interactions for common scattering/spectroscopy techniques.

Particle	Mass, kg	Charge, C	Spin	Magnetic moment (10^{-27} J · T^{-1})	Interaction	Depth
Electron	9.11×10^{-31}	-1.602×10^{-19}	½	−9284.764	Electrostatic	≪ μm
Neutron	1.67×10^{-27}	0	½	−9.662 36	Nuclear Dipole–dipole (magnetic)	~ cm (Al)
X-ray/photon	—	No fixed charge	—	—	Electromagnetic	~ mm (Al)
Muon	$\sim 1.88 \times 10^{-29}$	-1.602×10^{-19}	½	−44.904 478	Electrostatic	~ cm

Spectroscopy is the study of matter interacting with radiation. Such techniques can be used to probe the dynamics of a molecule such as lattice vibrations (phonons) or magnetic excitations (spin waves, magnons). The choice of radiation (or matter) for such measurements will, of course, influence the type of interaction that occurs. For example, electromagnetic radiation (x-rays, visible light etc) has no fixed charge and a penetration depth of the order of a few mm. This means that it is typically limited to probing energy levels within an atom (absorbing and emitting photons). Electrons on the other hand, have a fixed negative charge that will interact with the electron cloud surrounding a typical atom (the electrostatic interaction), and given

Table 1.2. List of spectroscopy techniques and the information accessed by them.

Technique	Frequency range, Hz	Information accessed
X-ray spectroscopy	10^{16}–10^{19}	Deep electronic states
UV spectroscopy	10^{14}–10^{16}	Energy levels of the valence band
IR spectroscopy	10^{11}–10^{14}	Molecular vibrations
Microwave spectroscopy	10^{6}–10^{11}	Excitations of molecular rotations
Raman spectroscopy	10^{14}	Molecular vibrations
Nuclear magnetic resonance	10^{6}	Intramolecular magnetic field
Electron spin resonance	10^{6}–10^{11}	Electronic structure of materials with unpaired electrons [11]
Inelastic neutron scattering	—	Molecular vibrations and magnetic excitations
(Gamma ray) Mössbauer spectroscopy	10^{20}	Deep electronic states (chemical, structural, magnetic and time-dependent properties of a material)

their small size and mass typically have larger penetration depths. Neutrons have larger mass, no charge, and a magnetic moment, which means that their interaction with matter will not be directly linked to the atomic number of the element being investigated and will show magnetic contrast. When coupled with other techniques the wide variety of interactions that can arise from different types of particles (photons, electrons, neutrons, muons, protons) can provide a more complete picture of the material. A summary of common spectroscopy techniques and key figures has been given in figure 1.1 and tables 1.1 and 1.2.

1.3 Introductory optics

Optics describes the study of how light interacts with matter, usually in the context of lenses, fibre optics, diffraction gratings or prisms, where it can be guided, or scattered in order to manipulate the electromagnetic wave for some purpose. In this section we will cover key elements of basic optics that will be useful in understanding the fundamental mechanisms of the characterisation techniques discussed later in this book.

1.3.1 Definitions of a wave

A wave can be defined as a periodic 'event'. For sound waves, for example, it would describe the periodic movement of a collection of particles, where we assign this periodicity a frequency (or pitch), f, that is directly proportional to the number of oscillations per second at a given point. The size of the wave—in this case, the distance moved by these particles—is described by the waves' amplitude, and the distance (in space) between identical points on the wave (i.e. where the particles are at maximum displacement) by the wavelength, λ. Whilst sound waves are longitudinal, which means that the direction in which they travel is the same as that of the particles that carry it, this is not the case for electromagnetic waves such as x-rays or visible light. Electromagnetic radiation, on the other hand, is a transverse wave, where the direction of propagation is at right angles to the electric and magnetic fields that vary periodically in order to make up the 'wave'. We can describe an electromagnetic plane wave mathematically by considering the electric and magnetic fields that it is comprised of, as shown in figure 1.4. In this context, the electric field along the x-axis can be described by:

$$E(x, t) = E_{max} \cos(kx - \omega t + \phi) \tag{1.2}$$

where E_{max} is the amplitude of the electric field, k is the wavevector ($k = 2\pi/\lambda$), ω is the angular frequency ($\omega = 2\pi f$), ϕ is the phase, and x and t are spatial and temporal coordinates. The phase is useful to describe the relative position of one wave with respect to another. For example, at $x = 0$, $t = 0$, the magnitude of the electric field of one wave may be 0, and the other E_{max}. In this case, as shown in figure 1.4(c), the phase difference between the two waves is 90 degrees, or $\pi/2$. In practice, when two waves coexist at the same point, then superposition will occur and the average of the two waves will be seen. In such a case, if the phase difference were 180 degrees, as shown in figure 1.4(d), then the amplitude of wave 1 would be $+E_{max}$, and wave 2 $-E_{max}$. The resultant sum of these two electric fields would be 0, and total destructive

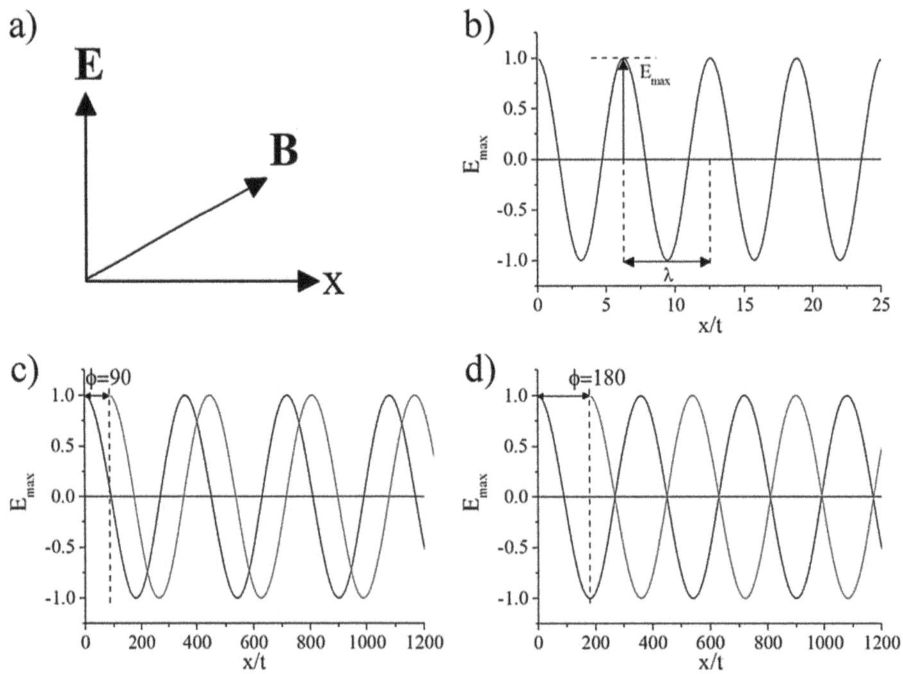

Figure 1.4. Conventions used to describe a plane wave. (a) An electromagnetic wave consists of electric and magnetic fields oscillating at right angles to one another, and to the direction of propagation, x. (b) Amplitude, E_{max}, wavelength, λ, and period, T (= $1/f$). (c) and (d) Two plane waves with a phase difference of 90° and 180°, respectively.

interference occurs. If the two waves were, instead *in phase*, i.e. $\phi = 0$, then the sum would be $2E_{max}$ and constructive interference occurs. Such superposition of waves forms the basis of classic interference patterns, as will be discussed in section 1.3.3.

1.3.2 Refraction, reflection, retardation

Retardation describes the change in speed (of propagation) of a wave, when it travels from one medium (such as air) to a more dense medium (such as glass). This results in a change in the wavelength of the wave, as shown in figure 1.5, whilst the frequency remains constant. The refractive index, n, describes this change in wave speed for an electromagnetic wave:

$$n = \frac{c}{v} \tag{1.3}$$

where c is the speed of light in a vacuum (3×10^8 m s^{-1}), and v is the speed of light in the material with refractive index n.

This retardation of the wave, can be used to explain the refraction of a wave that enters a material at an angle, θ_1, as described by Snell's law:

$$n_1 \sin \theta_1 = n_2 \sin \theta_2 \tag{1.4}$$

where n_1 is the refractive index of the medium in which the wave was initially travelling, n_2 is the refractive index of the new medium in which the wave is refracted, and θ_2 is the angle with respect to normal incidence, of the refracted ray. In addition to refraction, the incident ray can also be reflected by the surface of the new material, at an angle θ_1 to normal incidence.

Finally, a useful object in some optical experiments—the prism-works on the principle of retardation, where the refractive index of visible light within the prism varies with wavelength, as shown in figure 1.6. Whilst there are several different types of prism, the dispersive prism is the most familiar, as an object that splits an incoming light source into its component colours. Other types of prism, such as the

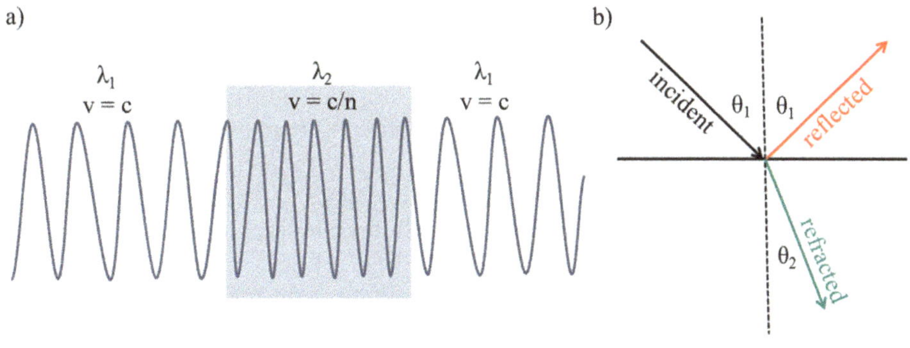

Figure 1.5. (a) Retardation of an electromagnetic wave travelling from a vacuum ($v = c$) to a dense medium ($v = c/n$) to a vacuum again ($v = c$). Whilst the wavelength, λ, will change, the frequency will remain constant. (b) Reflection/refraction of an electromagnetic wave off/through a material with different refractive index, n.

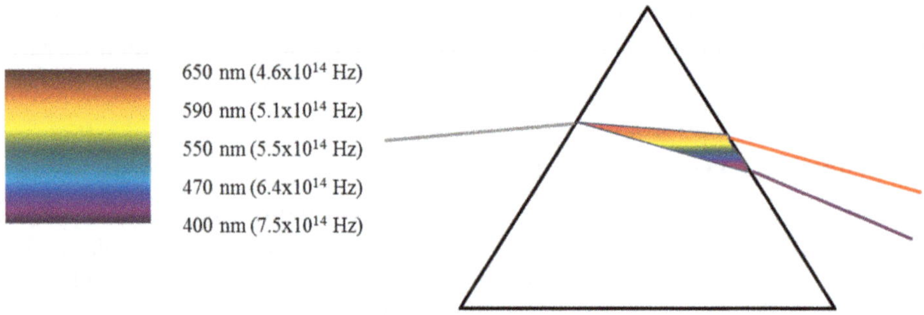

Figure 1.6. Example of a dispersive prism, in which white light can be split into its component wavelengths due to the differing refractive index within the prism.

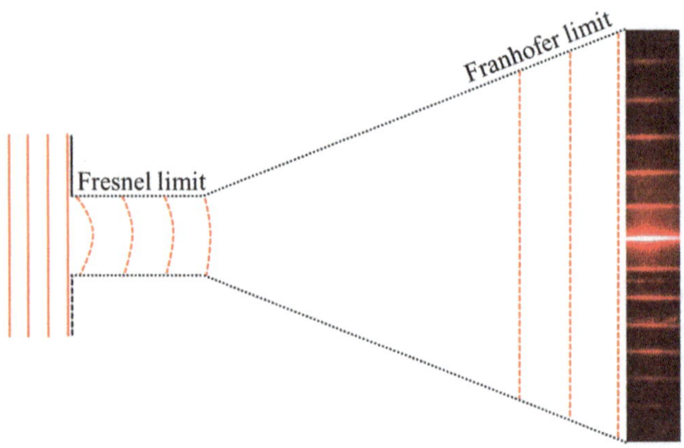

Figure 1.7. Fresnel and Fraunhofer diffraction limits for collimated light incident on an aperture. The red lines indicate the wavefront of the incident and diffracted light.

polarising Wollaston prism, are discussed in more detail in *Optical Properties of Solids* [12].

1.3.3 Diffraction

Standard optical microscopy is limited by the fundamental principles of light, in particular, diffraction: the divergence of light incident on an aperture or opening, as shown in figure 1.7. This diffraction of light is what ultimately limits the resolution of an optical microscope and it is useful at this stage to consider some common examples such as the circular aperture, single slit, double slit, and diffraction grating, as demonstrated by figure 1.8.

With regards to diffraction, there are two limiting cases that are often discussed: Fraunhofer and Fresnel diffraction. These are otherwise referred to as the far-field and near-field cases, respectively, and are shown schematically in figure 1.7. Fraunhofer diffraction requires a 'uniform beam' to be incident on the object which diffracts the light, and for the screen at which the diffraction pattern is observed to

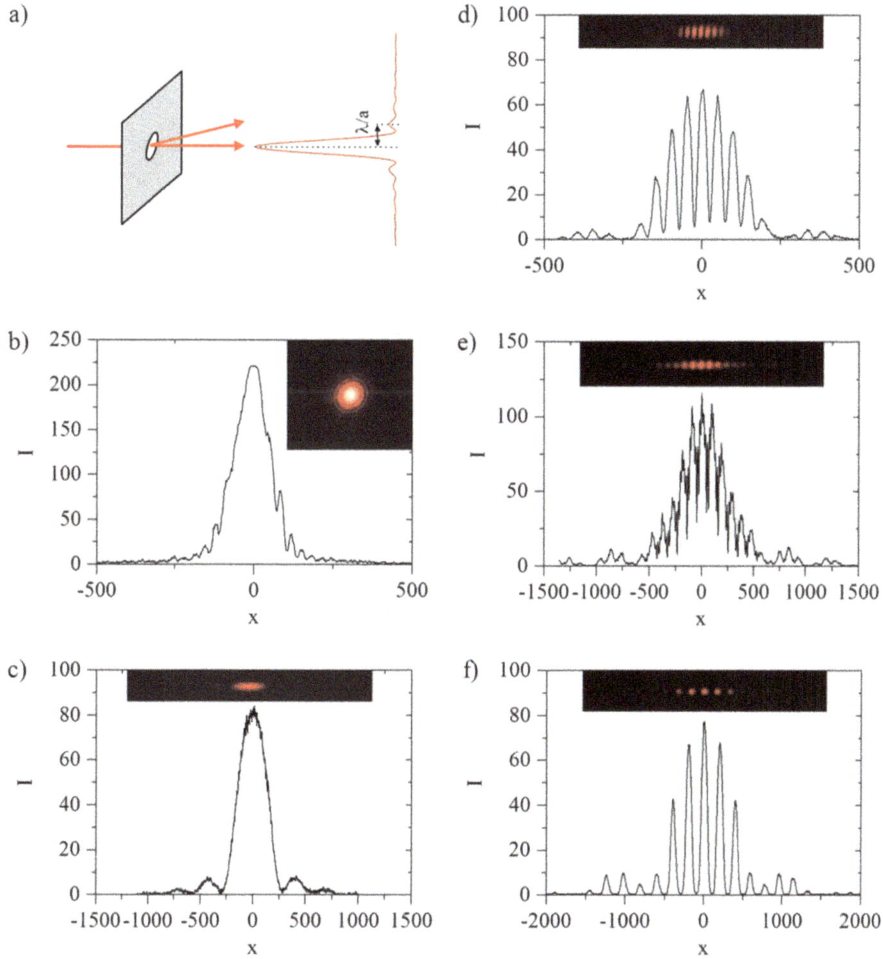

Figure 1.8. Diffraction of light through an aperture. (a) Schematic of the experiment, where a coherent light source (same wavelength and phase) passes through a slit and the diffraction pattern is observed some distance, D away. (b) circular aperture, where the image is resolution limited by the camera (c) single slit, (d) double slit, (e) 6 slit, (f) coarse diffraction grating.

be sufficiently far away for the wavefronts to be considered planar (parallel ray approximation), such that the diffraction pattern will not change further (when defined in terms of the angle subtended by the aperture) as the distance is increased. Fresnel diffraction, on the other hand, occurs if the diffracted image is observed close to the aperture. In this case the diffraction pattern will vary as the screen on which the diffraction image is projected is moved with respect to the aperture. For simplicity, most examples of diffraction given here will be in the far-field limit.

Practically, this means that the beam of light is comprised of waves with the same wavelength, λ, frequency, f, and to be in phase (i.e. $\phi = 0$ for all waves). In addition, the direction of propagation, should be uniform. For example, for a plane wave, the waves could all be travelling along the x-axis. This condition is often met

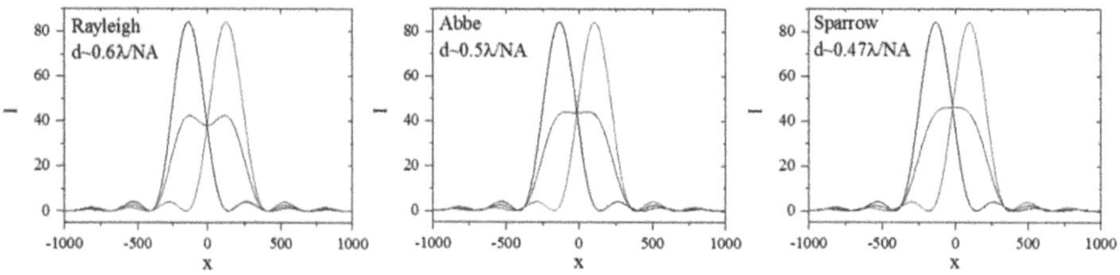

Figure 1.9. Definition of the resolution limit and the smallest resolvable object, *d*, in microscope with numerical aperture, NA.

experimentally by producing what is called a 'collimated beam', where plane waves travelling in the same direction are incident on an object. An example of how to obtain collimated light is given in section 1.3.5.

Some examples of diffraction of a helium neon laser (red, λ = 632.8 nm) by different objects is shown in figure 1.8. As can be seen, the diffraction of this uniform beam of light through an aperture, i.e. the degree to which it will diverge, will be roughly proportional to λ/a, where a is the diameter (in the case of a circular aperture) or width (in the case of a slit) of the aperture. This can be seen in figure 1.8 (a)–(c), where the majority of light (approximately 90% of the intensity) is contained in the central maxima, surrounded by 'fringes' of diffracted light about this point. The inset of figure 1.8(b), for example, shows the diffraction pattern formed by a circular aperture where the central spot is referred to as the 'Airy disc'.

As we move from a single slit to a diffraction grating, as shown in figure 1.8(c)–(f), the diffraction pattern becomes more complex due to interference from multiple slits. Initially, the single slit will result in a similar diffraction pattern to the circular aperture (a central bright peak surrounded by dimmer fringes, with periodicity proportional to λ/a). As another slit is added, the diffraction pattern will split into several narrower peaks, where the intensity no longer dies off as quickly. Now the periodicity of the diffraction fringes is controlled by the slit separation, d. As more slits are added (see the 6-slit example in figure 1.8(e)) these peaks will continue to become narrower, until eventually with a diffraction grating you have well-defined maxima and minima. These diffraction patterns will be explored further in chapter 2, where we use Fourier transforms to derive the mathematical expression for them in the far-field limit.

1.3.4 Resolution

The resolution of an optical system, describes the limit at which two objects could be reliably identified. There are several definitions of resolution, as shown in figure 1.9. The Rayleigh criterion, for example, describes the resolution of an optical system by considering two point sources. In this case, the point sources are considered resolved if the maxima of the airy disc produced by the first object coincides with the minima of the airy disc produced by a second object. The Abbe criterion, on the other hand, considers the object resolved if the two airy disc maxima can *just* be resolved (resulting in a dip in the observed intensity profile). Finally, the Sparrow criterion considers the objects resolved if the sum of the two Airy patterns produces a flat profile.

With regards to microscopy, the diffraction limit was until recently the controlling factor in determining the resolution of a given microscope [13, 14]. This is quantified by the Abbe diffraction limit, which can be used to describe the smallest resolvable object (d) in terms of the wavelength of light used (λ), the refractive index of the medium between the objective lens and the specimen shown in figure 1.2 (n), and the one-half angular aperture (α):

$$d = \frac{\lambda}{2n \sin \alpha} \tag{1.5}$$

where the denominator contains a representation of the numerical aperture, NA, a property of the microscope that is defined by:

$$NA = n \sin(\alpha) \tag{1.6}$$

Given that the highest NA is of the order of 1–1.5, and that optical light has a wavelength of 400–650 nm, this would suggest that the smallest resolvable object would be of the order of 3 μm. It also suggests that there are two experimental conditions that can be controlled to improve the resolution: (1) decreasing the wavelength by moving, for example, to x-rays, which can be expensive and destructive (biological samples, for example, will not withstand x-ray or gamma radiation for too long), or (2) increasing the NA. There have been considerable efforts to improve resolution in this way, by, for example, imaging objects in a medium with a higher refractive index (such as oil), however the numerical aperture is currently limited to ~1.5.

This is not the end of the story, however, as electrons can be used in place of light[1], or, as recognised by the 2014 Nobel Prize in Chemistry, flouresence microsopy can be used to work beyond the diffraction limit, with super-resolution microscopy [2, 3].

1.3.5 Lenses

The standard optical lens is an essential 'object' in optical measurements. It can be employed to focus or magnify an image, so that it can be easily manipulated, however, it requires some knowledge of the basic limitations. There are two common types of lenses: convex and concave, as shown in figure 1.10. When the surface of the lens is convex (or concave) on both sides it is termed 'bi'-convex, when it is convex (or concave) on one side only, with a flat edge on the other side it is termed 'plano'-convex.

In order to determine the size, location and orientation of an image formed by a lens, ray diagrams are employed. For the convex and concave lenses shown in figure 1.10, the method for constructing the ray diagram is the same:

[1] The DeBroglie wavelength of electrons, $\lambda = h/p$, where h is Planck's constant and p is the momentum of the electron. For electron energies of 200 keV the associated wavelength would be 2.75 pm.

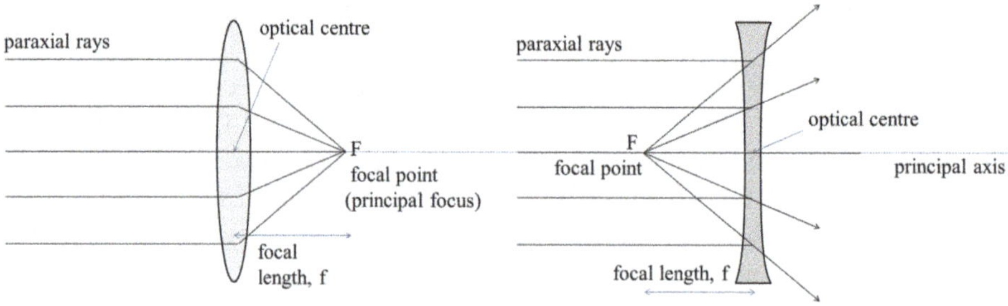

Figure 1.10. Examples of (left) bi-convex and (right) bi-concave lenses. Paraxial rays are defined as rays close to the optical axis.

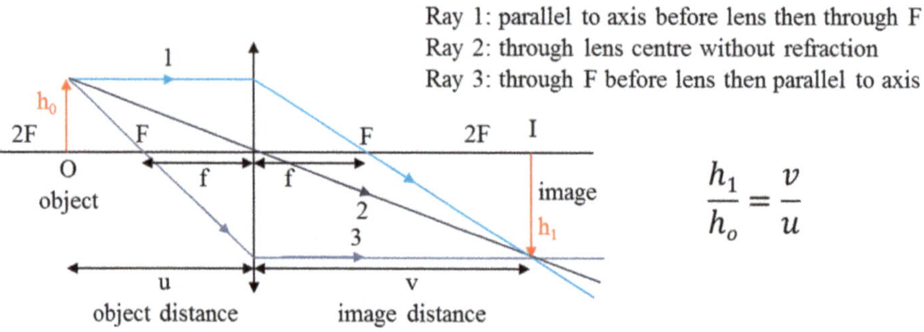

Figure 1.11. Constructing a ray diagram for a bi-convex lens, with the object at $F < u < 2F$, where F is the focal point.

1. From the object draw a line parallel to the principle axis until it reaches the centre of the lens, then draw the line from that point through the focal length of the lens.
2. From the object draw a line that passes through the optical centre of the lens without refracting.
3. From the object (if possible) draw a line that passes through the focal point before the lens. On the other side draw a line parallel to the principle axis.

By following these steps, you would in most cases obtain a point where the three 'rays' cross. This will be where the image is formed, as shown in figure 1.11. If the image is formed on the other side of the lens then it will be 'real' and as such could be projected onto a screen; otherwise it is virtual. From the ray diagram you can also determine if the image is inverted (i.e. it has been flipped vertically with respect to the object), at which point the image has been formed, and whether it has been magnified.

A summary of the types of images that can be formed by a simple convex lens is found in table 1.3. For example, the simple camera shown in figure 1.12 uses a lens where the object being imaged is at a distance greater than two focal lengths away. If this condition is satisfied, then the image can be projected onto a CCD or film. There are several other features that are employed by the camera to control the quality of the image: an aperture is used to control how much light is incident on the film/

Table 1.3. Applications of a single convex lens.

Object position	Image position	Nature of the image	Image size	Application
Beyond 2*F*	Between *F* and 2*F* on the opposite side	Real and inverted	Smaller	Camera, eye
At 2*F*	At 2*F* on the opposite side	Real and inverted	Same size	Inverter
Between *F* and 2*F*	Beyond 2*F* on the opposite side	Real and inverted	Bigger	Projector
At *F*	No image formed	—		To produce a parallel beam
Inside *F*	Further away on the same side	Virtual and upright	Bigger	Magnifying glass

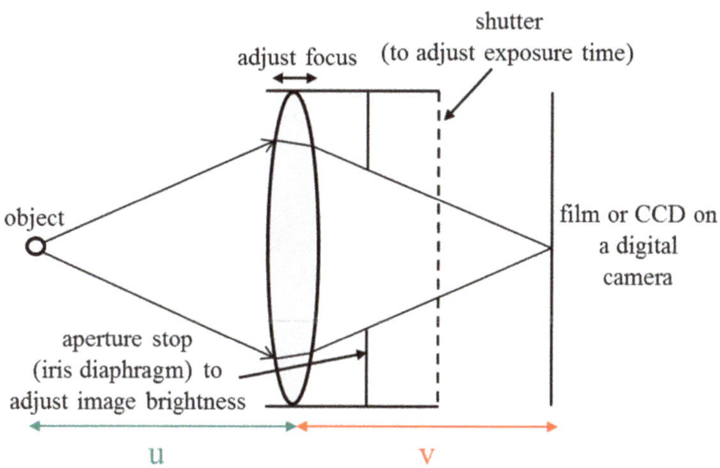

Figure 1.12. Example of the lens arrangement in a simple camera.

CCD; a shutter is used to control the exposure time; lastly, there will be some movement of the lens in order to adjust the focus of the image onto the film/CCD.

The position at which an image formed by a lens occurs can be determined using the simple lens equation:

$$\frac{1}{f} = \frac{1}{u} + \frac{1}{v} \tag{1.7}$$

where f is the focal length of the lens, u is the distance from the object to the lens and v is the distance from the lens to the image. Convention for more complex (i.e. multiple lenses) systems, requires this equation to be rewritten as:

$$\frac{1}{f}+\frac{1}{u}=\frac{1}{v} \tag{1.8}$$

where the distance between the lens and the object/image is now measured relative to the lens. In this case, a distance to the left of the lens is now negative (i.e. u) and a distance to the right is positive (i.e. v). Equation (1.8) is known as the Cartesian lens equation. Other conventions in this representation are that the light usually travels from left to right, converging lenses are positive, heights above the optical axis are positive, and angles measured clockwise form the optical axis are negative.

Finally, there are aspects of the lens arrangement that can lead to small imperfections in the image that is formed. This includes:

- Astigmatism: where the image formed is skewed with respect to the x- or y-plane.
- Chromatic aberrations: where the focal length varies with wavelength (due to wavelength dependence of the refractive index), as shown in figure 1.13. This can be mitigated by using a lens with large focal length, or an achromatic lens (usually comprised of two or more different materials in order to compensate for chromatic dispersion).
- Spherical aberrations: where slight variations in the focal length as a function of incident angle arise. This is due to small imperfections of the lens curvature and results in a loss of image definition as different rays will be focussed at different points (as shown in figure 1.13).

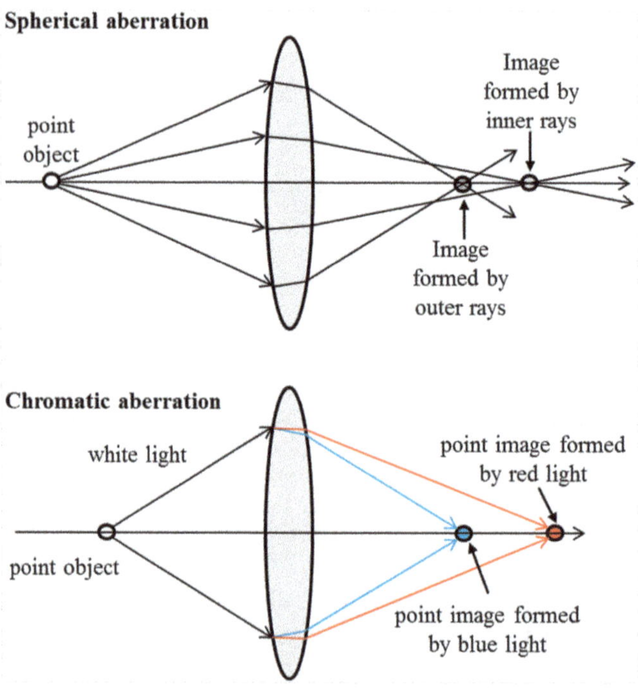

Figure 1.13. Examples of images formed with spherical and chromatic aberration.

1.4 Introduction to atomic physics

A general understanding of atomic theory is necessary when interpreting spectroscopy measurements. We start here with an overview of the periodic table, and follow with some key concepts such as the photoelectric effect and the Bohr model of the atom. Following on from this, the idea of atomic orbitals (or shells), expected energy levels and splitting with respect to degeneracy or applied magnetic fields will be presented.

1.4.1 The periodic table

The periodic table is an incredibly successful classification system that has the advantage of grouping similar elements, enabling prediction of missing elements, and simplifying discussion of the features of elements in nature [15]. Whilst it is now a universally accepted method for organising elements into groups that exhibit similar physical properties, with an atomic number that conveys the number of electrons (or protons) present, this was not always an obvious connection to make.

Historically, elements were first grouped by similar properties or their atomic weight. For example, in 1817 Döbereiner showed that with calcium, strontium and barium—three elements exhibiting similar properties—the atomic weight of strontium is the average of the other two. This was later picked up by De Chancourtois in 1862, who recognised that the properties of elements was a function of their atomic weight, and Newlands in 1864, who recognised a repetition of elemental properties after each series of seven elements (the 'law of octaves') [4]. The breakthrough that Mendeleev made in 1869 was to group elements together based on trends or similarities in properties such as density, atomic weight, atomic volume, melting point, and valency. Most importantly, he left room for undiscovered elements, such as gallium (discovered 1875), scandium (discovered 1879) and germanium (discovered 1886).

We now understand the patterns observed in the periodic table to be the result of the composition of an atom—a nuclear shell comprised of protons and neutrons, surrounded by a cloud of electrons. The periodic table is arranged by atomic number, Z—the number of electrons (or protons) in the atom. The atomic weight of each element will generally increase with Z, but vary slightly due to the existence of different isotopes, where the number of neutrons in the nucleus varies. The common features of a periodic group, are now also understood to be a result of the occupancy of their outer shell of electrons, where electrons that surround the nucleus arrange themselves in these 'shells' with periodic sizes denoted by the corresponding principal quantum number, n (table 1.4). Within each shell is a subset of stable orbitals that have corresponding orbital angular momentum denoted by the quantum number, $l (0 \leqslant l \leqslant n - 1)$. Finally, each orbital has a unique number of stable states with corresponding magnetic quantum number, m_l ($|m_l| \leqslant l$), and electron spin quantum number, m_s ($\pm 1/2$), which denotes the direction of the electron spin with magnitude 1/2. A fundamental principle of quantum physics, referred to as the Pauli exclusion principle, results in the requirement that 'no two

Table 1.4. Quantum states in the first four shells: principle quantum number, n, denotes the energy level; orbital quantum number, l, denotes the type of atomic orbital; magnetic quantum number, m_l, denotes the orbitals available within a subshell. As each electron can have spin ±1/2, there are two possible number of states for each unique combination of n, l and m_l.

n	l	m_l	Spectroscopic notation	Number of states		Shell
1	0	0	1s	2		K
2	0	0	2s	2	= 8	L
2	1	−1, 0, 1	2p	6		
3	0	0	3s	2	= 18	M
3	1	−1, 0, 1	3p	6		
3	2	−2, −1, 0, 1, 2	3d	10		
4	0	0	4s	2	= 32	N
4	1	−1, 0, 1	4p	6		
4	2	−2, −1, 0, 1, 2	4d	10		
4	3	−3, −2, −1, 0, 1, 2, 3	4f	14		

electrons can have the same set of quantum numbers $\{n, 1, m_l, m_s\}$'. The result of this is the gradual filling of these shells as seen in tables 1.4 and 1.5.

Note that the atomic weight is the mass of an element measured in atomic mass units, where it is equivalent to the mass in grams of 1 mole of atoms as defined by Avogadro's constant, $N_A = 6.02 \times 10^{23}$. Thus 1 mole of copper, with an atomic mass of 63.546 a.m.u. would describe a 63.546 gram sample with 6.02×10^{23} individual copper atoms.

1.4.2 The photoelectric effect

The photoelectric effect is described as a phenomena whereby a metal illuminated by light of a suitable frequency will emit electrons. It is often credited as a seminal piece of work with regards to establishing quantum physics—the particle nature of light—for which Einstein won the Nobel prize in 1922. What is perhaps more interesting about this effect, however, is the controversy it generated with regards to the theory: even though Einstein won the Nobel in 1922, it was not widely accepted in the physics community until Compton's scattering experiment of 1926 [16].

The photoelectric effect was first observed by Heinrich Hertz when experimenting with the generation and transmission of radio waves. A series of measurements performed by a student of Hertz—Hallwachs—showed that the dissipation of negative/positive charge on a zinc plate was accelerated/decelerated when UV light was incident on it. Additional pieces of the puzzle were provided by Thomson in 1899 who demonstrated that the charge emitted by Hertz' experiment exhibited particle characteristics, and Lenard in 1902, who demonstrated that these electrons had energy that was not affected by the intensity of the light incident, but the wavelength. Thus it was found that a lower wavelength led to faster electrons, and the hypothesis that 'light triggers release' and that the resulting energy originates

Table 1.5. Electronic configuration of the first 21 elements in the periodic table.

Element	Atomic number	Electronic configuration
Hydrogen, H	1	1s
Helium, He	2	$1s^2$
Lithium, Li	3	$1s^22s$
Beryllium, Be	4	$1s^22s^2$
Boron, B	5	$1s^22s^22p$
Carbon, C	6	$1s^22s^22p^2$
Nitrogen, N	7	$1s^22s^22p^3$
Oxygen, O	8	$1s^22s^22p^4$
Fluorine, F	9	$1s^22s^22p^5$
Neon, Ne	10	$1s^22s^22p^6$
Sodium, Na	11	$1s^22s^22p^63s$
Magnesium, Mg	12	$1s^22s^22p^63s^2$
Aluminium, Al	13	$1s^22s^22p^63s^23p$
Silicon, Si	14	$1s^22s^22p^63s^23p^2$
Phosphorous, P	15	$1s^22s^22p^63s^23p^3$
Sulphur, S	16	$1s^22s^22p^63s^23p^4$
Chlorine, Cl	17	$1s^22s^22p^63s^23p^5$
Argon, Ar	18	$1s^22s^22p^63s^23p^6$
Potassium, K	19	$1s^22s^22p^63s^23p^64s$
Calcium, Ca	20	$1s^22s^22p^63s^23p^64s^2$
Scandium, Sc	21	$1s^22s^22p^63s^23p^64s^23d$

from inside the atom. Given that the atomic model was not put forward by Neils Bohr until 1913, this was a reasonable assumption.

The step that Einstein's seminal 1905 paper made, was to describe 'light quanta', now known as photons, which interacted with the metal to release charged particles. He showed that the energy of the light quanta, E, was proportional to the frequency of the light, f, which is now commonly written as:

$$E = hf \tag{1.9}$$

where h is Planck's constant ($6.626\ 069 \times 10^{-34}$ J s^{-1}). The photoelectric effect can then be expressed in terms of the maximum energy, E, of emitted electrons (i.e. the kinetic energy, $1/2mv^2$), and the work function, ϕ, of the metal:

$$E = hf - \phi \tag{1.10}$$

This equation relates back to the hypothesis that 'the resulting energy originates from inside the atom'. Or, more precisely, the resulting energy of the emitted electrons is the difference between the available energy from a photon, and the energy required for the electron to escape (ϕ). With regards to measurement of the photoelectric effect, the combined work function of the circuit, ϕ_c, can be measured

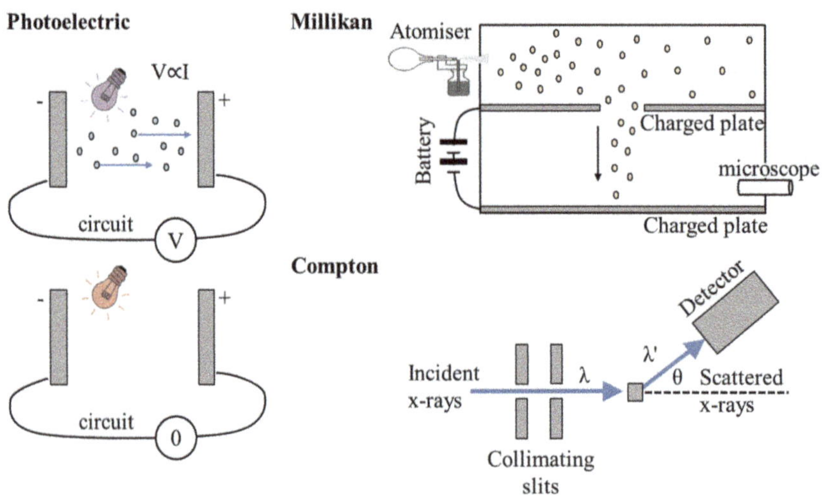

Figure 1.14. Schematic of the photoelectric, Millikan oil drop and Compton scattering experiments.

by the voltage required to stop ejection of electrons when light is incident on the surface of the metal to be measured (the stopping voltage, V):

$$eV = hf - \phi_c \tag{1.11}$$

where e is the charge on the electron.

To really verify the equations that Einstein had proposed for the photoelectric effect required an accurate knowledge of both the charge on an electron and Planck's constant. This was achieved by Millikan in two elegant experiments, the first of which was the oil drop experiment (see figure 1.14). In this, he determined the electron charge by measuring the terminal velocity of oil drops falling between two charged plates (an electric potential) and used this to determine the mass of the oil drops. Once the mass was known, the gravitational potential energy could be equated with the electric potential:

$$qE = mg \tag{1.12}$$

where q is the charge, E is the electric field, m is the mass of the drop, and g is the acceleration due to gravity. This enabled measurement of the electron charge, and it was the accuracy with which he determined this that then allowed him to measure Planck's constant and thus verify Einstein's model of the photoelectric effect. This was achieved by measuring the voltage required to stop ejection of electrons by the photoelectric effect, when a freshly cleaned surface of metal under vacuum was exposed to a monochromatic light source.

Finally, an observation made by Compton in 1923 of an unexpected decrease in the absorption factor of x-rays incident on aluminium led to the final acceptance of the light quanta theory. In his experiment he found that x-rays scattered off of an aluminium target exhibited a decrease in energy that varied with angle as:

$$\lambda' - \lambda_0 = \frac{h}{m_e c}(1 - \cos\theta) \tag{1.13}$$

where λ' and λ_0 are the wavelengths of the scattered and incident x-rays, respectively, m_e is the mass of the electron, h is Planck's constant, c is the speed of light, and θ is the angle at which the x-ray is scattered. This observation was explained by particle-like scattering of an x-ray quantum with electrons in the outer shell of the atom. This assumed that energy and momentum were conserved and thus provided further proof for the particle nature of light.

1.4.3 The atomic model, and energy levels

The periodic table indicated that there were regular patterns in elements, when arranged by atomic weight (for the most part), and this was followed by the determination of the particle nature of light, behaving as packets of energy which can excite an electron enough that it can escape a metal via the photoelectric effect.

In 1913, the Rutherford–Bohr model was put forward to describe the structure of an atom, where electrons were described to exist is stable orbitals (much like the Earth orbits the Sun). There were four key features of this model:

1. Electrons existed in stable orbits about the nucleus.
2. The electron's angular momentum is quantised ($nh/2\pi$).
3. The orbitals are stable despite acceleration (i.e. the electron does not emit radiation).
4. There are definite energy levels, where a photon with energy, $hf = E_i - E_f$, is emitted/absorbed by the transition of an electron between energy levels.

The concept of stable orbits was required in order to explain why the electrons did not simply spiral towards the charged nucleus as otherwise, according to classical electromagnetism, the constant acceleration of the electrons would result in energy being lost by radiating photons. As such, the angular momentum can be described by:

$$m_e vr = n\hbar \tag{1.14}$$

where m_e is the mass of the electron, v is the velocity, r is the radius of orbit, n is the principle quantum number, and $\hbar$ is the reduced Planck's constant ($= h/2\pi$). Given that the orbits are stable, Bohr postulated that the centripetal force would counteract the Coloumb force exhibited by charged particles such that:

$$F_{\text{Coulomb}} = F_{\text{radial}} \tag{1.15}$$

$$\frac{e^2}{4\pi\varepsilon_0 r^2} = \frac{m_e v^2}{r} \tag{1.16}$$

where e is the charge of the electron, and ε_0 is the permittivity due to free space. Rearranging this, we can obtain the radius of the orbit as:

$$r = \frac{n^2 4\pi\varepsilon_0 \hbar^2}{m_e e^2} = n^2 a_0 \qquad (1.17)$$

where $a_0 \sim 5.3 \times 10^{-11}$ m is the Bohr radius, and describes mean radius of the orbit of the electron. The energy of the orbits can then be calculated by adding the kinetic ($1/2mv^2$) and Coulomb $\left(\frac{e^2}{4\pi\varepsilon_0 r}\right)$ energies and substituting v (obtained from 1.14 and 1.17) and r, to obtain:

$$E = -\frac{1}{2}\frac{m_e e^4}{(4\pi\varepsilon_0)^2 \hbar^2 n^2} = -R_\infty \frac{hc}{n^2} \qquad (1.18)$$

where R_∞ is the Rydberg constant. Whilst the Bohr model is known to be a simplification, it was able to describe the classic emission spectra of hydrogen, seen in figure 1.15, using fundamental constants such as the electron mass, charge, and Planck's constant. For example, the emission spectra corresponding to a transition from a higher energy level ($n > 1$) to the ground state ($n = 1$), would result in a photon emitted with energy equal to the difference between these levels:

$$E_\gamma = R_\infty \frac{hc}{n^2} - R_\infty \frac{hc}{m^2} \qquad (1.19)$$

where m is the principal quantum number associated with the higher energy level. Given that the energy of a photon is related to the wavelength by:

$$\lambda = \frac{hc}{E} \qquad (1.20)$$

equation (1.19) can be rewritten as:

$$\frac{1}{\lambda} = R_\infty \left(\frac{1}{n^2} - \frac{1}{m^2}\right) \qquad (1.21)$$

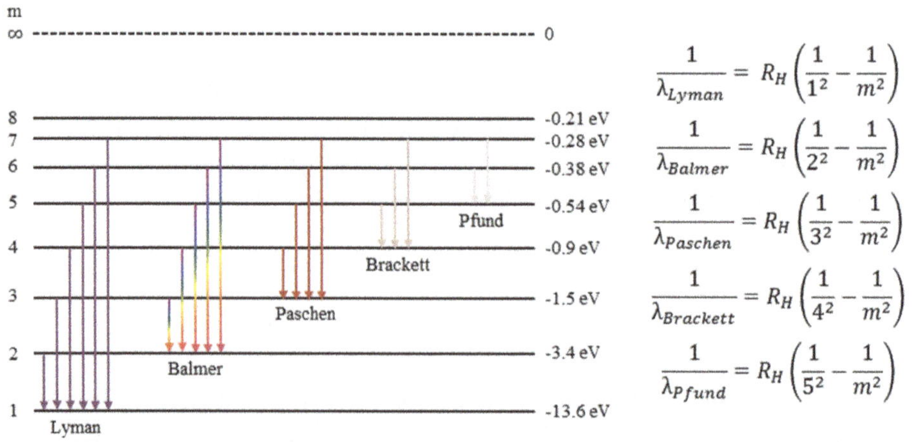

Figure 1.15. Emission lines for hydrogen atom, where R_H is the Rydberg constant for hydrogen and m indicates the energy level number, with $m = 1$ as the ground state.

Table 1.6. Balmer series: emission lines of hydrogen to the l shell (n = 2).

Colour	Wavelength (nm)	Transition
Violet	383.5384	9–2
Violet	388.9049	8–2
Violet	397.0072	7–2
Violet	410.174	6–2
Violet	434.047	5–2
Cyan	486.133	4–2
Red	656.272	3–2
Red	656.2852	3–2

which corresponds to the phenomological relationship observed for the Lyman, Balmer (see table 1.6), Paschen, Brackett and Pfund series, as seen in figure 1.15.

The Bohr model works well for single electron systems such as the Hydrogen atom, but there are additional effects to consider when electrons are added (to the system). This includes screening of the nuclear charge by electrons in lower energy states, and additional emission lines not expected from the Bohr model. In particular the electron spin (characterised by quantum number m_s) can lead to subtle splitting of the energy levels and thus the emission spectra. This is categorised as fine and hyperfine structure:

- *Fine structure—for example, the closely-spaced doublet in H (^2P splits to $^2P_{3/2}$ and $^2P_{1/2}$). Arises due to the interaction between electron spin (**S**) and orbital angular momentum (**L**) (spin–orbit interaction).*
- *Hyperfine structure—for example, subtle ground state splitting observed in hydrogen. Arises due to the interaction between electron spin (**S**) and nuclear spin, where anti-aligned states are lower in energy.*

Whilst the quantum numbers, n, l, m_l, m_s describe the quantisation of energy levels and momenta, they can also be used to determine the magnitude of orbital (**L**), spin (**S**) and total angular momentum (**J**). The magnitude of the orbital angular momentum can be determined by:

$$|\mathbf{L}| = \sqrt{l(1+1)}\,\hbar \tag{1.22}$$

where l can have integer values between 0 and $n - 1$ and the orbital angular momentum projected along the z-axis is:

$$L_z = m_l \hbar \tag{1.23}$$

where m_l can have integer values between 0 and ±l. The magnitude of the spin angular momentum, on the other hand, is determined by:

$$|\mathbf{S}| = \sqrt{\frac{1}{2}\left(\frac{1}{2}+1\right)}\,\hbar = \sqrt{\frac{3}{4}}\,\hbar \tag{1.24}$$

Table 1.7. Examples of atomic and x-ray notation. For atomic notation, $^2P_{3/2}$, 2 corresponds to the number of possible spin orientations, P corresponds to $l = 1$, and 3/2 corresponds to value of j.

Quantum numbers					
n	l	s	j	Atomic notation $^n\{\text{orbital}\}_j$	X-ray notation $\{\text{shell}\}_\#$
1	0	±0.5	0.5	$^1s_{1/2}$	K_1
2	0	±0.5	0.5	$^2s_{1/2}$	L_1
2	1	−0.5	0.5	$^2p_{1/2}$	L_2
2	1	0.5	1.5	$^2p_{3/2}$	L_3
3	0	±0.5	0.5	^3s	M_1
3	1	−0.5	0.5	$^3p_{1/2}$	M_2
3	1	0.5	1.5	$^3p_{3/2}$	M_3
3	2	−0.5	1.5	$^3d_{3/2}$	M_4
3	2	0.5	2.5	$^3d_{5/2}$	M_5

and the spin angular momentum projected along the z-axis is:

$$S_z = \pm m_s \hbar = \pm \frac{1}{2} \hbar \tag{1.25}$$

where m_s only has values ±1/2. Finally, the total angular momentum is a vector sum of **L** and **S** (**J** = **L** + **S**), where the magnitude of the total angular momentum, **J** is:

$$|\mathbf{J}| = \sqrt{j(j+1)}\,\hbar \tag{1.26}$$

and j can have values of $|l \pm 1/2|$. To account for this additional splitting of the energy levels, spectroscopic notation requires some acknowledgement of n, 1 and j, with the generic form $\{^n l_j\}$, where n and j are the quantum numbers, and the orbital label for 1 is used. For example $^2s_{1/2}$ corresponds to $n = 2$, $l = 1$, $j = 1/2$, and $^3d_{5/2}$ corresponds to $n = 3$, $l = 3$, $j = 5/2$, as shown in table 1.7.

In general, similar notation will be used to describe spectral lines from emission or absorption of a photon. By convention, an electron decay to a lower energy level will be indicated by the shell it decays to (K, L, M, N…), followed by a Greek letter (α, β, γ…) identifying which level it decayed from (α from neighbouring shell, β for next neighbour), as summarised in table 1.7 and figure 1.16.

It should be noted that there is a shift of notation away from Siegbahn to IUPAC (International Union of Pure and Applied Chemistry) [17], where the high and low energy levels are explicitly referenced. For example, the Siegbahn nomenclature would change in the IUPAC notation as follows:

- $K\alpha_1$—decay to K_1 (^1s), from L_3 would now read K-L_3.
- $K\alpha_2$—decay to K_1 (^1s), from L_2 would now read K-L_2.
- $K\beta_1$—decay to K_1 (^1s), from M_3 would now read K-M_3.
- $L\beta_1$—decay to L_2 ($^2p_{1/2}$), from M_4 would now read L_2-M_4.

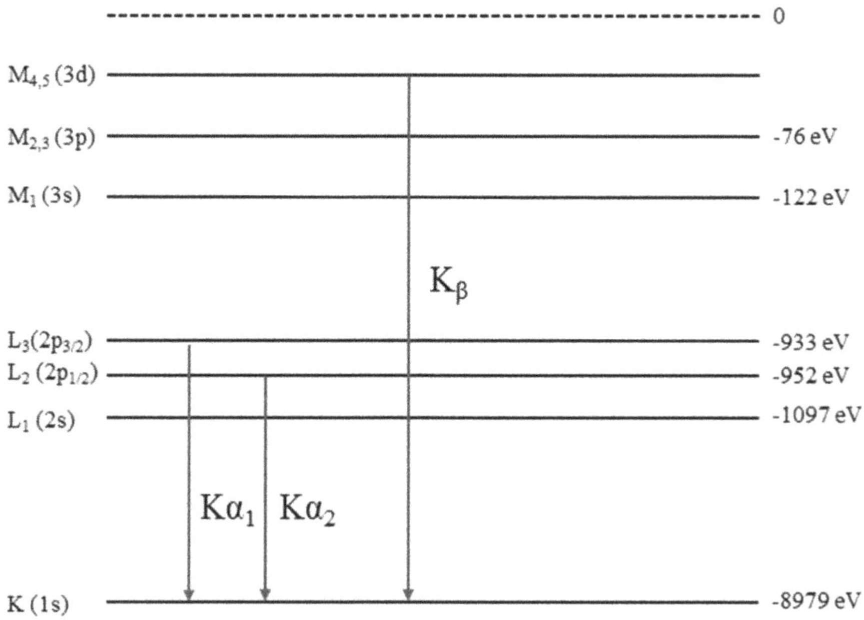

Figure 1.16. Energy level transitions in copper. The $K\alpha_1$, $K\alpha_2$ and K_β are the characteristic peaks observed when using a copper based cathode ray tube for x-ray diffraction.

Moseley's law

Moseley's law describes the empirical relationship between the energy of characteristic emission lines with their atomic number, Z. It briefly states that $E \alpha Z^2$. This is particularly useful when trying to categorise a series of electron spectroscopy measurements.

1.5 Introductory solid state physics

A significant part of this course will require you to understand the basic controlling factors of diffraction and resonance techniques. This will require a general understanding of the terminology used to describe crystal structures: symmetry groups, translation vectors, lattice parameters, and extinction conditions. Whilst these are fundamental to any solid state physics course, they will be briefly reiterated here.

1.5.1 The crystal lattice

You will be familiar with various types of matter in your everyday lives, such as polymers, plastics, glass, metals, alloys, salts, and rocks. The macroscopic properties of such materials (e.g. the thermal conductivity; electric resistivity; Young's modulus; state of matter—liquid, solid, gas) will depend on the arrangement of individual atoms within that material. Amorphous materials such as polymers and glass typically do not have any periodic arrangement of the elements on a macroscopic scale. There may be ordered microscopic chains, but on a macroscopic scale these chains may be organised in an irregular way. Crystalline solids, on the other

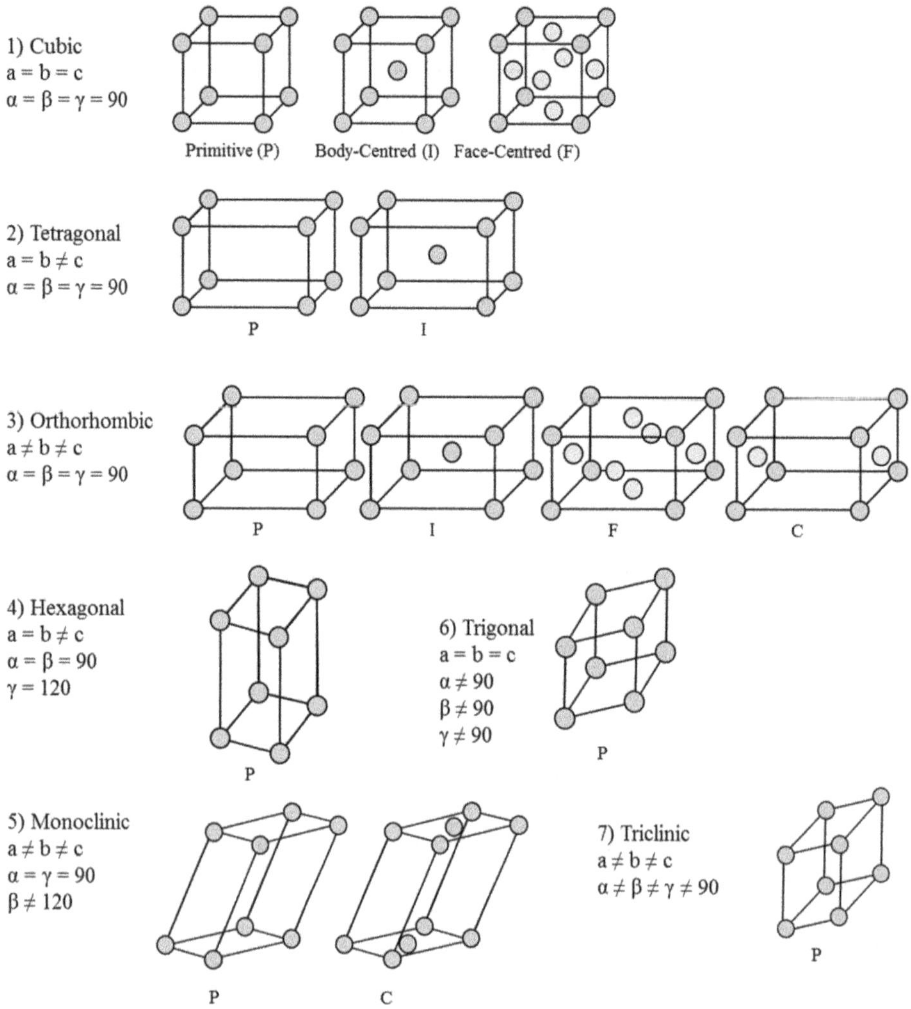

Figure 1.17. Definition of the seven crystal classes and 14 Bravais lattices.

hand, express a regularly arranged 'lattice' of atoms that can be described by its smallest repeating unit—the unit cell. In a perfectly ordered crystal the view from one specific atom of a unit cell will be indistinguishable from its corresponding atom in another unit cell.

There are seven crystal classes (cubic, tetragonal, orthorhombic, hexagonal, monoclinic, triclinic and trigonal) that are used to describe the general arrangement of atoms within a unit cell. These combine to form a total of 14 Bravais lattices, as summarised in figure 1.17. In general, there are three primary axes a,b,c that are not necessarily equal to the Cartesian x,y,z (i.e. not, by definition at right angles to one another), and thus, translation vectors, **r**, are used to describe a crystal structure rather than Cartesian coordinates.

1.5.2 Translation vectors

Translation vectors, **r**, are used to indicate the trajectory (e.g. from the (0,0,0) corner of a unit cell) to a particular atom/space in a periodic lattice, as shown in figure 1.18. For example:

$$\mathbf{r} = n_1\mathbf{a}_1 + n_2\mathbf{a}_2 + n_3\mathbf{a}_3 \tag{1.27}$$

where ($\mathbf{a}_1$,$\mathbf{a}_2$,$\mathbf{a}_3$) are the non co-planar vectors of the chosen lattice and (n_1,n_2,n_3) are numbers which repeat in a regular fashion. Used correctly, translation vectors can be used to describe the lattice (i.e. an infinite array of periodic points).

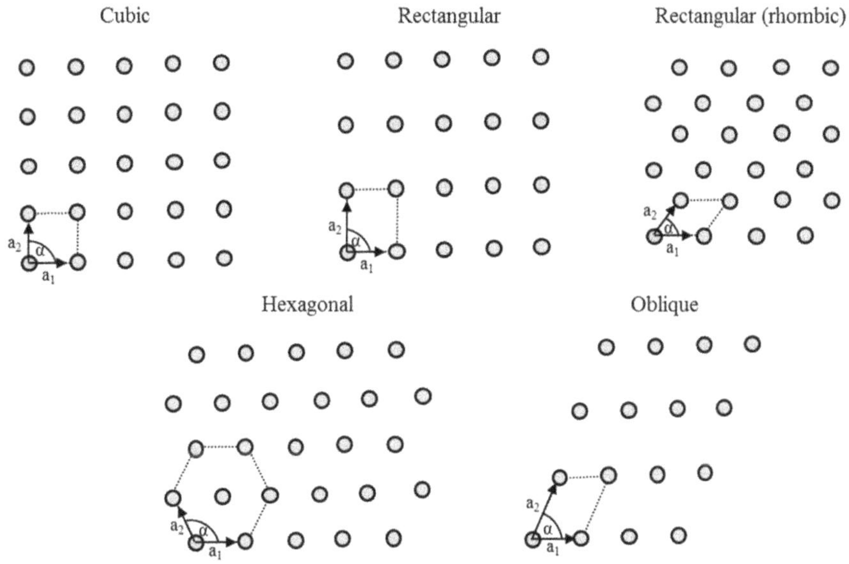

Figure 1.18. Types of translation vectors in different crystal classes.

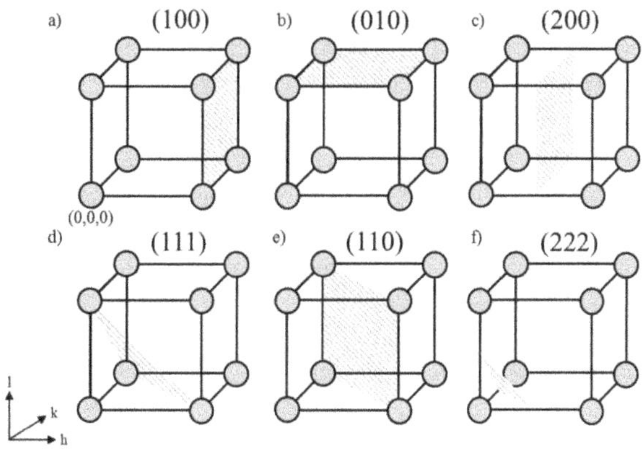

Figure 1.19. Example use of Miller indices to describe a crystal plane in a cubic lattice.

1.5.3 Miller indices and notation

Miller indices are a notation system used to describe planes in a crystal lattice. For a given crystal plane, the corresponding Miller indices can be determined by the following steps:

1) Find the intercepts on each of the axes in terms of the non co-planar vectors $\mathbf{a}_1$, $\mathbf{a}_2$, $\mathbf{a}_3$.
2) Take the reciprocal of these intercepts and reduce to three integers (maintaining the ratio between each).

For example, in figure 1.19(c) the (200) plane cuts the x-axis at ½, the y-axis at ∞, the z-axis at ∞ (i.e. does not cut these axes). This corresponds to $(\frac{1}{2}\mathbf{a}_1, \infty\mathbf{a}_2, \infty\mathbf{a}_3)$. The reciprocal of this is (2,0,0). In figure 1.19(d) the (111) plane cuts the x-axis at 1, the y-axis at 1, and the z-axis at 1. This corresponds to $(\mathbf{a}_1, \mathbf{a}_2, \mathbf{a}_3)$, the reciprocal of which is (1,1,1). In figure 1.19(f) the (222) plane cuts the x-, y-, z-axes at ½.

Note—for an intercept at infinity the corresponding index is zero.

When discussing crystal planes or directions, Miller indices (hkl) are used in conjunction with a specific bracket to describe:

- (hkl)—a specific plane. The plane that cuts the reciprocal (b_1,b_2,b_3) axis at $(1/h,1/k,1/l)$.
- {hkl}—planes equivalent by symmetry.
- [hkl]—a specific direction. The direction in the basis of the direct lattice vectors with respect to $(\mathbf{a}_1,\mathbf{a}_2,\mathbf{a}_3)$; perpendicular to ($hkl$) plane.
- <hkl>—a direction equivalent by symmetry.

For example, the (hkl) indicates a plane that cuts the $(\mathbf{a}_1,\mathbf{a}_2,\mathbf{a}_3)$ axis at $(1/h,1/k,1/l)$ such that in a cubic system (200), (020) and (002) indicate planes that cut $\mathbf{a}_1 = 1/2$,

Table 1.8. List of unit cells and their corresponding lattice and angular constraints. The lattice spacing, d, for a given plane is related to the lattice constants a, b, c by the formulas given in $1/d^2$.

Unit cell	Lattice constraints	Angular constraints	$1/d^2$
Cubic	$a = b = c$	$\alpha = \beta = \gamma = 90$	$\frac{h^2+k^2+l^2}{a^2}$
Tetragonal	$a = b$	$\alpha = \beta = \gamma = 90$	$\frac{h^2+k^2}{a^2} + \frac{l^2}{c^2}$
Orthorhombic		$\alpha = \beta = \gamma = 90$	$\frac{h^2}{a^2} + \frac{k^2}{b^2} + \frac{l^2}{c^2}$
Hexagonal	$a = b$	$\alpha = \beta = 90, \gamma = 120$	$\frac{4(h^2+hk+k^2)}{3a^2} + \frac{l^2}{c^2}$
Trigonal/ rhombohedral	$a = b = c$	$\alpha = \beta = \gamma$	$\frac{(h^2+k^2+l^2)\sin^2\alpha + 2(hk+hl+kl)(\cos^2\alpha - \cos\alpha)}{a^2(1-3\cos^2\alpha + 2\cos^2\alpha)}$
Monoclinic		$\alpha = \gamma = 90$	$\frac{h^2}{a^2\sin^2\beta} + \frac{k^2}{b^2} + \frac{l^2}{c^2\sin^2\beta} - \frac{2hl\cos\beta}{ac\sin^2\beta}$

$\mathbf{a}_2$ = 1/2, and $\mathbf{a}_3$ = 1/2, respectively. These three planes could be considered equivalent by symmetry (cubic system) and could thus be represented in general by {200}. Finally, the $\mathbf{a}_1$ axis is in the [100] direction, where again, for cubic systems [*hkl*] will be perpendicular to (*hkl*), and in general could be represented by <100>.

This notation is designed to be adaptable to the different Bravais lattices, where, for example, in the trigonal or hexagonal lattices the [*hkl*] may not be perpendicular to (*hkl*). Of particular note—the direction noted by [*hkl*] will not necessarily be perpendicular to the (*hkl*) plane for non cubic systems. For more, see table 1.8, or [18, 19].

1.5.4 Wyckoff sites and space groups

As the crystal lattice is the manifestation of a regular arrangement of atoms it is natural to try to describe this in a standardised way. As will be seen later, the diffraction of x-rays from such a solid results in a diffraction pattern that can be

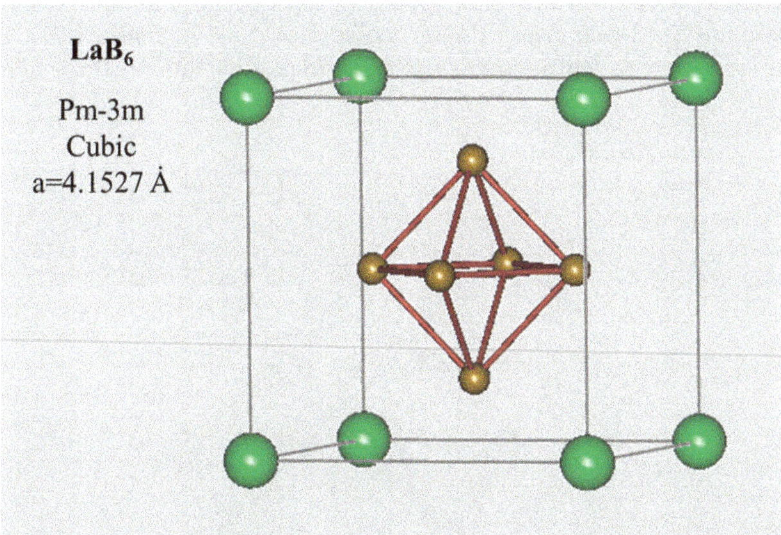

Wyckoff site		**x**	**y**	**z**	**B**	**Occ.**
La	1a	0	0	0	0.00157	1
		Multiplicity	**Wyckoff letter**	**Site Symmetry**	**Coordinates**	
		1	a	m-3m	(0,0,0)	
B	6f	0.1993	0.5	0.5	0.0027	1
		Multiplicity	**Wyckoff letter**	**Site Symmetry**	**Coordinates**	
		6	f	4m.m	(x,1/2,1/2), (-x,1/2,1/2), (1/2,x,1/2), (1/2,-x,1/2), (1/2,1/2,x), (1/2,1/2,-x)	

Figure 1.20. LaB_6 Wyckoff sites and corresponding coordinates.

analysed to determine quantities such as: the crystal structure, atomic disorder and strain. The complexity of this pattern, however, essentially boils down to the degree of symmetry within the unit cell (with regards to rotations, translations, reflections and a combination of these) as defined by:

- Translations—as described by the 14 possible Bravais lattices. Translations are defined as a motion (of the face of a plane) from one point to another.
- Glide planes—reflection in a plane followed by a translation.
- Screw axes—rotation about an axis followed by a translation.

For any given crystal structure, its corresponding space group describes the type of symmetry present. There are 230 space groups in total (listed in appendix C), each of which typically belong to a given crystal class (cubic, hexagonal…), and which have specific 'sites' within the unit cell that an atom can occupy and still fulfil the symmetry requirements. These are known as the Wyckoff sites (or Wyckoff positions).

Classic examples of this are LaB_6 (space group Pm-3m, cubic, a = 4.1527 A; figure 1.20) and Si (Space group Fd-3m, cubic, a = 5.43 A; figure 1.21), both of which are used for x-ray diffraction calibration measurements. In these figures the

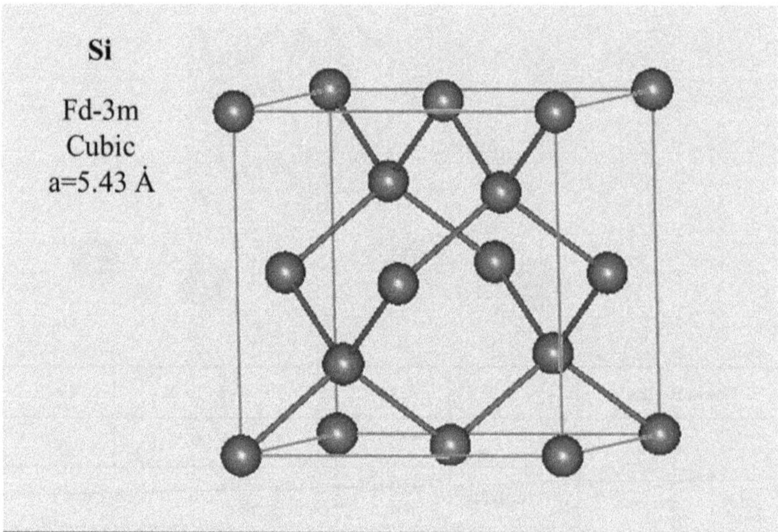

Wyckoff site		x	y	z	B	Occ.
Si	16c	0	0	0	1.39966	1
		Multiplicity	Wyckoff letter	Site Symmetry	Coordinates	
		16	c	-3m	(0,0,0),(3/4,1/4,1/2),(1/4,1/2,3/4),(1/2,3/4,1/4)	

Figure 1.21. Si Wyckoff sites and corresponding coordinates.

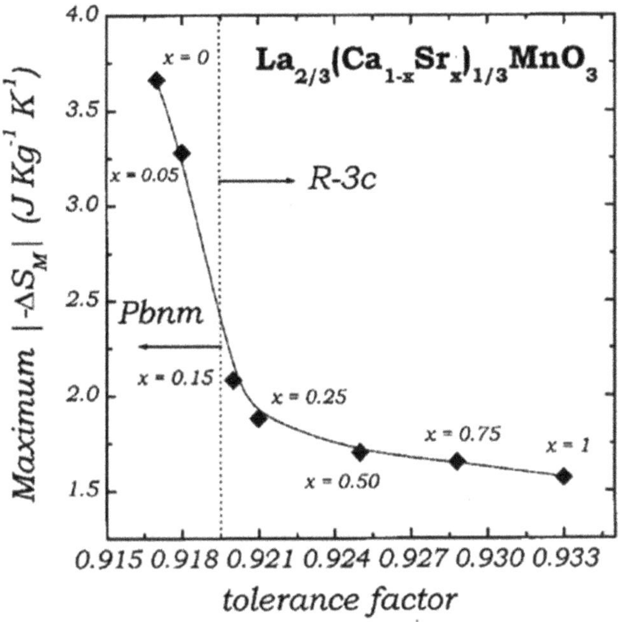

Figure 1.22. Crystallographic phase diagram of $La_{2/3}(Ca_{1-x},Sr_x)_{1/3}MnO_3$ as a function of tolerance factor. Reprinted from [20], with the permission of AIP Publishing.

unit cell is shown, and all the information required to build it from a 'Basis' of 1 or 2 atoms is given.

As an extended example, the manganites—$AMnO_3$ (where A is typically a combination of rare earths and alkalis such as La, Sr and Ca)—can form in cubic (Imma), orthorhombic (Pnma, or its non-standard variant Pbnm[2]) or rhombohedral (R-3c) lattices depending on the ratio of the ionic radii r_A, r_B (due to A and Mn) and r_O (due to O atoms). This is often discussed in terms of tolerance factor, t, a parameter that was devised in order to predict the crystallographic state of different manganites (i.e. when the Ca:Sr ratio is altered):

$$t = \frac{r_A + r_O}{\sqrt{2}(r_B + r_O)} \tag{1.28}$$

And which is demonstrated in figure 1.22 for $La_{1-x}Ca_xMnO_3$, In this example, the sample is orthorhombic (Pbnm) for high Ca content, and becomes rhombohedral (R-3c) as Ca is substituted for Sr and the tolerance factor, t, decreases below 1.

[2] With regards to the primitive axes (a_p):
In Pbnm $a_0 = \sqrt{2}a_p$, $b_0 = 2a_p$, $c_0 = \sqrt{2}a_p$.

Table 1.9. Wyckoff positions and corresponding coordinates for spacegroups Pnma (Pbnm) and R-3c. $La_{1-x}Sr_xMnO_3$ forms in R-3c with La on the 6a site, Mn on the 6b site and O on the 18e site, or in Pnma with La on the 4c site, Mn on the 4a site and O on the 4c and 8d site.

R-3c			
Multiplicity	Wyckoff letter	Site symmetry	Coordinates
36	f	1	(x,y,z), $(-y,x-y,z)$, $(-x+y,-x,z)$, $(y,x,-z+1/2)$, $(x-y,-y,-z+1/2)$, $(-x,-x+y,-z+1/2)$, $(-x,-y,-z)$, $(y,-x+y,-z)$, $(x-y,x,-z)$, $(-y,-x,z+1/2)$, $(-x+y,y,z+1/2)$, $(x,x-y,z+1/2)$
18	e	.2	$(x,0,1/4)$, $(0,x,1/4)$, $(-x,-x,1/4)$, $(-x,0,3/4)$, $(0,-x,3/4)$, $(x,x,3/4)$
18	d	−1	(1/2,0,0), (0,1/2,0), (1/2,1/2,0), (0,1/2,1/2), (1/2,0,1/2), (1/2,1/2,1/2)
12	c	3.	$(0,0,z)$, $(0,0,-z+1/2)$, $(0,0,-z)$, $(0,0,z+1/2)$
6	b	−3.	(0,0,0), (0,0,1/2)
6	a	32	(0,0,1/4), (0,0,3/4)
Pnma			
Multiplicity	Wyckoff letter	Site symmetry	Coordinates
8	d	1	(x,y,z), $(-x+1/2,-y,z+1/2)$, $(-x,y+1/2,-z)$, $(x+1/2,-y+1/2,-z+1/2)$, $(-x,-y,-z)$, $(x+1/2,y,-z+1/2)$, $(x,-y+1/2,z)$, $(-x+1/2,y+1/2,z+1/2)$
4	c	.m.	$(x,1/4,z)$, $(-x+1/2,3/4,z+1/2)$, $(-x,3/4,-z)$, $(x+1/2,1/4,-z+1/2)$
4	b	−1	(0,0,1/2), (1/2,0,0), (0,1/2,1/2), (1/2,1/2,0)
4	a	−1	(0,0,0), (1/2,0,1/2), (0,1/2,0), (1/2,1/2,1/2)

In this example, where there is little strain in the lattice due to r_A:r_B, it will form in a cubic lattice ($t \sim 1$). As the degree of strain is increased by replacing more of A with larger radii elements, the Mn-O octahedral will tilt, manifesting in a distortion of the Mn-O bond angles as the compound forms in the rhombohedral ($t < 1$), and then orthorhombic lattices ($t < 0.92$). A useful exercise is trying to reproduce the manganite crystal structures from the information given in table 1.9 and figure 1.23.

See appendix C for a full list of space group symbols [22] and the Bilbao Crystallographic Server for their corresponding Wyckoff positions.

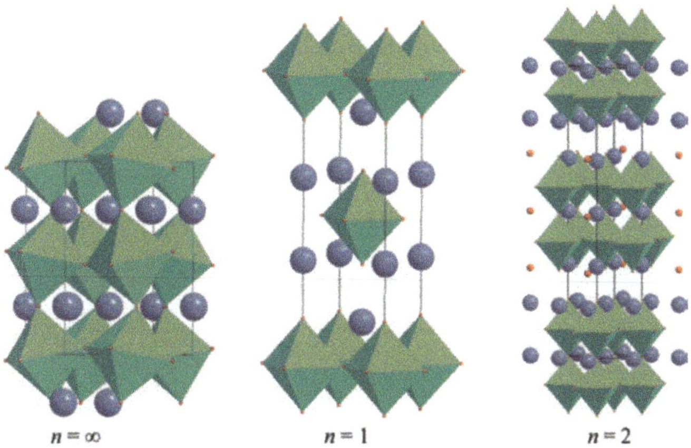

Figure 1.23. Schematic drawing of the arrangement of MnO_6 octahedra (green), cations (light blue) and oxygen (red) in orthorhombic crystal structuer (left), and n = 1, n = 2 Ruddelson Popper phases (middle and right). Dashed lines correspond to the unit cell of each lattice. Reproduced from [21] with permission from The Royal Society of Chemistry.

References

[1] Alberi K *et al* 2019 *J. Phys. D: Appl. Phys.* **52** 013001
[2] Wang B *et al* 2014 *Nanoscale* **6** 12287
[3] Nitta N *et al* 2015 *Mater. Today* **18** 252–64
[4] Franco V *et al* 2018 *Prog. In Mat. Sci.* **93** 112–232
[5] Waske A *et al* 2019 *Nature Energy* **4** 68–74
[6] Stadler A 2012 *Materials* **5** 661–83
[7] 2016 *Nat. Phys.* **12** 105
[8] Manna P K and Yusuf S M 2014 *Phys. Rep.* **535** 61–99
[9] Mas-Ballest R *et al* 2011 *Nanoscale* **3** 20
[10] Khare V *et al* 2017 *Int. J. of Precis. Eng. and Manuf.-Green Tech.* **4** 291
[11] Weckhuysen B M *et al* 2004 *Mol. Sieves* **4** 295–335
[12] Fox M 2010 *Optical Properties of Solids, Oxford Master Series in Condensed Matter Physics* (Oxford: Oxford University Press)
[13] Leung B O and Chou K C 2011 *Appl. Spectrosc.* **65** 967–80
[14] So S *et al* 2018 *Appl. Spectrosc. Rev.* **53** 290–312
[15] Brito A, Rodriguez M A and Niaz M 2005 *J. Res. Sci. Teaching* **42** 84–111
[16] Klaasen S 2011 *Sci. & Educ.* **20** 719–31
[17] Jenkins R, Manne R, Robin R and Senemaud C 1991 *Pure Appl. Chem.* **63** 735–46
[18] Singleton J 2001 *Band Theory and Electronic Properties of Solids, Oxford Master Series in Physics* (Oxford: Oxford University Press)
[19] Kittel C 2005 *Introduction to Solid State Physics* 8th edn (Chichester: Wiley)
[20] Mira J *et al* 2002 *J. Appl. Phys.* **91** 8903
[21] Malavasi L 2008 *J. Mat. Chem.* **18** 3295–308
[22] Hahn T 2002 ed T Hahn International Tables for Crystallography, Volume A: Space Group Symmetry *Int. Tables for Crystallography, A* 5th edn (Berlin, New York: Springer)

Kelly Morrison

Chapter 2

Fourier series, transforms and their relevance in diffraction and microscopy

In this chapter some of the key concepts of Fourier series, transforms and their application to real problems will be presented. This will start with a brief overview of the core equations followed by several key examples that can be used to develop an intuitive understanding of the varying contributions to typical experimental data such as the resolution function determined for microscopy or the expected diffraction patterns from periodic structures.

2.1 Introduction

The ability to spot and quantify patterns can be incredibly useful when trying to simplify a complex numerical problem. In diffraction, for example, there can be multiple contributions to the shape of a diffraction maxima (i.e. the Bragg 'peak' seen with x-ray diffraction), such as: sample size and quality, instrumental broadening, thermal broadening, and strain effects. The advantage of Fourier series and transforms are twofold: (1) they provide a way to separate these contributions to the final diffraction pattern in a quantifiable way, and (2) they enable the experimenter to develop an intuitive understanding of the relationship between an observed diffraction pattern and the object being imaged. Whilst the derivation and use of Fourier series is perhaps the bane of most undergraduate physics students, the general observations that they yield are incredibly powerful. As such, the focus of this chapter will not be an in-depth derivation of the relevant equations. Instead, the information presented will be sufficient to draw useful comparisons from classic experiments such as Young's double slit experiment, as well as to apply to later examples such as the Fourier treatment of x-ray reflectivity found in chapter 3.

doi:10.1088/2053-2563/ab2df5ch2

2.2 Fourier series

A Fourier series can be used to describe a periodic function by decomposing it into an infinite series of odd (sine; $f(x) = -f(-x)$) and even (cosine; $f(x) = f(-x)$) functions (as defined in figure 2.1):

$$f(x) = \frac{a_0}{2} + \sum_{n=1}^{\infty} a_n \cos\left(\frac{2\pi nx}{x_0}\right) + \sum_{n=1}^{\infty} b_n \sin\left(\frac{2\pi nx}{x_0}\right) \tag{2.1}$$

where a_0, a_n, b_n are coefficients that can be determined if the function is known. For a full derivation of equation (2.1), and examples of its use, the reader is referred to *Mathematical Methods for Physics and Engineering* [1] and chapter 7 of Boas' *Mathematical Methods in the Physical Sciences* [2].

Given that sin(x) and cos(x) can also be represented in terms of exponentials:

$$\sin(x) = \frac{e^{ix} - e^{-ix}}{2i} \tag{2.2}$$

$$\cos(x) = \frac{e^{ix} + e^{-ix}}{2} \tag{2.3}$$

where

$$e^{ix} = \cos x + i \sin x \tag{2.4}$$

And given that exponentials are often easier to integrate, it is common to rewrite equation (2.1) as:

$$f(x) = \sum_{n=-\infty}^{\infty} c_n e^{\frac{i2\pi nx}{x_0}} \tag{2.5}$$

where

$$c_0 = a_0/2,$$

$$c_n = (a_n - ib_n)/2,$$

$$c_{-n} = (a_n + ib_n)/2.$$

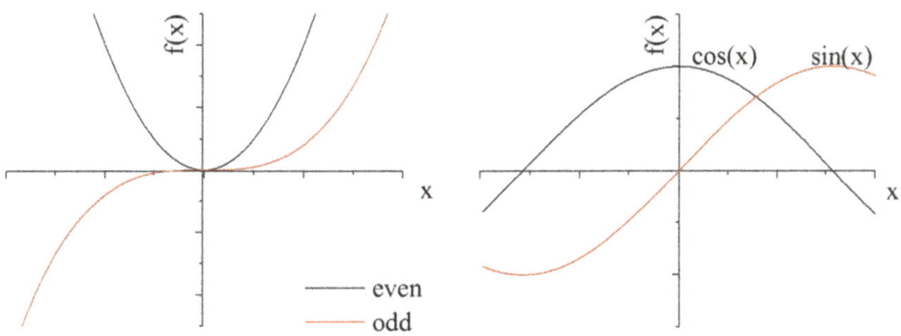

Figure 2.1. Examples of odd and even functions.

The idea that a function can be represented as a combination of odd and even functions is utilised whenever we perform a Fourier transform.

2.3 The Fourier transform

The Fourier transform is a mathematical operation that can be used to switch between what is described as 'real' and 'reciprocal' space. As the Fourier series identifies patterns in a given function, the Fourier transform enables the experimenter to switch between the function or signal being described, and a representation of the patterns identified in this function. For example, an electrical signal may exhibit periodic variations in amplitude that could be described by a Fourier series with common frequencies identified within the time varying signal (and amplitudes described by the corresponding coefficients a_n, b_n). Once we can identify patterns within a signal this way, then we can easily move from the time domain (our perception of the signal) to the frequency domain where the amplitude of the signal is a representation of the relative strength of each frequency component. This is essentially what we do when we perform a Fourier transform.

This is an obvious strength of the Fourier transform: taking a function in a given domain and transforming it into its reciprocal space (e.g. time to frequency [s → Hz], or distance to reciprocal space [m → m^{-1}]).

There are various ways to represent Fourier transforms, depending on how you define the lengthscale in reciprocal space. Here we shall use lowercase $f(x)$ for real space when referring to the 'spatial domain' (where $f(t)$ would represent the time domain), and the uppercase $F(u)$ to represent reciprocal space, where $f(x) \leftrightarrows F(u)$ and the lengthscale in reciprocal space, u, has been defined as $1/\lambda$ (cycles/length). This results in the following Fourier transform pair:

$$f(x) = \int_{-\infty}^{\infty} F(u)e^{i2\pi ux}\mathrm{d}u \tag{2.6}$$

$$F(u) = \int_{-\infty}^{\infty} f(x)e^{-i2\pi ux}\mathrm{d}x \tag{2.7}$$

Where the integral from $-\infty$ to ∞ is a reflection of extending the Fourier series to infinity (equation (2.5)). As we will be using Fourier transforms to describe diffraction, and it is common to represent the momentum transfer q (reciprocal space in diffraction measurements) as $2\pi/\lambda$ (radians/meter) we may also use:

$$f(x) = \frac{1}{\sqrt{2\pi}}\int_{-\infty}^{\infty} F(q)e^{iqx}\mathrm{d}q \tag{2.8}$$

$$F(q) = \frac{1}{\sqrt{2\pi}}\int_{-\infty}^{\infty} f(x)e^{-iqx}\mathrm{d}x \tag{2.9}$$

Notice that a prefactor of $\frac{1}{\sqrt{2\pi}}$ has appeared in equations (2.8) and (2.9). This is to ensure they are normalised, i.e. so that the Fourier transform of $F(q)$ returns the original function $f(x)$ (or the Fourier transform of the Fourier transform of $f(x)$

returns $f(x)$). Sometimes, this is achieved by including a prefactor of $\frac{1}{2\pi}$ for only one of the Fourier transform pairs; the result is the same. You should be able to convince yourself of this for both u-space ($1/\lambda$) and q-space ($2\pi/\lambda$), as suggested in section 2.9.2.

Aside: The Fourier transform of a Fourier transform is my friend...
By substituting equation (2.9) into (2.8) you can determine the value of the prefactor for the Fourier transform pair:

$$\begin{aligned} f(x) &\propto \int_{-\infty}^{\infty} F(q)e^{iqx}\mathrm{d}q = \int_{-\infty}^{\infty}\int_{-\infty}^{\infty} f(x')e^{-iq(x+x')}\mathrm{d}q\mathrm{d}x' \\ f(x) &\propto \int_{-\infty}^{\infty} F(q)e^{iqx}\mathrm{d}q = \int_{-\infty}^{\infty} f(x')\left[\int_{-\infty}^{\infty} e^{-iq(x+x')}\mathrm{d}q\right]\mathrm{d}x' \\ f(x) &\propto \int_{-\infty}^{\infty} F(q)e^{iqx}\mathrm{d}q = 2\pi\int_{-\infty}^{\infty} f(x')\delta(x+x')\mathrm{d}x' \\ f(x) &\propto \int_{-\infty}^{\infty} F(q)e^{iqx}\mathrm{d}q = 2\pi f(-x') \end{aligned} \tag{2.10}$$

As the Fourier transform yields a quantitative picture of the frequency content of a function this could be useful, for example, in quantifying the timbre (or mix of harmonic frequencies) of an instrument. It could also be used with an electric signal in order to identify the frequency of electrical noise (for example, you would expect a spike at the mains frequency—50 Hz in the UK).

For electromagnetic waves (e.g. visible light or x-rays) interacting with optical objects such as a diffraction grating or a lens, the Fourier transform can be used to determine the resultant diffraction pattern (image). This is the type of application that is often referred to as Fourier optics, and is the focus of an example labscript that can be found in the appendix. For more, see [1–3].

2.4 Key functions and their Fourier transforms

The calculation of the Fourier transform of the functions given here has been, for the most part, left for the student to determine, and is the focus of section 2.9.1. The aim of this section is to present the starting equations and key results, so that comparisons can be drawn.

2.4.1 The aperture function (single slit of finite width)

The aperture, or rect(x) function, as defined in equation (2.11) is often used to describe a single slit of finite width. This is because it represents a section that either lets the signal through ($|x| < a/2$) or doesn't ($|x| > a/2$), as shown in figure 2.2. Note, that this function could be complex (i.e. it lets some signal through), but that we use this binary representation here for simplicity.

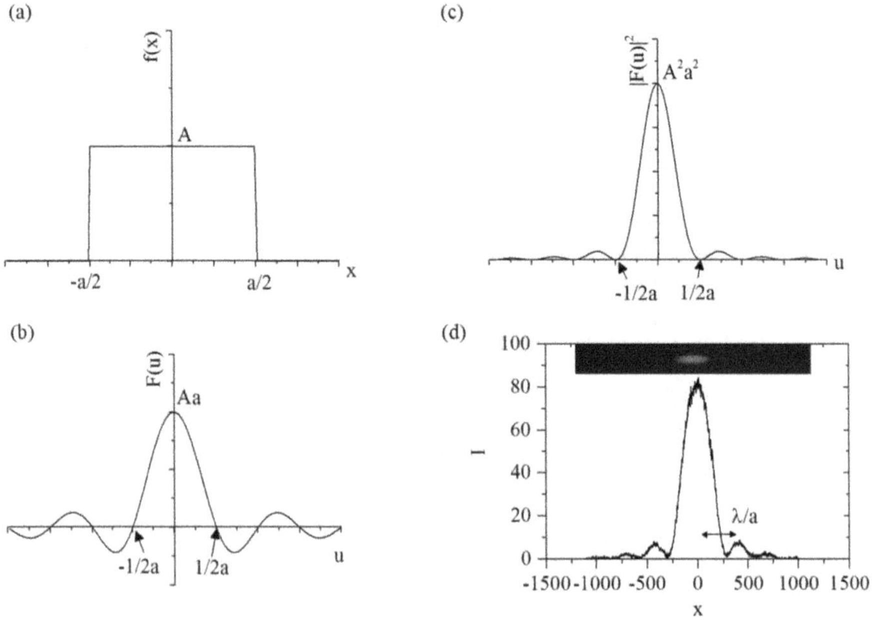

Figure 2.2. Fourier transform of the aperture function. (a) Aperture function, A rect (x/a). (b) Fourier transform of (a). (c) The intensity profile, $|F(u)|^2$, expected from (b). (d) Example of a single slit diffraction pattern.

$$A\ \mathrm{rect}\left(\frac{x}{a}\right) = \begin{matrix} A \text{ where } |x| < a/2 \\ 0 \text{ where } |x| > a/2 \end{matrix} \tag{2.11}$$

The Fourier transform of equation (2.11), can be calculated by considering how the integral reduces from infinity to a fixed range (where the aperture is 'open'), eventually yielding the solution given in equation (2.12).

$$F(u) = \int_{-\infty}^{\infty} A\ \mathrm{rect}\left(\frac{x}{a}\right) e^{-2\pi i u x} \mathrm{d}x = Aa\ \mathrm{sinc}\,(\pi u a) \tag{2.12}$$

where $\mathrm{sinc}\,(\pi u a) = \frac{\sin(\pi u a)}{\pi u a}$.

As in any optical experiment we do not observe the Fourier transform, but an intensity equal to the square modulus of $F(u)$ (where phase information is lost), this leads to the intensity distribution for a single slit of:

$$I = |F(u)|^2 = A^2 a^2\ \mathrm{sinc}^2\,(\pi u a) \tag{2.13}$$

2.4.2 The Dirac delta function (infinitely narrow slit)

The Dirac delta function, $\delta(x)$, is incredibly useful if we want to simplify an integration, as by definition it is zero everywhere except at a specified point:

$$f(x) = \delta(x - a) = \begin{cases} \infty & x = a \\ 0 & x \neq a \end{cases} \tag{2.14}$$

$$\int_{-\infty}^{\infty} \delta(x - a)\mathrm{d}x = 1 \tag{2.15}$$

With regards to diffraction measurements it can be used to describe an infinitely narrow slit at a given position. When integrating a function, $f(x)$, multiplied by $\delta(x)$ you simply substitute the x-position of the delta function as the values of x for the corresponding function, $f(x)$. For example:

$$F(u) = \int_{-\infty}^{\infty} \delta(x - a)e^{-2\pi iux}\mathrm{d}x = e^{-2\pi iua} \tag{2.16}$$

where:

$\delta(x)$ indicates a delta function at $x = 0$
$\delta(x + a)$ indicates a delta function at $x = -a$
$\delta(x - a)$ indicates a delta function at $x = +a$

Similarly, the Fourier transform of two delta functions separated by distance $2x_0$ could be written as:

$$F(u) = \int_{-\infty}^{\infty} (\delta(x - x_0) + \delta(x + x_0))e^{-2\pi iux}\mathrm{d}x = 2\cos(2\pi ux_0) \tag{2.17}$$

and for an infinite array of delta functions separated by distance, d:

$$F(q) = \frac{1}{\sqrt{2\pi}}\int_{-\infty}^{\infty} \sum_{-\infty}^{\infty} \delta(x - md)e^{-iqx}\mathrm{d}x \propto \sum_{-\infty}^{\infty} \delta\left(q - n\frac{2\pi}{d}\right) \tag{2.18}$$

2.4.3 The Lorentzian and Gaussian functions

The Lorentzian (equation (2.19)) and Gaussian (equation (2.20)) functions are often used to describe a general distribution of intensities, or more practically, broadening of a characteristic line profile due to experimental constraints (slit width, sample size and shape). They are defined as:

$$l(x) = \frac{1}{\pi}\frac{1/2\Gamma}{(x - x_0)^2 + (1/2\Gamma)^2} \tag{2.19}$$

$$g(x) = \frac{1}{\sqrt{2\pi}}\frac{e^{-(x-x_0)^2/2\sigma^2}}{\sigma} \tag{2.20}$$

where x_0 is the position of the functions with respect to the x-axis, and the full width half maximum (FWHM) describes the width of the function at half its maximum intensity. For these functions, Γ is by definition, the FWHM of the Lorentzian function, σ is the FWHM of the Gaussian function, as shown in figure 2.3.

Whilst determining the Fourier transform of the Gaussian function is fairly straightforward, to determine the Fourier transform of the Lorentzian function

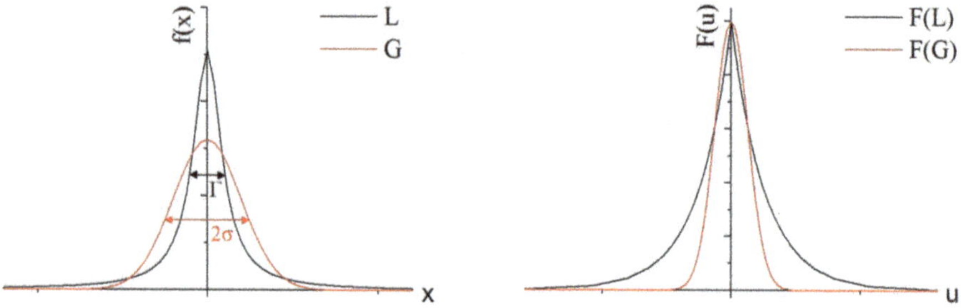

Figure 2.3. The Gaussian and Lorentzian functions and their Fourier transforms.

requires use of the Cauchy integral. The general results are given below, but you can see from figure 2.3, that the general trend is for the functions to be broader in reciprocal space. For example, the narrow Lorentzian function of figure 2.3 in real space, becomes broader than the Fourier transform of the Gaussian function in reciprocal space.

$$L(u) = e^{-2\pi i u x_0 - \Gamma\pi|k|} \tag{2.21}$$

$$G(u) = e^{-2\pi^2\sigma^2 u^2} \tag{2.22}$$

2.5 The convolution theorem

Another powerful application of Fourier transforms is the convolution of two functions. Before we move on, we first need to define what is meant by *convolution*, as it is a common concept that can be used to describe often complex contributions to an observed function or signal.

Convolution is a mathematical operation that maps one function onto another. If we wanted to define a function, $c(x)$, as the convolution ($\otimes$) of the functions $f(x)$ and $g(x)$, we could write:

$$c(x) = g(x) \otimes f(x) = \int_{-\infty}^{\infty} g(x')f(x - x')\mathrm{d}x' \tag{2.23}$$

This generally conveys the shape of one function onto another, as demonstrated in figure 2.4.

Convolution is useful for introducing more complex components of a function gradually. For example, the Dirac delta function could be used to map the position of multiple slits, however, they are rarely infinitely narrow (as is the case with the delta function). In order to describe the position of multiple slits with finite width we could convolve the aperture function (section 2.4.1) with this series of delta functions. If we wanted to take a complex function and tease apart the original functions that lead to its shape we could perform the inverse operation: deconvolution.

The convolution theorem is a useful result of Fourier transforms that simplifies what could otherwise be a complex mathematical operation. It states that the

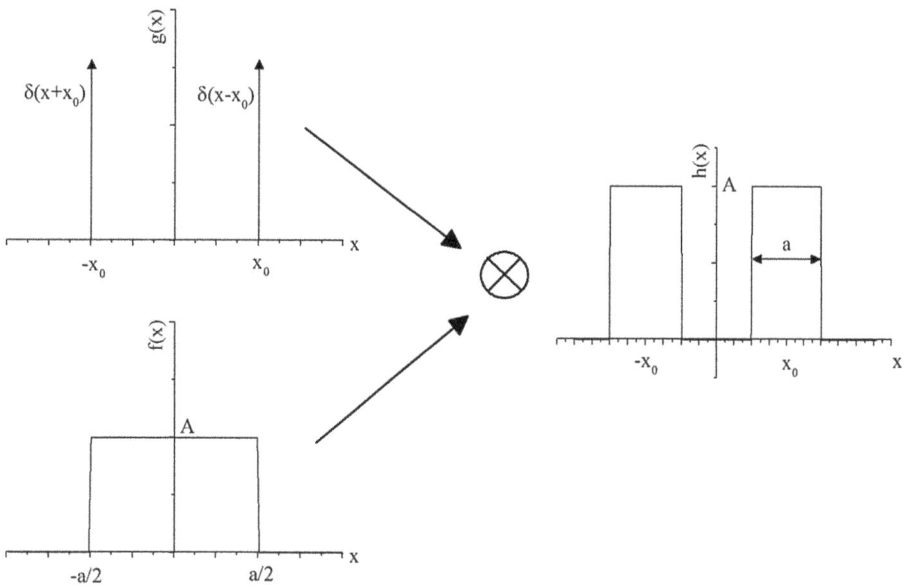

Figure 2.4. Schematic example of the convolution of two simple functions.

convolution of two functions, $f(x)$ and $g(x)$, in real space is equal to the multiplication of the two Fourier transforms of these functions, $F(u)$ and $G(u)$, in reciprocal space:

$$c(x) = f(x) \otimes g(x) \leftrightarrows \int_{-\infty}^{\infty}\int_{-\infty}^{\infty} g(x')f(x - x')e^{-2\pi iux}\mathrm{d}x\mathrm{d}x' = F(u)G(u) \quad (2.24)$$

So, for the example of multiple finite slits given above, rather than computing the function that results from convolution of a delta function and the aperture function followed by performing a Fourier transform on this new function, we can take the known Fourier transforms of these two independent functions and simply multiply them together. This is demonstrated with the Young's double slit example discussed in section 2.6.1.

Care should be taken when using this theorem, however, as it is important to identify whether convolution of two functions is appropriate in real or reciprocal space. For worked through examples of this, see chapter 2 of [3].

2.6 Classic examples of diffraction patterns

2.6.1 Young's double slit

We have already calculated the intensity distribution for a single slit of finite width using the aperture function (see equation (2.12)). Young's double slit experiment—the diffraction of light through two finite slits, separated by distance, $2x_0$ (see figure 2.5)—is an extension of this.

The double slit experiment was initially proof of the wave nature of light, but has since been used to demonstrate the wave–particle duality of waves and particles. If a

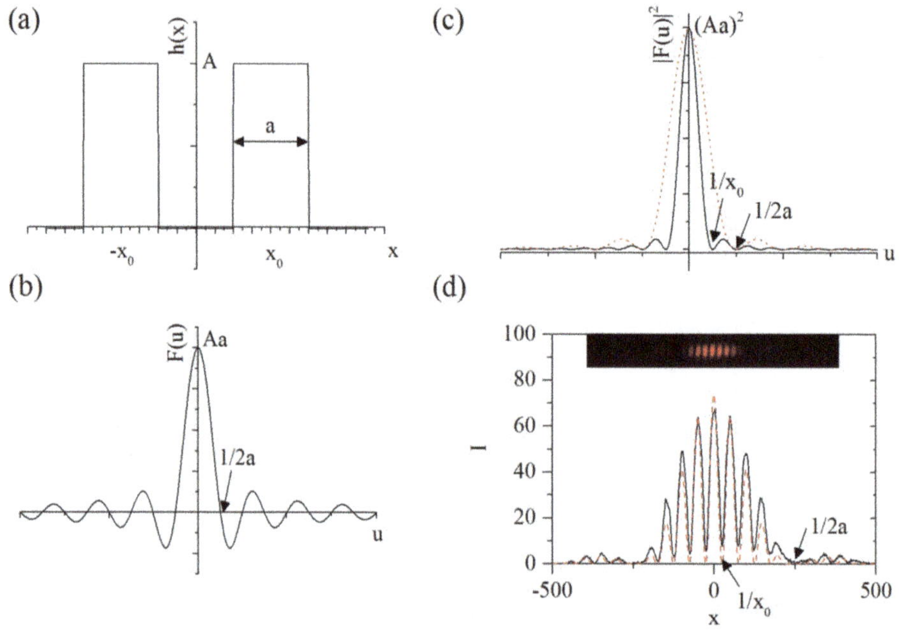

Figure 2.5. Deriving Young's double slit diffraction pattern using Fourier transforms and the convolution theorem. (a) The aperture function and (b) its Fourier transform. (c) The resultant intensity distribution for a finite single slit (dotted line) and double slit (solid line). (d) Example double slit measurement where the red dotted lines indicate the profile expected from equation (2.27).

monochromatic plane wave is incident on a sheet with two well defined slits (width, a) separated by some distance, $2x_0$, a diffraction pattern can be observed on a screen a distance, D, away from the slits. This diffraction pattern can be classically described by considering the constructive and destructive interference of two waves where a path difference is introduced by them travelling slightly different lengths, as well as diffraction from the slits themselves. It manifests as a bright central spot or 'fringe', surrounded by similarly thick, dimmer fringes a distance, s, apart (figure 1.8(d)). This fringe separation can be found to be:

$$s = \frac{D\lambda}{2x_0} \tag{2.25}$$

where λ is the wavelength of the light used.

We could initially approximate the Young's double slit experiment with two delta functions separated by distance, $2x_0$. The result of this would be a periodic function (equation 2.16), where the intensity oscillates between a maximum and minimum value. The decrease in intensity observed in a real double slit experiment such as seen in chapter 1 is because the slits have a finite width, a.

You can hopefully imagine that for a double slit experiment the function describing the experiment in real space should be the convolution of two delta functions with an aperture function of width, a, as demonstrated in figure 2.5. As we know that the Fourier transform of a delta function is $e^{-i2\pi ux_0}$, where x_0 is the position of the delta function and that the Fourier transform of a slit of finite width, a, is Aa sinc(πua), we can find the Fourier transform of these convolved functions to be:

$$F(u) = 2Aa\cos(2\pi ux_0)\text{sinc}(\pi ua) \tag{2.26}$$

which yields an intensity distribution of:

$$I = 4A^2a^2\cos^2(\pi ux_0)\text{sinc}^2(\pi ua) \tag{2.27}$$

as shown in figure 2.5. This is a fairly trivial application of the convolution theorem, but is a useful example of how it might simplify the calculation of $F(u)$. To convince yourself of this, calculate $F(u)$ using both methods outlined in Q1.6 of section 2.9.

2.6.2 The diffraction grating

An infinite diffraction grating can be represented by a series of delta functions (Dirac comb), which was shown in equation (2.18) to be proportional to $\sum_{-\infty}^{\infty}\delta\left(q - n\frac{2\pi}{d}\right)$, i.e. a series of delta functions separated by distance $2\pi/d$. As the intensity observed is equal to $|F(q)|^2$, this leads to a diffraction pattern of:

$$I \propto \left|\sum_{-\infty}^{\infty}\delta\left(q - n\frac{2\pi}{d}\right)\right|^2 \tag{2.28}$$

Again, in reality, diffraction gratings have a finite width, and this results in broadening of the fringes observed (or the convolution of a sinc function with the Dirac comb, as seen in figure 2.6(f)). In this sense, the finite diffraction grating is similar to the Young's double slit in that the calculation of the convolution of two functions could be simplified by using the convolution theorem. In this case, however, as it is the convolution of two diffraction patterns, the aperture function (representing a finite aperture width) and the infinite grating should be multiplied in real space. The Fourier transform of this function would then result in the expected diffraction pattern. This is summarised in figure 2.6.

2.7 Autocorrelation and the loss of phase information in measurements

One last concept to discuss is the autocorrelation function, ACF. This describes the degree of structure in a given pattern or function based on the amplitude and separation of its components and is defined by equation (2.29).

$$\text{ACF}(x) = \int_{-\infty}^{\infty} f(t)^* f(x+t)\text{d}t = \int_{-\infty}^{\infty} |F(k)|^2 e^{ikx}\text{d}k \tag{2.29}$$

where $f(t)^*$ is the complex conjugate. In essence, autocorrelation describes the distance between distributions of structures in $f(x)$, or in other words, the similarity between a signal and itself delayed by time t (figure 2.7). There are a couple of things

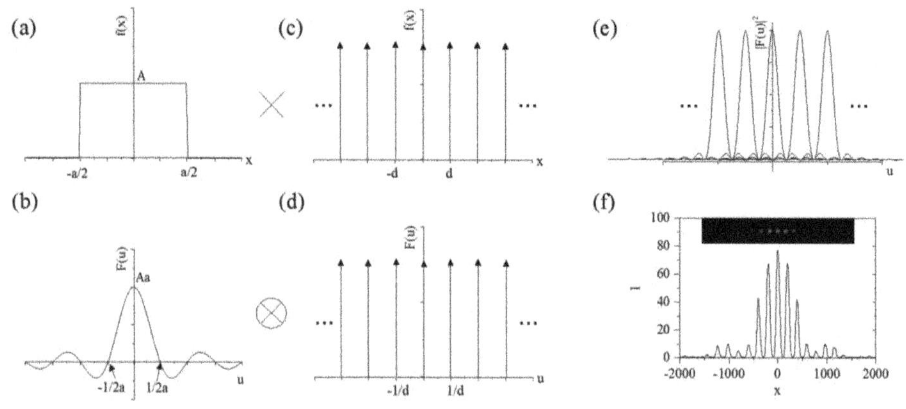

Figure 2.6. The infinite diffraction grating. The convolution of the diffraction from a finite aperture (a) with an infinite diffracting grating (c) results in the diffraction pattern shown in (e) (where (b) and (d) show the Fourier transforms of (a) and (b), respectively). For a finite grating, the diffraction pattern is modulated by the Fourier transform of a second aperture function leading to the pattern seen in (f).

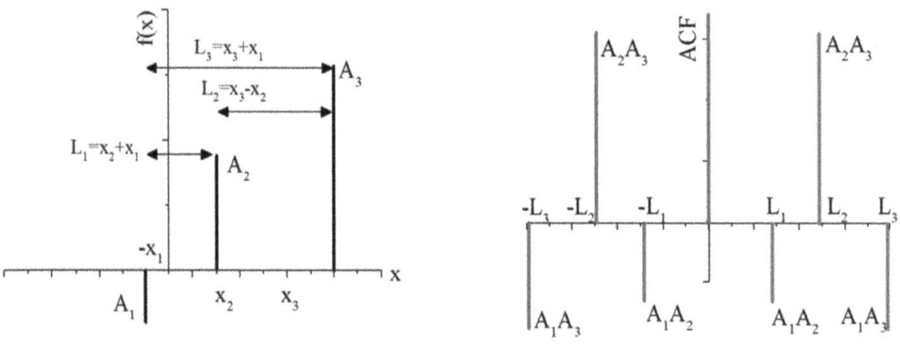

Figure 2.7. Example of the corresponding autocorrelation function ACF for $f(x)$.

to note from this, the first being that the ACF is maximum at zero (everything correlates with itself), the second being that it is proportional to the Fourier transform of the function $f(x)$ multiplied by a phase component, e^{ikx}. As in practice we observe the intensity of a diffraction pattern $|F(q)|^2$, and not the phase, information contained in e^{ikx} is lost. As a result of this loss of information determination of the original structure can be difficult.

2.8 Aside: some useful theorems

Finally, some useful theorems to remember when dealing with Fourier transforms or trying to develop a qualitative understanding of general results are presented here. Some of these will be tested in the Fourier Optics labscript (appendix H) that you are encouraged to try.

2.8.1 Linearity theorem

The Fourier transform of a sum of two functions multiplied by a constant is the sum of their Fourier transforms multiplied by the same constant(s).

$$af(x) + bg \leftrightarrows aF(u) + bG(u) \tag{2.30}$$

2.8.2 Scaling theorem

The Fourier transform of a function, $f(x)$, where the variable is multiplied by a constant, a, will be rescaled by the reciprocal of this constant.

$$f(ax) \leftrightarrows \frac{1}{|a|}F\left(\frac{u}{a}\right) \tag{2.31}$$

2.8.3 Shifting theorem

A shift in real space of a function, $f(x)$, will manifest as a phase shift in the Fourier transform.

$$f(x - x_0) \leftrightarrows e^{-2\pi i u x_0}F(u) \tag{2.32}$$

2.8.4 Parseval's theorem

The Fourier transform is unitary.

$$\int_{-\infty}^{\infty} |f(x)|^2 \mathrm{d}x = \int_{-\infty}^{\infty} F(u)F^*(u)\mathrm{d}u \tag{2.33}$$

2.8.5 Correlation theorem

If

$$p(x) = \int_{-\infty}^{\infty} f^*(x')g(x + x')\mathrm{d}x' \tag{2.34}$$

$$p(x) \leftrightarrows F^*(u)G(u) \tag{2.35}$$

where $f(x) = g(x)$ signifies autocorrelation.

2.8.6 Convolution theorem

The Fourier transform of the convolution, $\otimes$, of two functions in real space is equal to the multiple of their reciprocal functions in reciprocal space.

$$c(t) = g(t) \otimes f(t) = \int_{-\infty}^{\infty} g(t')f(t - t')\mathrm{d}t' \tag{2.36}$$

$$c(t) \leftrightarrows \int_{-\infty}^{\infty}\int_{-\infty}^{\infty} g(t')f(t - t')e^{-2\pi i v t}\mathrm{d}t\mathrm{d}t' = F(v)G(v) \tag{2.37}$$

where $x \leftrightarrows u$ and $t \leftrightarrows v$.

2.9 Questions

2.9.1 Some standard Fourier transforms

In this set of questions we will derive several of the Fourier transforms discussed earlier in this chapter and start to consider experimental data.

$$f(x) = \frac{1}{\sqrt{2\pi}}\int_{-\infty}^{\infty} F(q)e^{iqx}\mathrm{d}q$$

$$F(q) = \frac{1}{\sqrt{2\pi}}\int_{-\infty}^{\infty} f(x)e^{-iqx}\mathrm{d}x$$

1.1 *The aperture function*

Show that the aperture function:

$$A\,\mathrm{rect}\left(\frac{x}{a}\right) = \begin{matrix} A \text{ where } |x| < a/2 \\ 0 \text{ where } |x| > a/2 \end{matrix}$$

leads to the intensity distribution for a single slit of:

$$I = \frac{2A^2\sin^2(qa/2)}{\pi q^2}$$

1.2 *The delta function*

The Fourier transform of the delta function gives some insight into the differences between the atomic form factors for x-rays (scattering from a cloud of electrons) and neutrons (scattering from the nucleus).

Show that the Fourier transform of the delta function $\delta(x)$ is equal to a constant in reciprocal space.

1.3 *Multiple delta functions*

Derive the Fourier transform of:

$$\delta(x - x_0) + \delta(x + x_0)$$

1.4 *The infinite grating (HARDER)*

An infinite diffraction grating can be represented by a series of delta functions (Dirac comb):

$$\sum_{n=-\infty}^{\infty}\delta(x - nx_0)$$

Show that this would lead to a diffraction pattern of:

$$I \propto \left| \sum_{n=-\infty}^{\infty} \delta\left(q - \frac{2\pi n}{x_0}\right) \right|^2$$

1.5 *The Gaussian function*

Show that the Fourier transform of the Gaussian function:

$$g(x) = \frac{1}{\sqrt{2\pi}} \frac{e^{-x^2/2\sigma^2}}{\sigma}$$

is:

$$G(q) = \frac{1}{\sqrt{2\pi}} e^{-\sigma^2 q^2/2}$$

For help refer to *Mathematical Methods for Physics and Engineering* [1] p 436.

1.6 *Young's double slit*

The diffraction pattern for two slits at $x = x_0$ and $x = -x_0$, with finite width, a, can be determined by considering the Fourier transform of the delta function(s) and the aperture function. Show that for Young's double slit:

$$I = \frac{4A^2 \cos^2(qx_0)\sin^2(qa/2)}{\pi^2 q^2}$$

As outlined in the text, this can be attempted one of two ways:

1) Calculate the Fourier transform of:

$$A\,\mathrm{rect}\left(\frac{x - x_0}{a}\right) + A\,\mathrm{rect}\left(\frac{x + x_0}{a}\right)$$

2) Calculate the Fourier transform of:

$$\delta(x - x_0) + \delta(x + x_0) \text{ and } A\,\mathrm{rect}\left(\frac{x}{a}\right)$$

2.9.2 Normalisation of the Fourier transform pair

2.1 *Defining reciprocal space in terms of q (2π/λ)*

If the Fourier transform pair were described as:

$$f(x) \propto \int_{-\infty}^{\infty} F(q) e^{iqx} \mathrm{d}q$$

$$F(q) \propto \int_{-\infty}^{\infty} f(x) e^{-iqx} \mathrm{d}x$$

Show that $f(x) = 2\pi(\boldsymbol{FT}(F(q))) = 2\pi(\boldsymbol{FT}(\boldsymbol{FT}(f(x))))$, where $\boldsymbol{FT}$ is shorthand for the Fourier transform. Hence convince yourself that the following Fourier transform pairs would be self consistent:

$$f(x) = \frac{1}{\sqrt{2\pi}} \int_{-\infty}^{\infty} F(q) e^{iqx} \mathrm{d}q$$
$$F(q) = \frac{1}{\sqrt{2\pi}} \int_{-\infty}^{\infty} f(x) e^{-iqx} \mathrm{d}x$$

Or

$$f(x) = \int_{-\infty}^{\infty} F(q) e^{iqx} \mathrm{d}q$$
$$F(q) = \frac{1}{2\pi} \int_{-\infty}^{\infty} f(x) e^{-iqx} \mathrm{d}x$$

2.2 *Defining reciprocal space in terms of u ($1/\lambda$)*
Determine the normalisation constant for the following Fourier transform pair:

$$f(x) \propto \int_{-\infty}^{\infty} F(u) e^{i2\pi ux} \mathrm{d}u$$
$$F(u) \propto \int_{-\infty}^{\infty} f(x) e^{-i2\pi ux} \mathrm{d}x$$

2.3 *Defining reciprocal space in terms of ω ($2\pi/t$)*
Determine the normalisation constant for the following Fourier transform pair:

$$f(x) \propto \int_{-\infty}^{\infty} F(\omega) e^{i\omega x} \mathrm{d}\omega$$
$$F(\omega) \propto \int_{-\infty}^{\infty} f(x) e^{-i\omega x} \mathrm{d}x$$

2.9.3 Further questions

3.1 *A broadened aperture function*
How would you determine the Fourier transform of an aperture function described by the convolution of a Gaussian function, $f(x)$, and $A\,\mathrm{rect}(x/a)$, $g(x)$?

3.2 *Shifted delta functions*

What would you expect to happen if you shifted the delta functions defined in Q1.3 by distance d:

$$\delta(x - x_0 + d) + \delta(x + x_0 + d)$$

3.3 *Intensity incident on the double slit*

How would you expect the intensity calculated in Q1.6 to vary if you doubled the width of the slits:

$$A\,\mathrm{rect}\left(\frac{x - x_0}{2a}\right) + A\,\mathrm{rect}\left(\frac{x + x_0}{2a}\right)$$

References

[1] Riley K F, Hobson M P and Bence S J *Mathematical Methods for Physics and Engineering* 3rd edn. (Cambridge: Cambridge University Press)

[2] Boas M L *Mathematical Methods in the Physical Sciences* 2nd edn. (New York: Wiley)

[3] Sivia D S *Elementary Scattering for X-ray and Neutron Users* (Oxford: Oxford University Press)

Kelly Morrison

Chapter 3

Diffraction techniques

In this chapter a broad overview of diffraction techniques for the study of crystal structure and thin films will be given. This will start with a discussion of the theory of elastic scattering, including the cross section, and structural and magnetic form factors for specific elements. Key details on the general methods, such as sources and detectors will then be presented, followed by more detailed treatment of each experimental technique. The aim will be to provide the reader with a broad understanding of the limiting factors in such measurements, with enough detail to meaningfully compare the advantages of using x-rays or neutrons (for example) for a particular material system. Finally, some examples of these approaches will be highlighted, followed by questions specific to data analysis.

3.1 Elastic scattering

X-ray diffraction is a commonly used tool for identification of different compounds in a powder (phase analysis), or to extract detailed information about crystal structure from single crystal or powder measurements. X-ray diffraction itself was first observed by von Laue in 1912, for which he was awarded the 1914 Nobel Prize for the 'discovery of diffraction of x-rays by crystals'. This experiment was significant in that it demonstrated both the wave–particle duality of light as well as the space-lattice structure of crystals. It was not until the seminal work of William Lawrence Bragg, however, that this observation could be used to really interrogate a crystal structure. With the help of his father, he built the first x-ray spectrometer, earning Sir William Lawrence Bragg and (father) Sir William Henry Bragg the 1915 Nobel prize for *'their services in the analysis of crystal structure by means of x-ray'*.

doi:10.1088/2053-2563/ab2df5ch3

The breakthrough that Lawrence Bragg made was to treat diffraction mathematically as the reflection of waves from successive parallel planes, separated by distance, d, and to relate the wavelength of x-rays used, λ, to the angle of reflection, θ, as:

$$n\lambda = 2d \sin \theta \tag{3.1}$$

One way to derive this classically is to consider the path difference between two rays of incident light reflected off of successive planes, as demonstrated in figure 3.1. If there is constructive interference (path difference = $n\lambda$) then a peak will be observed. How sharp the peak is will depend on the variation in d across the sample measured, as well as other factors (such as instrumental broadening) that will be discussed later in this chapter.

Of course, this relation assumes that the waves are scattered elastically, where there is no energy loss and the incident wavelength is equal to the scattered wavelength. As >99% of scattering events will be elastic, any inelastic scattering where there is energy loss (i.e. a frequency shift) is merely a nuisance to the spectrographer. There is, however, useful information to be obtained from inelastic scattering, as will be discussed in chapter 5.

Elastic scattering is defined as an interaction where energy is conserved, and there is a change in direction of the wave (or particle). It is useful, therefore, to consider how we might express these aspects of the incident and final waves.

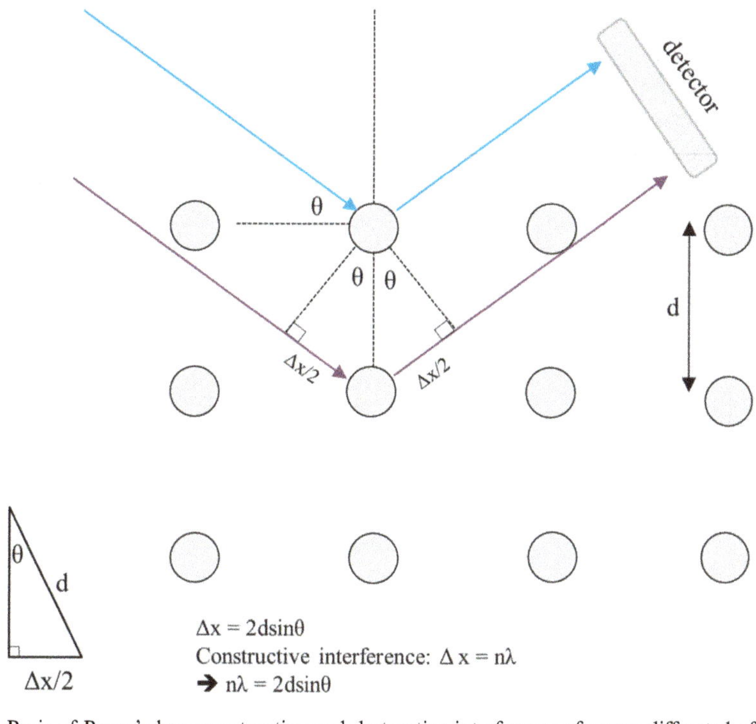

Figure 3.1. Basis of Bragg's law: constructive and destructive interference of waves diffracted off crystalline planes in a cubic lattice separated by distance, d.

The energy, E [unit of Joules = kg m^2 s^{-2}], of an electromagnetic wave can be defined as:

$$E = hf = \frac{hc}{\lambda} \tag{3.2}$$

where h is Planck's constant [m^2 kg s^{-1}], f is the frequency [Hz = s^{-1}], λ is the wavelength [m], and c is the speed of light [m s^{-1}]. We know, from quantum mechanics (de Broglie), that the wavelength and momentum, p, are related in the following way:

$$\lambda = \frac{h}{p} \tag{3.3}$$

Therefore, for an elastic scattering event where the energy is conserved:

$$E_i = E_f \tag{3.4}$$

Substituting equation (3.2):

$$\frac{hc}{\lambda_i} = \frac{hc}{\lambda_f} \tag{3.5}$$

Substituting equation (3.3):

$$\frac{1}{p_i} = \frac{1}{p_f} \tag{3.6}$$

In other words, the absolute momentum, p, is also conserved. This should make sense: as no energy is lost, the wave or particle is merely redirected by the elastic collision. As the change here is not the velocity but the direction, in this case we need to start considering the momentum as *vectors*.

Typically, when discussing diffraction measurements, the momentum transfer, $\boldsymbol{p}$, or wavevectors, $\boldsymbol{k}_\mathbf{i}$ and $\boldsymbol{k}_\mathbf{f}$, will be used to convey the change in direction of the scattered wave, where:

$$\boldsymbol{p} = \hbar \boldsymbol{k}_i - \hbar \boldsymbol{k}_f = \hbar \boldsymbol{Q} \tag{3.7}$$

$$\boldsymbol{Q} = \boldsymbol{k}_i - \boldsymbol{k}_f \tag{3.8}$$

and:

$$|\boldsymbol{k}| = \frac{2\pi}{\lambda} \tag{3.9}$$

Notice that the wavevector here is defined in a similar way to reciprocal space in chapter 2. If we consider the trigonometry of a typical scattering experiment in 2D, as shown in figure 3.2, then we can show that the momentum transfer, Q, is defined as:

$$Q = |\boldsymbol{Q}| = \frac{4\pi \sin\theta}{\lambda} \tag{3.10}$$

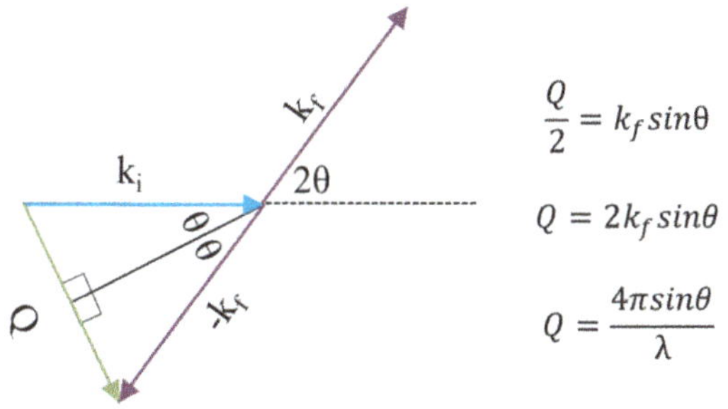

Figure 3.2. Vector diagram for an elastic scattering event. Initial wavevector, $\boldsymbol{k}_i$, is scattered to form final wavevector, $\boldsymbol{k}_f$. The momentum transfer, $\boldsymbol{Q}$, is the difference between these vectors, $\boldsymbol{k}_i - \boldsymbol{k}_f$.

This is useful as diffraction experiments are commonly expressed either in terms of θ (for monochromatic radiation) or in terms of Q (for time of flight).

To extend this to 3D (e.g. for single crystal diffraction), we would need to define a 3D system in spherical polar coordinates, resulting in:

$$\boldsymbol{k}_i = \left(0,\, 0,\, \frac{2\pi}{\lambda}\right) \tag{3.11}$$

where we have set the incoming wave to be along the z-axis. The scattered wave could then be described by:

$$\boldsymbol{k}_f = \frac{2\pi}{\lambda}(\sin 2\theta \cos \varphi,\, \sin 2\theta \sin \varphi,\, \cos 2\theta) \tag{3.12}$$

where the wave is now travelling in a direction determined by the scattering event. Typically, what happens at a scattering centre (such as an atom) is that the wave will be scattered isotropically (in all directions). For this reason, it is simpler to use spherical polar coordinates to describe scattering events, as defined in appendix E. For example, using the trigonometric relations—$\sin 2\theta = 2 \sin \theta \cos \theta$, $\cos 2\theta = 1 - 2 \sin^2 \theta$—and the combination of equations (3.11) and (3.12) we can describe the momentum transfer, $\boldsymbol{Q}$, as follows:

$$\boldsymbol{Q} = \frac{4\pi \sin \theta}{\lambda}(-\cos \theta \cos \varphi - \cos \theta \sin \varphi,\, \sin \theta) \tag{3.13}$$

3.1.1 X-rays versus neutrons

Until now, we have considered the interaction of x-rays with a crystal lattice, but what happens if we perform diffraction with a particle, such as the neutron? Again, we can draw on wave–particle duality, in other words the definition of the de Broglie wavelength given in equation (3.3). This can be used to describe the wavelength associated with particles such as the electron or neutron, which can also be used in diffraction measurements. The difference here is that we define the momentum and energy in terms of the particle's speed, v, and mass, m. For a particle, the energy can be expressed as its kinetic energy:

$$E = \frac{mv^2}{2} \tag{3.14}$$

and the momentum:

$$p = mv \tag{3.15}$$

If we substitute (3.14) and (3.15) in a similar way to equations (3.4) to (3.6) we obtain:

$$\Delta E = \frac{|\hbar \boldsymbol{k}_i|^2}{2m} - \frac{|\hbar \boldsymbol{k}_f|^2}{2m} = 0 \tag{3.16}$$

Unlike the electromagnetic wave, we no longer have a linear relationship between the momentum transfer and the energy gain/loss. However, by definition of elastic scattering, the difference in energies, ΔE, will be zero, which yields a simple relationship between the incident and final wavevectors ($\boldsymbol{k}_i = \boldsymbol{k}_f$).

So, what are the advantages of neutron diffraction?

- Different contrast (to x-ray diffraction);
- No charge;
- Magnetic interaction.

To understand this, this we need to go back one step and consider the interaction that leads to diffraction of x-rays and neutrons. In table 1.2 of chapter 1 the key features of particle interactions were summarised for x-rays, neutrons, electrons and muons. For electrons, the key interaction is electromagnetic: they will be easily diffracted by the electron cloud surrounding an atom. Neutrons, on the other hand will interact with the nucleus, or via the magnetic dipole–dipole interaction. To describe the strength of these interactions we refer to the nuclear scattering length and the atomic form factor (x-rays).

3.1.2 Nuclear scattering length and form factors

To describe the possible scattering interaction in any given diffraction experiment requires knowledge of the wavelength of the incident wave as well as the scattering angle itself. It can also depend on the magnetic, electromagnetic and nuclear interactions of the incident wave and the diffraction medium. When quantifying this, the function $f(\lambda, \theta)$ is used, which describes the probability of scattering for a given wavelength and in a direction defined by θ.

3.1.2.1 X-rays

For x-rays, as the strength of the interaction will depend on the number of electrons present (electrostatic interaction), it is simple enough to see that it will increase with Z. Less obvious, perhaps, is how it will vary with λ and θ. This is what the atomic form factor describes and the functional form for Si is shown in figure 3.3 alongside

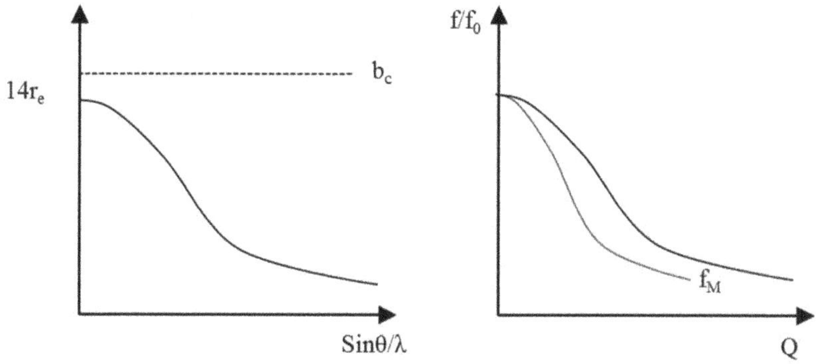

Figure 3.3. Left: schematic of the atomic form factor of Si for neutrons, and x-rays, respectively. Right: illustration of the difference between the normalised atomic form factor for x-rays and the magnetic form factor for neutrons, f_M (r_e = 0.2818 fm, b_c = 4.106 fm).

its corresponding value for neutrons. Notice how the atomic form factor decreases with increasing $\sin\theta/\lambda$. If you recall from chapter 2, we described the Fourier transform of a delta function and a finite aperture, and found that the modulus squared of these functions, $|F|^2$, was a constant and a $\text{sinc}^2\,\theta$ function, respectively. With the example of x-ray and neutron scattering we have a similar scenario: the size of the nucleus is of the order of fm [10^{-15} m], whereas the size of the electron cloud is of the order of Å [10^{-10} m]. Therefore, the diffraction of neutrons could be considered analogous to the delta function, whereas the x-ray interaction could be considered analogous to the aperture function. Because of the differing lengthscales of these interactions, the resulting form factors are very different.

3.1.2.2 Neutrons

The nuclear scattering length, b, defines the strength of interaction for neutrons with a given element. As it describes the interaction with the atomic nucleus it will be isotope dependent, however, beyond this, there is no obvious trend in the value of the scattering length. In addition, the magnetic moment of the neutron means that the scattering length will also have different values for an atom with non zero spin.

In general:

$$f(\lambda, \theta) = -b_c \tag{3.17}$$

where b_c is the bound coherent scattering length (in femtometres). For an atom with non-zero spin, S, this will be described by:

$$b_c = g_+ b_+ + g_- b_- \tag{3.18}$$

where b_+ and b_- describe the scattering length of the symmetric and antisymmetric states, respectively, and g_+ and g_- describe the multiplicity of these states. For example, a neutron of spin $S_n = 1/2$ could interact with an atom of spin $S_a = 1/2$ to form a combined spin of either $S = 1$ (triplet, +) or $S = 0$ (singlet, −). The ratio of the b_+ and b_- scattering lengths is thus described by equations (3.19) and (3.20), where

Table 3.1. X-ray (f) and neutron (b_c) scattering lengths for common elements [1]. Where S_a is the average spin value for the listed element; b_+ and b_- are the spin dependent scattering length for $I = +1/2$ and $I = -1/2$, respectively; and σ_{abs} is the thermal absorption cross-section for 0.0253 eV.

	Z	f(1.54 Å,0) [fm]	b_c [fm]	S_a	b_+ [fm]	b_- [fm]	σ_{abs} [Barns]
H	1	1 r_e	−3.7423	1/2	10.817	−47.420	0.3326
D	1	1 r_e	6.674	1	9.53	0.975	0.000 5197
O	8	8 r_e	5.805	0			0.000 10
Si	14	14 r_e	4.106	0			0.177
Mn	25	25 r_e	−3.75	5/2	−4.93	−1.46	13.3
Fe	26	26 r_e	4.2	0			2.25
Co	27	27 r_e	2.49	7/2	−9.21	3.58	37.18
Ni	28	28 r_e	14.4	0			4.6
Cu	29	29 r_e	6.477	3/2			4.5

S_a is the combined spin of the atom. Some examples of different scattering lengths are given in table 3.1.

$$g_+ = \frac{S_a + 1}{2S_a + 1} \tag{3.19}$$

$$g_- = \frac{S_a}{2S_a + 1} \tag{3.20}$$

Lastly, the variance in the scattering length, b_i, is described by the root mean square:

$$b_i^2 = g_+ g_- (b_+ b_-)^2 \tag{3.21}$$

3.1.2.3 Magnetic contrast

The **magnetic form factor** on the other hand, quantifies the magnetic scattering that a neutron (which itself acts as a bar magnet) can experience. This is due to the magnetic interaction between the neutron and the diffraction medium. In this case, we are utilising the dipole–dipole interaction of the atom with the neutron, which will occur on a larger scale than the nuclear (short-ranged) interaction. This results in a variation with sin θ/λ that drops off sharper than the corresponding atomic form factor for x-rays. For a more detailed discussion, see pp 69–75 of *Elemental Scattering Theory* by D Sivia [2] or section 2.5 of the *Neutron Data Booklet* [1].

3.1.3 Scattering cross-section

The scattering cross-section, σ, is an effective area that describes scattering. It can be thought of as the area required to achieve a chosen scattering rate for a given material and is defined by:

$$\sigma = \frac{R}{|\psi_0|^2} \tag{3.22}$$

where $|\psi_0|^2$ is the amplitude of the incident flux [s^{-1} m^{-2}], and R is the scattering rate for a given atom [s^{-1}]. It is worth remembering at this point that the scattering rate will depend on $f(\lambda, \theta)$, that elastic scattering will be isotropic (so useful to work in spherical polar coordinates, as seen in figure 3.4 and defined in appendix E), and that the flux will decrease with r^2. Thus, in spherical polar coordinates R can be defined as:

$$R = \int_{2\theta=0}^{\pi} \int_{2\varphi=0}^{2\pi} |\psi_f|^2 \mathrm{d}A \tag{3.23}$$

$$R = \int_{2\theta=0}^{\pi} \int_{2\varphi=0}^{2\pi} |\psi_f|^2 r^2 \sin 2\theta \, \mathrm{d}\varphi \mathrm{d}2\theta \tag{3.24}$$

where ψ_f is the wavefunction of the final state and:

$$|\psi_f|^2 = \frac{|\psi_0|^2}{r^2} |f(\lambda, \theta)|^2 \tag{3.25}$$

All of these concepts are key to describing the overall rate of scattering for a diffraction measurement if the elements involved, wavelength of radiation, and incident flux are known. If we substitute (3.24) and (3.25) into (3.22) then we obtain:

$$\sigma = 2\pi \int_{2\theta=0}^{\pi} |f(\lambda, \theta)|^2 \sin 2\theta \, \mathrm{d}2\theta \tag{3.26}$$

In other words, the scattering cross-section, σ, will depend on the function, $f(\lambda, \theta)$. *It is also worth mentioning that sometimes reference will be made to the differential cross-section, dσ/dΩ: the fraction of scattered particles that emerge in different directions, where dΩ is the solid angle.*

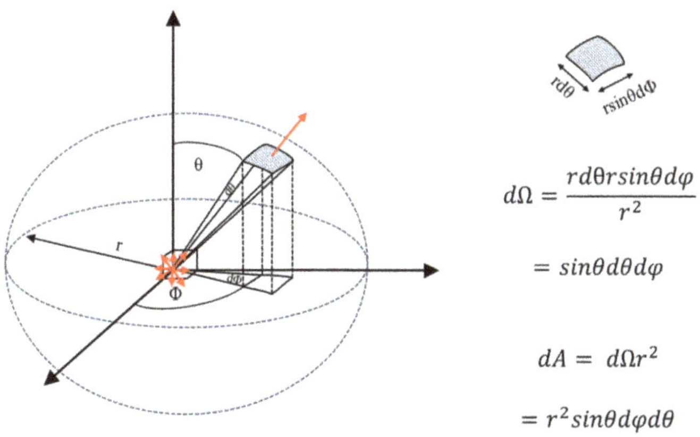

Figure 3.4. Scattering cross-section: the rate of scattered particles that pass through solid angle, dΩ, can be used to determine the rate of scattering if the incident flux, $|\psi_f|^2$ is known. As the scattering length function, $f(\lambda, \theta)$, describes the strength of interaction of an x-ray or neutron with a given element, it can be used to convey the fraction of incident particles that scatter off a given atom. Spherical polar coordinates are used to simplify the mathematical treatment, where the scattering centre is indicated by the red arrows.

3.1.4 Some final notes

So far we have mostly discussed elastic **coherent** scattering: this is all of the scattering events that will (coherently) interfere with one another to produce characteristic Bragg peaks. There will, however, also be **incoherent** contributions to the total diffraction pattern. This can be due to non-ordered material (such as polymers or glass), or as in the case of neutrons, a random distribution of different isotopes with different nuclear scattering lengths. In general, this will result in a general background, as will be seen later when Rietveld refinement is discussed.

Secondly, we have also assumed that all scattering is **kinematic**: that there is only one scattering event per particle before it reaches the detector. This is not necessarily true, and the result of multiple scattering events (**dynamic** scattering) is that the intensity of the scattered particles will decrease.

3.2 Methods

3.2.1 X-ray sources

We have discussed the basics of x-ray diffraction, but how are the x-rays first produced? In a standard lab, an x-ray generator consists of a simple cathode ray tube (CRT), as shown in figure 3.5. In this example, a beam of electrons is generated at a filament (cathode) and then accelerated towards an anode, where the interaction between the electrons and the material of the anode (usually copper) produces a steady stream of x-rays. If we look at the intensity of the x-rays produced, shown in figure 3.5, then we see a broad background due to bremsstrahlung (breaking radiation), and characteristic peaks K_{α_1}, K_{α_2} and K_β, that are related to energy level transitions within the anode. As discussed in chapter 1, this happens as electrons are excited by bombardment of electrons on the copper anode, and then relax to the ground state. For copper, the wavelength corresponding to $K_{\alpha_1} = 1.540\ 56$ Å.

For a given measurement, more often than not we will want a monochromatic source as this simplifies detection. The reason for this is that you would only need to know the angle and intensity of the observed, scattered, x-rays in order to implement Bragg's law (equation (3.1)). For a copper CRT this can be achieved with an absorbing material such as a Ni filter, which will absorb the K_β and some of the bremsstrahlung radiation. It can also be achieved using a single crystal monochromator, where diffraction of the x-rays off of a crystal surface is utilised to produce Bragg peaks (due to diffraction) at a specific angle. This will spatially separate the wavelengths enabling the user to select a specific wavelength from the alignment of the monochromator and slits to block out higher order reflections. The limitation of this is the mosaic spread of the crystals: slight misalignment or defects (in the crystal) that lead to broadening of the Bragg peaks.

For example, the characteristic peaks of Cu (K_{α_1}, K_{α_2} and K_β), each have a specific wavelength (1.540 56 Å, 1.544 39 Å, 1.392 22 Å). By Bragg's law, if these wavelengths were scattered off of a crystal surface with lattice separation $d = 5$ Å, it would produce Bragg peaks at 8.862°, 8.884°, and 8.003°, respectively. As long as

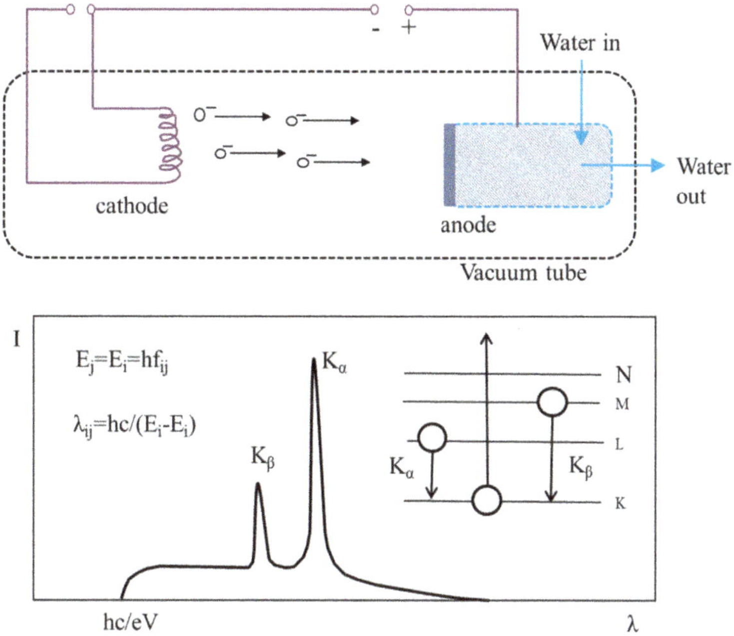

Figure 3.5. Schematic of a cathode ray tube (CRT) and spectrum of the x-rays produced.

the mosaic spread of the crystal(s) in the monochromator is small enough, these peaks could be separately resolved.

Crystals that are commonly used in monochromators are silicon, germanium, graphite, and quartz. They must have a suitable lattice separation in order to produce Bragg peaks at small values of 2θ (limiting loss of signal), a large structure factor for the reflection to be used, low absorption, be easy to cut, and a low coefficient of thermal expansion (so they are stable for small temperature fluctuations). Commonly used monochromators in the lab are graphite and silicon, which differ by their typical mosaic spread (graphite is typically broader than silicon). Whilst a narrow linewidth (the result of small mosaic spread) is advantageous for high resolution measurements, it will also require precise alignment of the monochromator.

The crystal monochromator can also be designed to focus the monochromated x-rays to a point, by bending or polishing the crystal such that it has a curved surface (much like a lens) [3].

It should be noted as well that while copper is the most common target for x-ray generation, it is not the only possibility. Alternative targets such as Mo or W might be used in order to access a different wavelength. This is useful if the sample is a strong absorber for Cu wavelengths, or if there is significant fluorescence of the sample to be studied (at that wavelength). The limitation of this type of generator, however, is the maximum intensity of x-rays that can be produced. Given that a continuous stream of electrons is colliding with the anode, there will be energy transfer. This would ordinarily warm up the anode, if not for water cooling built into

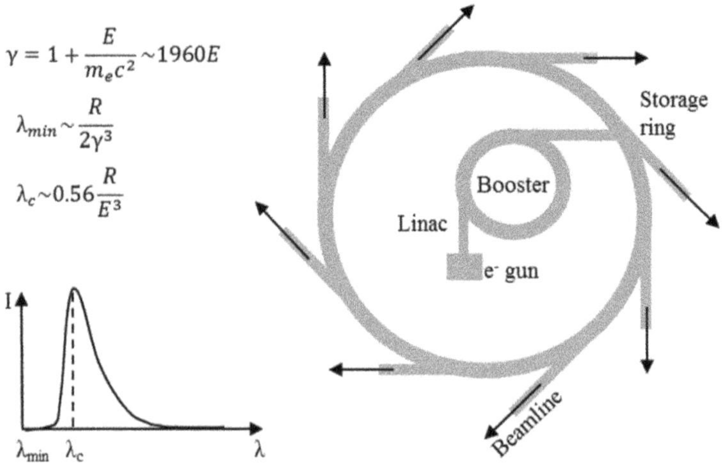

Figure 3.6. Simplified synchrotron schematic. A continuous beam of photons is produced with a minimum wavelength, λ_{min}, dependent on the radius of the ring and kinetic energy of the electrons, E.

the CRT. The limitation of x-ray generation, therefore, is how quickly we can cool the target.

Higher intensity x-ray sources, typically require access to a synchrotron, as shown in figure 3.6. Examples of this include the Diamond light source in Oxfordshire, or the European Synchrotron Radiation Facility (ESRF) in Grenoble. In such cases, a charged beam of particles is accelerated close to the speed of light in a circular ring much like the particle colliders used by CERN in the Large Hadron Collider (LHC). The difference here is the radius of the ring used: whereas the LHC has a circumference of 27 km, the Diamond Light source, for example, is only 0.5 km. The radius of the ring controls how much energy is wasted accelerating or decelerating the electrons through the curvature of the ring: as the curvature is increased more of this energy is needed and is expelled as photons.

Much like the CRT, in a synchrotron source, electrons are generated in an electron gun then accelerated. Initially, this occurs in a linear accelerator (linac), followed by a booster synchrotron; at which point the electrons will be travelling close to the speed of light. Once they enter the storage ring of the synchrotron, their path is controlled by a series of electromagnets that steer the electrons around the ring. For example, at Diamond, the storage ring is actually comprised of 48 linear sections arranged in a tetracontakaioctagon, with an electromagnet at the end of each linear section in order to redirect it to the next. At each point where the path of electrons is redirected, energy will be lost as photons, thus producing x-rays.

Highly collimated beams of x-rays are produced due to a combination of relativistic effects and the acceleration of the particles themselves. In addition, insertion devices (wigglers) might be added to further increase the intensity of light produced, by causing the electrons to wiggle around. Typically, this approach will produce a spectrum of wavelengths without the characteristic spikes seen with the CRT sources. The experimenter then faces a choice: (a) to use a monochromator in order to select an incident wavelength, λ_i, or (b) to use another method of measurement (such as the time of flight technique) in order to determine the momentum transfer from an unknown λ_i.

As x-ray generation no longer relies on smashing particles into a target there is less of a problem with cooling, which means higher intensities are possible but with higher operation costs and problems with control. For example, a problem with just one of the electromagnets in the storage ring would require the whole synchrotron to be shut down for hours, days, or as was the case with the LHC in 2008, up to a year.

3.2.2 Neutron sources

The two main sources of neutrons are by spallation or nuclear fission. The production of neutrons by spallation also involves the use of a synchrotron: a beam of charged particles is produced, accelerated, and then collided into a target such as tungsten (W), where neutrons are emitted. The neutrons which are produced can then be slowed down with moderators such as 300 K water, 100 K methane, or 20 K hydrogen.

Examples of spallation source facilities are the ISIS Neutron and Muon Source, the Spallation Neutron Source at Oakridge National Laboratory, the Spallation Neutron Source at the Paul Scherrer Institute, and the European Spallation Source in Lund, Sweden (from 2020). The neutron source at ISIS, for example, produces a beam of negative hydrogen ions and accelerates them to 665 keV in an ion source, before they are accelerated to 70 MeV in the linac of the synchrotron. The electrons are then stripped from the hydrogen ions by a thin Al_2O_3 foil, leaving a beam of protons which is subsequently accelerated to 800 MeV. This proton beam is shared between two target stations: 4 out of 5 pulses will be sent to target station 1; the remaining pulses are sent to target station 2, which is optimised for longer wavelengths. At each target station a W target will produce approximately 20 neutrons per incoming proton (or 2×10^{16} per second).

The alternative to spallation source production of neutrons is the nuclear fission of radioactive material, for example ^{235}U. This has the advantage of producing a higher flux of neutrons (e.g. the ILL at Grenoble will produce on average 58 MW), but relies on sourcing the necessary fuel elements.

There are many possible decays of ^{235}U, a couple of which are outlined below:

$$^{235}U + n \rightarrow {}^{236}U \rightarrow {}^{90}Rb + {}^{143}Cs + 3n$$

$$^{235}U + n \rightarrow {}^{236}U \rightarrow {}^{134}Xe + {}^{100}Sr + 2n$$

The key aspects of this fission are that:

1) It requires slow neutrons for the initial absorption stage (to produce ^{236}U).
2) On average 2.5 neutrons will be produced per event.
3) The neutrons produced from the fission will be high energy. These will need to be slowed down by a moderator for (1), and will then be directed to a beamline for experiment.

Due to the significantly different energies and implications of neutron energy for detection, neutrons produced in these ways are referred to as thermal (1 meV < E < 1 eV) and fast (~MeV) neutrons, respectively.

3.2.3 Time of flight

Finally, whilst we have concentrated on standard spectrometers that utilise the diffraction angle for a given monochromatic incident wave, this is not the only method that can be employed. The time of flight technique, for example, is commonly used for detection of neutrons from a wide beam source (i.e. a range of incident wavelengths, λ_i). For elastic scattering, if the distance between the source, material and detector(s) are known, and the time taken to travel that distance can be measured, then λ_i can be determined. This is commonly employed in neutron measurements, where, the advantage of using the entire spectrum of neutrons is a higher count rate. For simplicity, the diffraction pattern is then represented in terms of momentum transfer, Q (as the angle 2θ would shift with λ_i).

$$\lambda = \frac{h(t + t_0)}{m_n(L + L_0)} \tag{3.27}$$

One key aspect of the time of flight technique is that the bunches of neutrons are well separated in time, such that there is no overlap between the detection of counts in each pulse. This is demonstrated in figure 3.7, where the period of neutron production is denoted by τ. If the profile of neutron wavelengths was elongated, by, for example, moderation in 20 K H, then the proton bunches may start to overlap. At this point you start to see 'frame overlap' in the experiment due to ambiguity of the measured λ_i.

3.2.4 Detectors

You should now have a passing understanding of the basic principles of diffraction, and the production of x-rays and neutrons for experiment. It is worth at this point looking at the various types of detectors that are employed, and how this can affect the experiment. For example, whilst the time of flight technique offers parallel measurement of several wavelengths of incident radiation, the reason why frame overlap is a problem and why precise determination of the length of the beamline and time from source is required is due to the fact that the detectors used are not likely to have good energy resolution; the incident energy must be determined in another way. Detectors typically fall into one of two main categories—scintillators and gas filled chambers—but examples of solid state based detectors will also be given here.

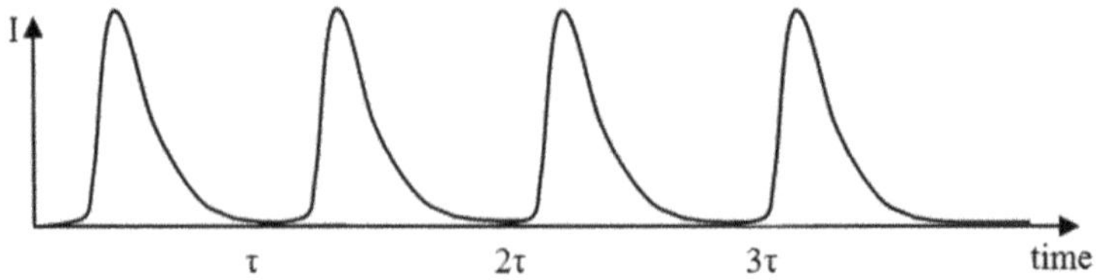

Figure 3.7. Time of flight technique. The incident wavelength, λ_i, can be determined if the time at which a neutron is detected, t, is measured for a given distance travelled, L. t_0 and L_0 are small offsets determined by calibration measurements.

3.2.4.1 A short note on dead time

Before we start discussing how different detectors work, it is worth taking some time to consider the impact of 'dead time' on the detection of x-rays or neutrons. Imagine a device that generates a measureable electric signal when hit by a particle. The mechanism driving this will more than likely involve excitation or ionisation of some medium, and will therefore have a characteristic time associated with it for the excited state to decay. Dead time is a term used to describe this effect, i.e. the time it takes a detector to reset to its original unexcited state. In reality, one of two situations often occur:

Paralysed dead time: where the detector will not detect another particle until the excited state has decayed. This can be complicated if several detections continue to 'excite' the detector—i.e. the dead time is prolonged by the existence of more particles.

Unparalysed dead time: where the detector simply won't detect (or respond) to further particles until it has decayed to its original state. This means that saturation of the signal will not continue indefinitely.

The most common manifestation of dead time is an 'unexplained' saturation of the signal for high intensities. As it is not uncommon for a continued increase in intensity to damage the detector, absorbers or scatterers are often used to decrease the amplitude for strong scattering events. For example, a thin copper sheet could be used to block part of the straight on beam when aligning a sample for x-ray reflectivity, and removed once the lower intensity measurement starts. Alternatively, a scatterer such as plexi glass might be used when aligning an inelastic neutron instrument on an elastic Bragg peak, and removed for the much lower intensity inelastic measurements.

3.2.4.2 Scintillation counters

The scintillation counter is comprised of two elements: a scintillator material, and a photomultiplier tube. The aim of scintillators is to absorb gamma- or x-rays and convert them to photons that can be more readily observed. This is then passed to a photomultipler, which converts these photons to a measureable electrical signal.

Properties of a good scintillator material include:

- High detection (μ) and conversion efficiency (so that it is sensitive);
- Being transparent to the emitted light (so that it can reach the photomultiplier);
- Short decay times of excited states (so that the dead time is not too long);
- Being robust and inexpensive (so that it is affordable and behaves in a stable manner).

Typical scintillator materials include:

- Sodium iodide activated with thallium (NaI(Tl)). This material is often used for nuclear medicine applications, however it is fragile and hygroscopic;
- Bismuth gemanate (BGO). This is used in pulse mode as detectors for PET scanners.

The aim of photomultipliers, is to take photons incident on them and amplify this into an observable electrical signal (typically by a factor of the order of 10^6). This is achieved by applying a voltage bias to photocathode-electrode based devices contained in an evacuated glass tube, as shown in figure 3.8. The vacuum is required to reduce any interaction of the incident photons with air or gas particles (such as ionisation). The most common set-up is a single photocathode surrounded by 10–12 electrodes referred to as 'dynodes' followed by an anode.

The photocathode converts photons to electrons by the photoelectric effect and is usually comprised of a screen with a photoresponsive coating. The electrons created by the photocathode are then accelerated towards the dynodes. Amplification of the signal occurs at each dynode, where electrons are accelerated due to the potential difference between the dynode and the previous step (whether it is the photocathode entrance, or the previous dynode). The kinetic energy that each electron gains on this path is enough to eject multiple electrons from the dynode it collides with. The number of electrons ejected (per incident electron) will depend on the voltage difference between each pair (i.e. the kinetic energy gained).

To appreciate the scale of amplification that a PMT can achieve you need only consider the number of electrons emitted (amplification) per dynode. If each dynode only has an amplification of 3, and there are ten dynodes in the PMT, then the total amplification would be 3^{10} = 59 049 (if there were 12 dynodes, then the total amplification would be 3^{12} = 531 441). If, on the other hand, each dynode had an amplification of 5, then the total amplification after 10 dynodes would be 9 765 625.

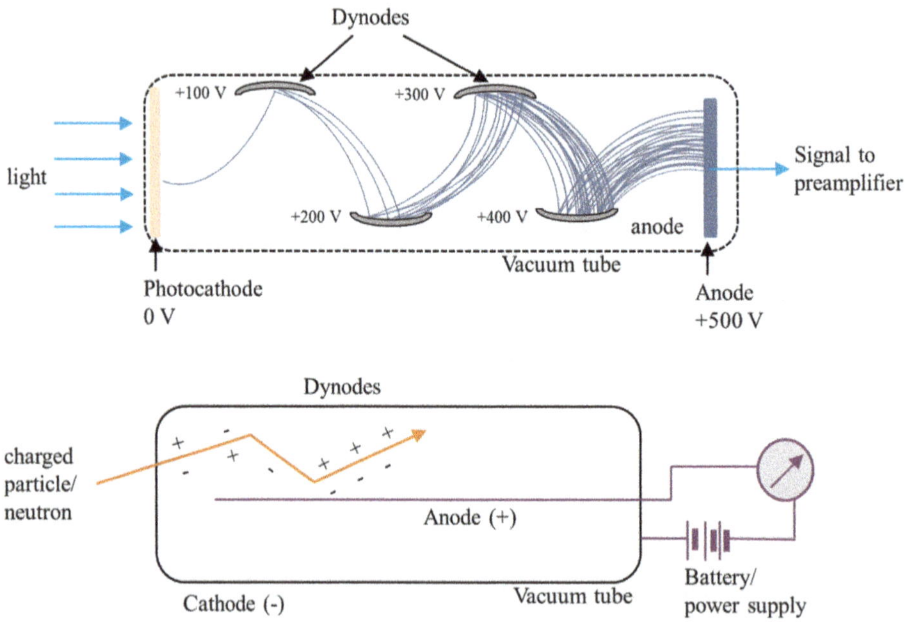

Figure 3.8. Schematic of the standard photomultiplier set-up (top) and gas filled detector (bottom).

3.2.4.3 Gas filled detectors

The gas filled detector relies on the ionisation of inert gas atoms when x-rays interact with them. As the energy associated with a single x-ray photon ($\lambda = 0.154$ nm) is:

$$E = \frac{hc}{\lambda} = 12.9 \times 10^{-16}\,\text{J} = 8.1\,\text{keV} \tag{3.28}$$

and compared to the ionisation energy of Ar (10–20 eV), it is easily seen that there is more than enough energy to ionise multiple gas particles. For example, one x-ray photon would in principle be capable of producing >500 electron–ion pairs.

Gas filled detectors are typically arranged as tubes of inert gas such as Ar, often in an outer shell of steel or Al, with an inner gold plated tungsten shell (for strength and conductivity). A thin wire is housed in the centre, and kept at a potential of around 1000 V. This potential will accelerate the electron–ion pair produced by the x-ray interaction with Ar towards the wire and thus cause further ionisation (gas amplification). Once the electron–ion pair hits the wire, it will yield an electrical signal (voltage pulse) that can then be shaped and counted. To minimise dead time, the gas is often mixed with a quenching gas such as methane (e.g. 10% CH_4:90% Ar).

A slight variant on the simple gas filled detector is the position sensitive detector (PSD). Here a voltage is applied to the two ends of the wire, and as the electron hits, a signal will travel to both ends. The difference between timings of the signal at the top and the bottom of the wire gives information on the position at which the electron hit the detector. For this to work, the anode wire needs to be poorly conducting (so that there is variance in the high tension voltage along the wire), and the gas detector needs to be carefully calibrated. PSDs can record information over a wide range of scattering angles and are particularly useful for time sensitive measurements (such as time resolved powder diffraction or time of flight).

Other variants include:

- The size of the PSD: this will determine whether a curved or straight wire is required;
- Whether there is a sealed or continuous flow of gas.

The PSDs require careful calibration if they are to be used quantitatively to determine both scattering angle and intensity. This is determined from knowledge of the wire position and efficiency (of the wire). The wire efficiency is often determined by measuring an amorphous iron foil—high flat background with no Bragg peaks. For calibration of the 2θ position the detector would have to be scanned through the Bragg reflection of a single strong peak (such as Si (111)) or many Bragg peaks (e.g. for a curved detector). Signals from multiple PSDs are then fed to a multichannel analyser (MCA).

For the gas filled detector, the relationship between pulse height (magnitude of electrical signal) and intensity of incident x-rays is determined by the voltage bias between the anode and cathode. There are four regions determined by this voltage

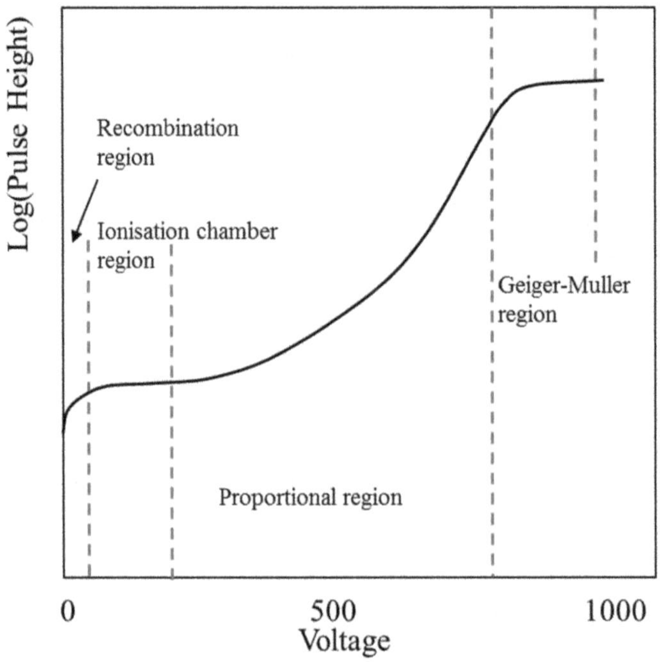

Figure 3.9. Schematic of the different regions of operation with gas filled detectors.

bias, as shown in figure 3.9, and the type of detector will depend on which region we are working in:

- Recombination region: the ionised particles are likely to recombine before they reach the anode;
- Ionisation chamber region: all the electrons are collected and the charge is proportional to the energy deposited;
- Proportional counter region: the voltage is strong enough for secondary ionisation to occur (avalanche ionisation). The charge will be linearly proportional to the energy deposited;
- Geiger–Müller counter region: the anode wire is now saturated with electrons and avalanche effects will occur over the entire wire. This detector type is not suitable for high count rates [4].

Proportional and Geiger–Müller counters consist of a thin wire anode, whereas the design of ionisation chambers can be more flexible.

3.2.4.4 Detection of neutrons

As neutrons are neutral, their detection is based on one of two types of indirect interactions: scattering, where a neutron is scattered by a nucleus that may then scatter off nuclei and thus ionise surrounding material (only effective with light elements such as helium or hydrogen); or nuclear reactions, where products of known reactions such as gamma, alpha, proton or fission fragments are detected.

The choice of technique falls down to the type of neutron source, and the corresponding cross-section for the energies involved. For example, figure 3.10 shows the interaction probability (cross-section) for commonly used elements He, B and Li. For energies greater than 10^6 eV the cross-section is too low for reliable absorption. As a result, fast neutrons (produced by nuclear fission, for example), would be better detected by scattering. For thermal neutrons (~meV), such as produced by the ISIS pulsed source, Oxfordshire, absorption reactions would be the preferred method.

The cross-section of some commonly used materials—^{3}He, ^{10}B, ^{7}Li, ^{6}Li, ^{235}U—in neutron gas detectors for thermal neutron absorption detection is given in figure 3.10. As the energies (of neutrons) involved with use of these types of detectors are low, it is typical for only a single ionisation pair to be produced. This will be detected as a discrete pulse rather than the broad continuum that would be associated with gamma ray detection. ^{3}He based gas detectors work by absorbing a neutron and producing ^{1}H and ^{3}H:

$$^{3}\mathrm{He} + \mathrm{n} \rightarrow {}^{3}\mathrm{H} + {}^{1}\mathrm{H} + 765\ \mathrm{keV}$$

Sensitivity to gamma radiation is negligible, so it makes a good detector for thermal neutrons. However, as the supply of ^{3}He is scarce (it is a byproduct of the decay of tritium produced from reactor operation or weapons programmes), it is expensive and difficult to source. The tube is wrapped in cadmium and boron layers to reduce contributions from unwanted thermal neutrons, with a lead shield to reduce contributions from photons.

Alternatively, enriched gaseous boron trifluoride, BF_3 (96% ^{10}B compared to naturally occurring 20%), or ^{10}B lined gas-filled proportional counters could be employed. In the latter case, the reaction will take place at the surface of the counter where one or two particles are released into the gas. For example:

$$^{10}\mathrm{B} + \mathrm{n} \rightarrow {}^{7}\mathrm{Li}^{*} + {}^{4}\mathrm{He} + 2310\ \mathrm{keV}$$

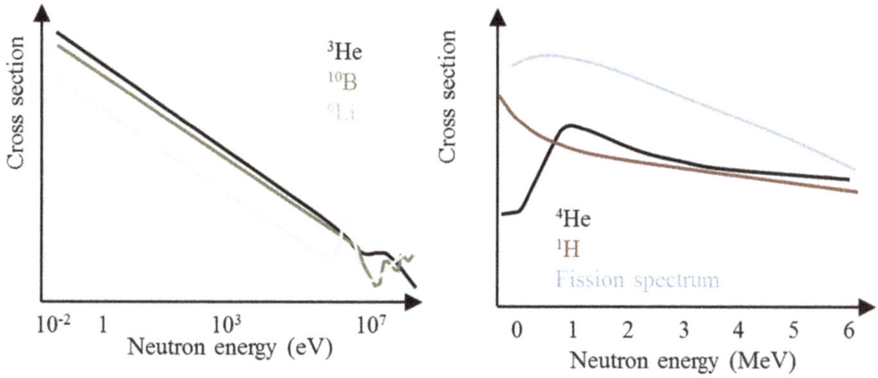

Figure 3.10. Sketch of the absorption (left) and scattering (right) cross-sections of commonly used detector materials.

Where the excited Li will eventually decay:

$$^{7}Li^{*} \rightarrow {}^{7}Li + 480\ keV$$

For fast neutrons (~MeV) a typical detection material is hydrogen as it is light, easily sourced, and has a high cross section for scattering. The scattering event of a neutron off a hydrogen nucleus will ionise the hydrogen, which can then be measured.

Scintillation neutron detectors work on a similar principle to their x-ray sensitive counterparts, except in this case a neutron sensitive material such as ^{6}Li couples with the photoactive ion Ce^{3+} to produce light that can be detected by a photomultiplier tube. A typical example is ^{6}Li, C^{3+} doped glass fibres as the scintillator. Interaction of a neutron with the ^{6}Li atom produces ionised particles that are attracted to the charged Ce^{3+} ion, transferring energy to excite the Ce^{3+} ion. As it returns to its ground state the Ce^{3+} ion releases photons in the range 390–600 nm. This is then easily detected with a PMT. The advantage of scintillation based neutron detectors is that they do not require ^{3}He, and even with a lower thermal cross section per atom, the higher concentration of ^{6}Li per unit volume (solid versus gas) results in a larger interaction probability overall. As an alternative, inorganic $LiCaAlF_6$ can be used. It has a higher interaction probability than ^{6}Li doped glass due to a higher concentration of ^{6}Li atoms, so will produce more light (i.e. is more sensitive). The disadvantage of this option is that it has a longer dead time compared to the doped glass fibres, so is not suitable for high radiation detection.

3.2.4.5 Semiconductor detectors

Semiconductors can also be used to detect neutrons if neutron sensitive dopants such as ^{6}Li are used. In a typical semiconductor detector a single neutron absorption can produce something of the order of 1.5 million electrons and holes, which is equivalent to 2.4×10^{13} C. This is easily observed, with electrical measurement (i.e. a multimeter). The disadvantage of such devices is low probabilities of absorption and complex wiring.

3.3 Diffraction: Rietveld refinement

Now that we have described the basic principles of diffraction measurements, including the production of x-rays and neutrons, and their detection, we can now bring all of this together in order to understand the contributing factors to a general diffraction pattern. One such way of achieving this is to outline the principles of Rietveld refinement, when applied to powder diffraction. This method requires some knowledge of the sample being studied as well as the resolution limiting aspects of the instrument and particle size, as will be outlined in this section.

In the first instance, powder x-ray diffraction can be used for simple phase analysis: presence of impurities, or degree of chemical reaction. There is, however, a wealth of additional information that can be extracted from the general shape of individual peaks, and this is achievable with Rietveld refinement. This was a technique that was pioneered by Rietveld before computer simulations of experimental data were commonplace and has since developed into a standard analysis tool for powder XRD.

An example of a standard Rietveld refinement pattern is given in figure 3.11. Here the experimental data (symbols) is compared to a simulated diffraction pattern (red line) and the difference between the two is shown (blue) alongside the positions of each Bragg reflection (green bars).

3.3.1 Least squares refinement

During the refinement, the program makes reference to the observed and expected pattern to see whether an improvement has been made by adjusting one of the variables. This is the least squares method, as defined by:

$$S_y = \sum_i w_i(y_i - y_{ci})^2 \tag{3.29}$$

where S_y is known as the residual, w_i is a weighting factor ($= 1/y_i$); y_i is the observed intensity at the ith step; and y_{ci} is the calculated intensity at the ith step.

This is, of course, the strength, and the weakness of the Rietveld method because the program can fall into false minima and the user needs to take care to avoid this. It can be indicated by a low value of S_y that does not accurately describe the crystal structure.

3.3.2 General equation

The Rietveld method can also be used to simulate a pattern. In order to do this we need to construct a general equation that can be adjusted in a logical way to account for changes in atomic structure, resolution, grain size, absorption, background, and any other effects that might affect the resulting diffraction pattern. The general equation for Rietveld refinement is given by:

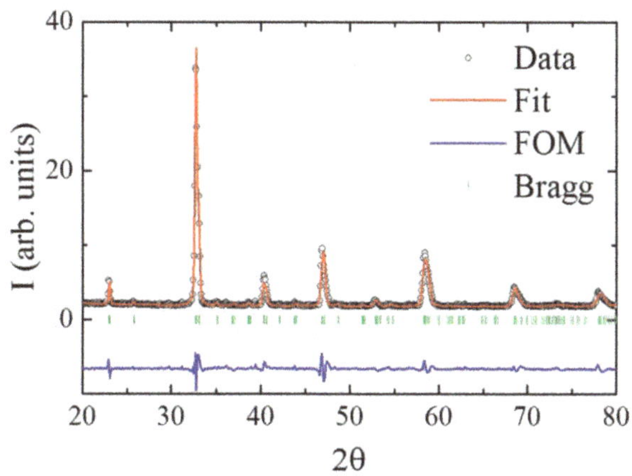

Figure 3.11. Example of indexed x-ray powder diffraction data for $La_{2/3}Ca_{1/3}MnO_3$ with corresponding Rietveld refinement. FOM indicates the figure of merit (for the refinement), and Bragg indicates the Bragg reflections for the model.

$$y_{ci} = s\sum_{K} L_K|F_K|^2 \varnothing(2\theta_i - 2\theta_K)P_K A + y_{bi} \tag{3.30}$$

where:

- s is a scale factor, which largely accounts for the intensity of incident radiation;
- K represents the Miller indices (h,k,l) that are used to index a Bragg reflection;
- L_K describes the Lorentz polarisation and multiplicity function, which varies from 1 to ½$(1 + \cos^2 2\theta)$ depending on whether the source is polarised or not;
- F_K is the structure factor, as described by equation (3.31);
- Φ is the reflection profile function: a function that describes the convolution of sample and instrument contributions to line broadening (broadening of the Bragg peaks);
- P_K is the preferred orientation function, which describes systematic distortions of reflection intensities often due to a non-polycrystalline sample;
- A is the absorption factor, which describes the amount of radiation that is simply absorbed by the sample and therefore does not contribute to the diffraction pattern;
- y_{bi} is the background intensity at the ith step due to incoherent scattering or amorphous samples and/or sample holders.

Note that the above makes the assumption that scattering is kinematic where $y_{ci} \sim |F_k|^2$ rather than dynamic ($y_{ci} \sim |F_k|$).

3.3.3 Structure factor

As a first approximation, the user needs to know the crystal structure of the powder that they are measuring so that a simple model of the expected Bragg peaks can be determined. This is the focus of the calculation of the structure factor, F_K:

$$F_K = \sum_j N_j f_j e^{-2\pi i(hx_j + hx_j + hx_j)} e^{-M_j} \tag{3.31}$$

where f_j is the function $f(\lambda, \theta)$, i.e. the atomic form factor or nuclear scattering length described earlier, $\{h,k,l\}$ are the Miller indices, $\{x_j,y_j,z_j\}$ are the positions of the jth atom in the unit cell, N_j is the site occupancy multiplier for the jth atom site (actual site occupancy divided by site multiplicity), and e^{-Mj} is the Debye–Waller term.

The Debye–Waller term describes the general line broadening of the diffraction pattern as the temperature is increased. This can be understood by considering the increased vibration of the individual atoms in the crystal lattice. As these vibrations increase it leads to some uncertainty over the position of the atoms and thus broadening of each Bragg peak. The exponent in this term is defined by:

$$M_j = \frac{8\pi^2 \overline{u_s^2} \sin^2\theta}{\lambda^2} \tag{3.32}$$

where u_s^2 is the root-mean-square (RMS) thermal displacement of the jth atom.

The classic example of a structure factor calculation is that of the simple cubic (SC), face centred cubic (FCC), and body centred cubic (BCC) structures, as shown in figure 3.12. In the questions at the end of this chapter you are walked through the corresponding calculations, where you use a simpler form of equation (3.31):

$$S_G = \sum_j f_j e^{-2\pi i(hx_j + hx_j + hx_j)} \tag{3.33}$$

Application of this to each example should show that there can be Bragg reflections which are not observed due to cancellation of the exponential terms in the summation. This is summarised in figure 3.12. Similarly, for some alloys, where different elements can occupy each of the sites in the crystal structure, superstructure lines can arise that are present when the alloy is ordered, but not when disordered, as demonstrated in the questions given in section 3.7.2.

3.3.4 Profile function

Typically, the profile function might be parametrised by the combination of Lorentzian and Gaussian profiles in order to describe sample and instrument contributions, respectively. This is achieved by using a pseudo-Voigt (pV) combination of the two functions $L(\theta_i, H_K)$ and $G(\theta_i, H_K)$, where a mixing parameter, η, determines the ratio of Lorentzian contribution, and H_K is the full width half maximum (FWHM) of the kth reflection, [$C_0 = 4 \ln 2$ and $C_1 = 4$]:

$$L = \frac{C_1^{0.5}}{\pi H_K \left(1 + \frac{C_1(2\theta_i - 2\theta_k)^2}{H_K^2}\right)} \tag{3.34}$$

$$G = \frac{C_0^{0.5}}{\pi^{0.5} H_K} e^{-C_0(2\theta_i - 2\theta_K)^2 / H_K^2} \tag{3.35}$$

$$pV = \eta L + (1 - \eta)G \tag{3.36}$$

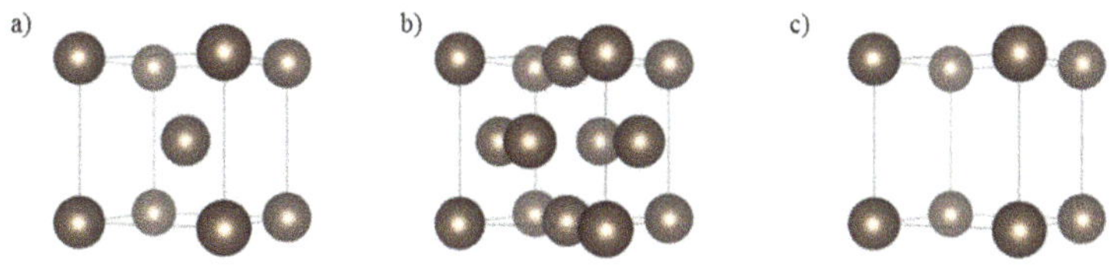

Structure	{100}	{110}	{111}	{200}	{210}	{220}	{221}	{300}	{310}	{320}	{100}	{311}
SC	✓	✓	✓	✓	✓	✓	✓	✓	✓	✓	✓	✓
FCC	×	×	✓	✓	×	✓	×	×	×	×	×	✓
BCC	×	✓	×	✓	×	✓	✓	×	×	✓	×	×

Figure 3.12. (a) Body centred cubic, BCC, (b) face centred cubic, FCC and (c) simple cubic SC unit cells. The expected Bragg peaks are listed for some example Miller indices. The observation (or not) of a given Bragg reflection, could therefore be used to determine if a cubic crystal structure is SC, FCC or BCC.

An alternative method is to use the instrument function described by Cagliotti *et al* [5, 6]:

$$H^2 = W + V\tan\theta + U\tan^2\theta \quad (3.37)$$

where the variable W is approximately the FWHM of the peak. This is often determined for a given diffractometer by measuring a calibration standard such as Si or LaB_6, where it can be assumed that line broadening due to sample size and/or strain is negligible. Further adjustments could then be made to include sample effects for any subsequent measurement.

3.3.5 Basic refinement strategy

When performing Rietveld refinement, you want to ensure that you avoid false minima. This can be achieved by varying the starting parameters and checking that you get the same result, or by comparing to complementary data such as neutron diffraction. For a given refinement it is often useful to resort to a simple refinement strategy as outlined below. The aim of this is to include parameters that start fairly simple and get more complex as you continue with the fit: start small and pin down key features, then perturb the fit to improve the figure of merit (FOM).

Strategy:

1. Scale factor, s;
2. Zero shift or specimen displacement (not both);
3. Linear background;
4. Lattice parameters;
5. More background (e.g. polynomial fit);
6. Peak width, w;
7. Atom positions;
8. Preferred orientation;
9. Isotropic temperature factor, B;
10. U, V and other profile parameters;
11. Anisotropic temperature factors.

Note also that each diffraction peak is an observable that should help to refine the parameters involved. For a suitable degree of confidence fitting N parameters, you ideally want $>N + 1$ peaks. A good discussion of the pitfalls in x-ray diffraction measurements and how to ensure data is suitable for Rietveld refinement is discussed by L B McCusker *et al* [7].

3.3.6 Figures of merit

The general aim of refinement is to reduce the residual in the least squares fitting method (see equation (3.29)), where methods of assessing the quality of fit include: the structure factor, R_F; the Bragg factor R_B; R-pattern, R_P; weighted pattern, R_{wp}; and expected weighted profile factor, R_{exp}, as defined by equations (3.38)–(3.42) [8].

$$R_F = \sum \frac{|(I_K(\text{obs}))^{0.5} - (I_K(\text{calc}))^{0.5}|}{I_K(\text{obs})^{0.5}} \tag{3.38}$$

$$R_B = \sum \frac{|I_K(\text{obs}) - I_K(\text{calc})|}{I_K(\text{obs})} \tag{3.39}$$

$$R_P = \sum \frac{|y_i - y_{c,i}|}{y_i} \tag{3.40}$$

$$R_{\text{wp}} = \sqrt{\frac{\sum w_i \, |y_i - y_{c,i}|^2}{\sum w_i y_i^2}} \tag{3.41}$$

$$R_{\text{exp}} = \sqrt{\frac{(N - P)}{\sum w_i y_i^2}} \tag{3.42}$$

where I_K(obs) and I_K(calc) are the observed and calculated integrated intensities, respectively; y_i, $y_{c,i}$ are the observed and calculated profiles, respectively; w_i is a weighting factor (to bias against a background signal); N is the total number of points used in the refinement; and P is the number of refined parameters ($N - P$ is the number of degrees of freedom). Commonly used goodness of fit indicators are χ^2 or S:

$$S = \frac{R_{\text{wp}}}{R_{\text{exp}}} \tag{3.43}$$

$$\chi^2 = \left[\frac{R_{\text{wp}}}{R_{\text{exp}}}\right]^2 \tag{3.44}$$

Where, as a general rule of thumb, $S_y < 1.3$ indicates a good fit ($\chi^2 = 1.69$), and $S_y < 1$ ($\chi^2 < 1$) probably indicates a false minimum.

3.3.7 Single crystal diffraction

Whilst polycrystalline x-ray diffraction measurements are commonplace, it requires some knowledge of the material being studied in order to fit the data sufficiently. Careful analysis of single crystal diffraction, on the other hand, can be used to determine the structure by comparing the symmetry and intensity of observed diffraction spots, as seen in figure 3.13. In this case, a collimated beam is incident on the single crystal to be studied, which can be rotated about 360° using a goniometer in order to align it with respect to the detector screen.

One limitation of single crystal diffraction is that it requires a large, high quality crystal, otherwise the diffraction spots will become distorted, or in the case of two crystals misaligned with respect to one another, result in overlapping diffraction patterns. For example, the mosaic in a single crystal [9] describes the misalignment of multiple crystallites with respect to one another. This can lead to a reduction in diffraction intensity.

3.3.8 Magnetic contrast

As discussed in section 3.1.2 the dipole–dipole interaction of the neutron with the atom results in magnetic contrast for neutron scattering. This manifests as Bragg peaks that can occur alongside the expected crystal Bragg reflections (for ferromagnets), or where the symmetry of the magnetic structure (for example, in an antiferromagnet) does not match the crystal symmetry, this can result in additional Bragg reflections. This makes neutron diffraction useful for identification of the specific magnetic structure in a crystal structure that could not otherwise be easily determined. A good example of this is given in chapter 8, section 8.5.5.

Whilst magnetic contrast is not normally observed for x-ray diffraction, there are methods which can result in magnetic contrast—referred to as resonant x-ray techniques. In such cases, the wavelength of the incident x-ray is tuned to match

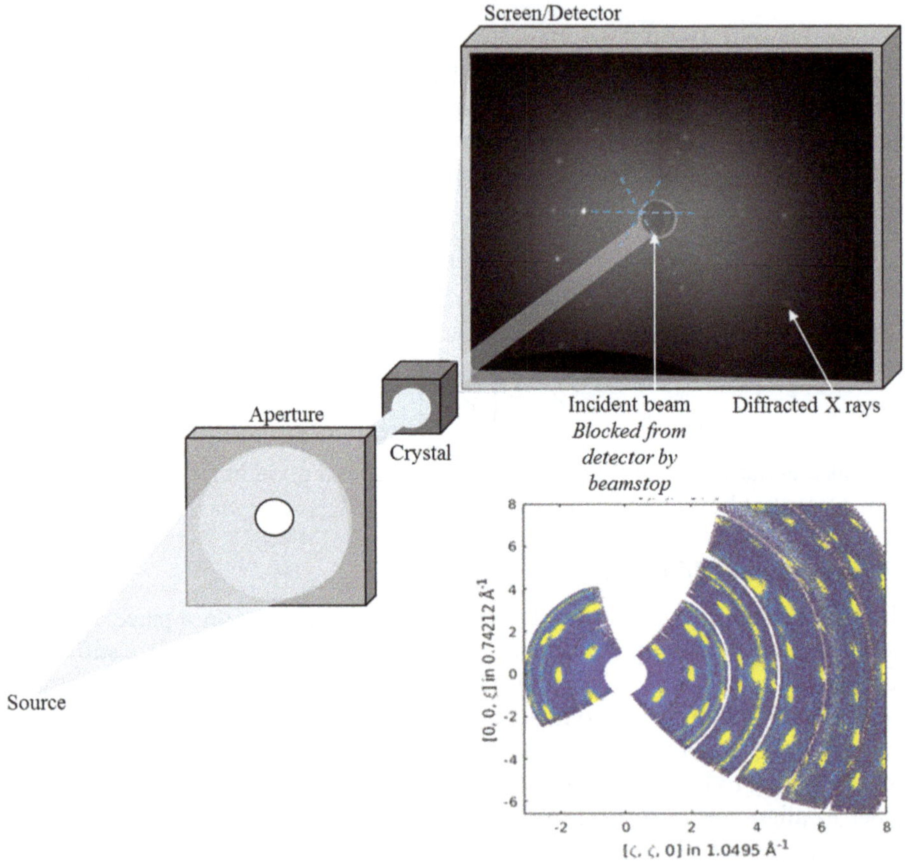

Figure 3.13. Schematic of a single crystal diffraction measurement. Example data from x-ray (black and white) and neutron (colour) diffraction off of a Fe_3O_4 cubic crystal are shown. When orientated in the (111) direction, threefold symmetry is observed, as indicated by the dashed blue lines in the upper figure.

the absorption edge of a specific element such that subtle differences in the energy levels and hence the absorption edge (see sections 1.4.3 and 5.5.3) will result in magnetic contrast. Examples of this include resonant x-ray diffraction (RXD), x-ray magnetic circular dichroism (XMCD), and x-ray magnetic reflectometry (XMR). The key advantage of resonant x-ray techniques is that it provides *element specific* magnetic information.

3.4 Grazing incidence diffraction

Grazing incidence diffraction is a method that is used to determine information on in-plane orientation of grains in a sample that may have preferential orientation. It is ideally suited to thin films or bulk materials, where this is most often the case.

The main difference between standard x-ray diffraction (XRD) and grazing incidence x-ray diffraction (GIXRD) is the relationship between the incident angle (source to sample surface) and the detector (sample surface to detector). Whilst in XRD the Bragg–Brentano geometry is used (θ:2θ), in GIXRD, the incident angle, θ, is kept fixed at an angle close to the critical angle for specular reflection (usually $< 3°$), and the detector angle is varied. To avoid confusion, the incident angle is often referred to as ω rather than θ in GIXRD geometry.

The advantage of this technique is that if there is any preferred orientation, such that a peak that might be expected from powder diffraction is not observed, it should become visible by varying ω. An example of the application of this technique and its associated geometry is given in figure 3.14.

3.5 X-ray and neutron reflectivity (XRR & NR)

3.5.1 Overview

X-ray and neutron reflectivity are surface sensitive techniques that rely on the reflection of x-rays or neutrons at an interface, i.e. it is dependent on critical reflection (such as employed with total internal reflection in fibre optic technology). At each interface, the change in refractive index can lead to reflection of an incident x-ray (or neutron). The combination of reflected rays off of multiple layers can then lead to an interference pattern similar to that observed for the classic air wedge experiment or Michelson interferometry, convoluted with the sharp drop in reflectivity expected as you move away from the critical angle. For thin films, they can be a useful technique for assessing:

- Roughness;
- Density;
- Layer thicknesses.

The refractive index of each layer will be directly proportional to the scattering length, β, of each layer. For x-rays this means the number of electrons per atom combined with density of atoms:

$$\beta = \frac{\rho N_A}{m} \langle b \rangle \tag{3.45}$$

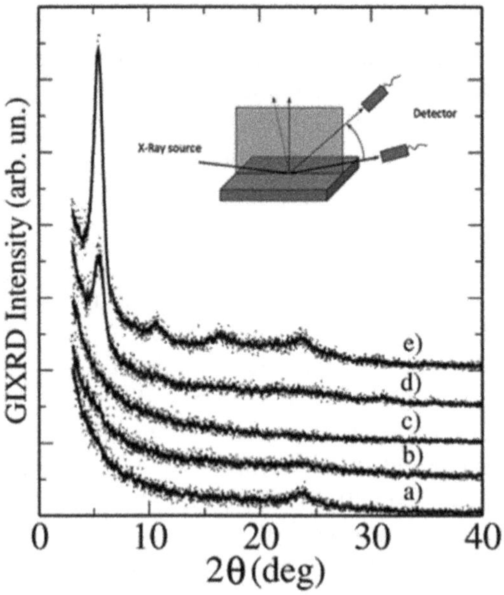

Figure 3.14. Grazing incidence XRD from P3HT films grown under different conditions, where the incident angle, ω, was 0.2°. In XRD only the 5° peak was observed. The inset shows the measurement geometry for GIXRD. Reprinted from [10], with the permission of AIP Publishing.

where:

$$\langle b \rangle = Z_j r_e \tag{3.46}$$

and Z is the number of electrons per atom whilst r_e is the Thomson scattering length = 2.818×10^{-5} m. For neutrons this will be proportional to the coherent scattering length ($\langle b \rangle = b_c$) defined earlier:

$$\beta = n \langle b \rangle \tag{3.47}$$

where n is the number of atoms per unit volume.

In a standard reflectivity measurement, the sample is aligned so that it cuts the incident flux of the x-ray or neutron source in half when the source and detector are aligned at 0°. The source and detector are then scanned in Bragg–Brentano geometry (i.e. coupled θ, 2θ), but for 2θ <15°. A typical XRR scan might, for example, span 0.4°–10°, demonstrating a critical edge (where the reflected x-ray starts to rapidly drop) at around 0.6°–1.2°, such as shown in figure 3.15.

3.5.2 Fourier derivation of the reflectivity curve

For a full description of an x-ray or neutron reflectivity curve, it would be necessary to solve the wave equation for this measurement geometry. It is more useful here, however, to walk through the Fourier derivation of this measurement as it allows the user to develop an intuitive understanding of what the different features of a reflectivity pattern represent.

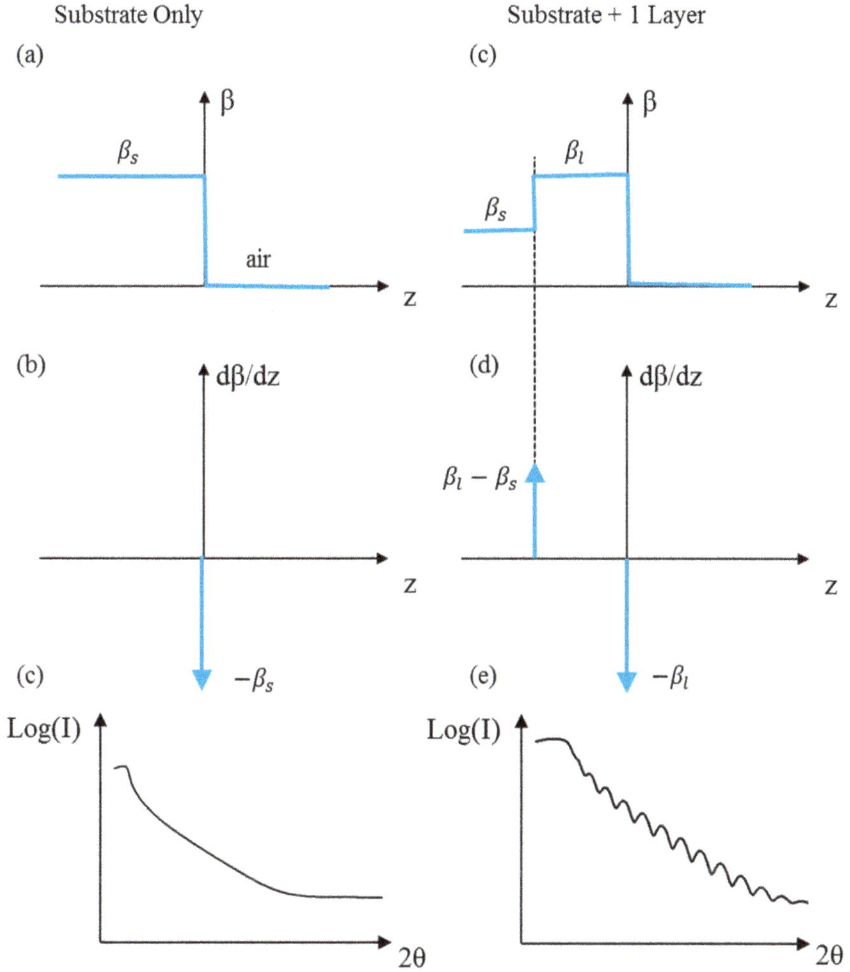

Figure 3.15. Example SLD functions (β and dβ/dz) and solutions (a)–(c) substrate only, (d)–(f) substrate + 1 layer.

There are several assumptions that have been made in order to simplify the derivation, and the reader is left to confirm the steps from one equation to the next. For a full treatment of this derivation, see [2].

Assumption 1—that the scattering length density, β, is invariant in the xy-plane:

$$\beta(\vec{r}) \rightarrow \beta(z) \text{ for } \left.\begin{matrix} |x| < L_x \\ |x| < L_y \end{matrix}\right\} \text{Instrument footprint} \quad (3.48)$$

Otherwise $\beta = 0$.

This is a reasonable assumption for thin films, where the thickness or density is not expected to vary significantly across the xy-plane.

Assumption 2—that we are working in the Fraunhofer limit (far-field), and that there is weak scattering (Born approximation). This means that the differential cross section can be written as:

$$\left(\frac{d\sigma}{d\Omega}\right)_{el} \propto F(\text{SLD function}, \beta(\vec{r})) \propto \left| \iiint_V \beta(\vec{r}) e^{i\vec{Q}.\vec{r}} d^3\vec{r} \right|^2 \tag{3.49}$$

$$\left(\frac{d\sigma}{d\Omega}\right)_{el} \propto \frac{16 \sin^2(L_x Q_x)}{Q_x^2} \frac{\sin(L_y Q_y)}{Q_y^2} \left| \int_{-\infty}^{\infty} \beta(z) e^{izQ} dz \right|^2 \tag{3.50}$$

Notice that $\sin^2(L_x Q_x / Q_x)$ can be written as $L_x^2 \operatorname{sinc}^2(L_x Q_x)$, i.e. a $\operatorname{sinc}^2(x)$ function, which will be maximum at $x = 0$. Equation (3.50) will therefore be a maximum when the specular condition is met:

$$\left.\begin{matrix} Q_x = 0 \\ Q_y = 0 \end{matrix}\right\} \frac{16 \sin^2(L_x Q_x)}{Q_x^2} \frac{\sin(L_y Q_y)}{Q_y^2} = 16 L_x^2 L_y^2 \tag{3.51}$$

where for the first order diffraction Q_x and Q_y are confined to:

$$|Q_x| < \frac{\pi}{L_x} \tag{3.52}$$

$$|Q_y| < \frac{\pi}{L_y} \tag{3.53}$$

Therefore, the differential cross-section could be described by:

$$\left(\frac{d\sigma}{d\Omega}\right)_{el} \propto 16 L_x^2 L_y^2 \left| \int_{-\infty}^{\infty} \beta(z) e^{-izQ_z} dz \right|^2 \tag{3.54}$$

Note that the convention, when it comes to reflectivity, is to define the z-axis such that the surface of the sample is at $z = 0$, and each subsequent interface is at $z < 0$. As a result, $Q_z = -Q$.

In order to describe the reflectivity curve, $R(Q)$, we refer to the differential cross section defined earlier:

$$\left(\frac{d\sigma}{d\Omega}\right) = \frac{\text{Number deflected per unit solid angle}}{\text{Number incident per unit area of beam}} \tag{3.55}$$

$$R = \frac{\text{Rate of specular scattering}}{\text{Rate of incidence}} \tag{3.56}$$

Note that the area of the sample illuminated—the sample footprint—is dependent on the angle of incidence, θ ($= 4L_xL_y \sin\theta$), therefore:

$$R(Q) = \frac{1}{4L_xL_y \sin\theta} \iint_{\Delta\Omega} \left(\frac{d\sigma}{d\Omega}\right)_{el} d\Omega \approx \frac{\Delta\Omega}{4}\left(\frac{d\sigma}{d\Omega}\right)_{el}\Bigg|_{Qx=0,Qy=0,Qz=-Q} \tag{3.57}$$

where the solid angle, $\Delta\Omega$, $= 16\pi^2 \sin\theta/L_xL_yQ^2$ (see p 96 of [2] for full derivation), therefore:

$$R(Q) \propto \frac{16\pi^2}{Q^2}\left|\int_{-\infty}^{\infty} \beta(z)e^{-izQ_z}\, dz\right|^2 \tag{3.58}$$

Or if we integrate by parts where β goes to zero as z goes to infinity:

$$R(Q) \propto \frac{16\pi^2}{Q^4}\left|\int_{-\infty}^{\infty} d\beta/dz e^{-izQ_z}\, dz\right|^2 \tag{3.59}$$

As can be seen in the following section, by considering the reflectivity measurement in this way, it is possible to intuitively determine the possible contributions to a given reflectivity curve.

3.5.3 Classic examples of reflectivity profiles

3.5.3.1 Substrate only

For a bare substrate, the main contribution to the reflectivity profile will be from the substrate surface itself (as it can be assumed that it is thick enough for most x-rays to either be absorbed or reflected from the surface). This means that $R(Q)$ will be heavily influenced by the density of the substrate (number density of scattering particles) and the surface roughness (probability of a coherently reflected ray).

We could choose to either describe a scattering length density, β, of the substrate as a function of z (the depth):

$$\beta(z) = \begin{cases} \beta_s & z < 0 \\ 0 & z > 0 \end{cases} \tag{3.60}$$

or in terms of the change at each interface, $d\beta/dz$:

$$\frac{d\beta}{dz} = -\beta_s\delta(z) \tag{3.61}$$

where β_s is the scattering length density of the substrate, and $\delta(z)$ is a delta function located at $z = 0$. This has been summarised in figures 3.15(a–c).

By inserting equation (3.61) into equation (3.59) it can easily be shown that:

$$R(Q) \propto \frac{16\pi^2}{Q^4}\left|\int_{-\infty}^{\infty} \beta_s\delta(z)e^{-izQ}\, dz\right|^2 \approx \frac{16\pi^2}{Q^4}\beta_s^2 \tag{3.62}$$

This Q^4 dependence results in a sharply decreasing intensity as Q is increased, and accounts for why most XRR data will be plotted as $\log(I)$ versus 2θ or Q.

3.5.3.2 Substrate + 1 layer

If we add a thin layer on top of the substrate then it will have a significant impact on the reflectivity profile, as we start to see interference between reflections off of two interfaces. As before, the scattering length density can be described as a function of z:

$$\beta(z) = \begin{cases} \beta_s & z < -L_1 \\ \beta_l & L_1 < z < 0 \\ 0 & z > 0 \end{cases} \tag{3.63}$$

where β_1 is the scattering length density of this new layer, and L_1 is its thickness. Or it can be described in terms of dβ/dz (which is easier to integrate):

$$\frac{\mathrm{d}\beta}{\mathrm{d}z} = (\beta_l - \beta_s)\delta(z + L_1) - \beta_1\delta(z) \tag{3.64}$$

In this case, equation (3.64), which has a delta function at $z = -L_1$ and 0, as shown in figure 3.15. The result of inserting this into equation (3.59) and integrating is:

$$R(q) \approx \frac{16\pi^2}{Q^4}\left(\beta_l^2 + (\beta_l - \beta_s)^2 - 2\beta_l(\beta_l - \beta_s)\cos(L_1 Q)\right) \tag{3.65}$$

Now, on top of the sharp decline in intensity with Q, we will observe an oscillating component, which has an amplitude proportional to the difference in scattering lengths (β_s and β_l) of the two layers.

3.5.3.3 The general case

You might notice that the result of the single layer case is the convolution of a standard interference pattern (dark/light fringes) with the reflectivity profile for the substrate. This is useful to remember when trying to understand more complex reflectivity profiles. For example, the development of the XRR pattern for additional layers can be appreciated qualitatively if you recall known Fourier transforms coupled with use of the convolution theorem. For two layers on top of a known substrate, you might then expect a more complex interference pattern to emerge (much like the combination of two sine waves compared to three sine waves), as shown in figure 3.16. Here, the total thickness could be identified from the highest frequency modulation, and the smallest feature from the overall lengthscale of the regular pattern.

For multilayer samples, as with the change from two slits to 3-, 4- and infinite gratings (figure 1.8), you might expect the contrast from the interference pattern to become more pronounced. It is at this stage that the reader is strongly encouraged to simulate such patterns, in order to identify these changes for themselves. (See Question 3.7.5 and the instructions in appendix H.)

Of course the derivation of the XRR reflectivity profile using Fourier analysis is only an estimate—there were several assumptions made in order to keep it relatively simple, and as such it breaks down for low values of Q (i.e. does not well describe the critical edge), and does not treat the impact of surface and interface roughness. For a full treatment of XRR you would need to solve the wave equation; a good treatment of which can be found in [11].

Some key things to note are:

1. The critical edge will be largely determined by the density of the topmost layer;
2. The speed with which oscillations decay with respect to Q or 2θ will depend on the surface roughness;
3. For multiple layers, the emergence of correlated or uncorrelated roughness will have a significant impact on the decay of oscillations;
4. Just as with a finite slit, for samples that have large curvature, or where the thickness is not uniform in the XY plane, the features in XRR will be smeared out;
5. Just as the interference pattern for a two slit experiment becomes sharper as more slits are added, a thin film sample that is comprised of repeats of multiple layer stacks will result in more defined peaks in the XRR data.

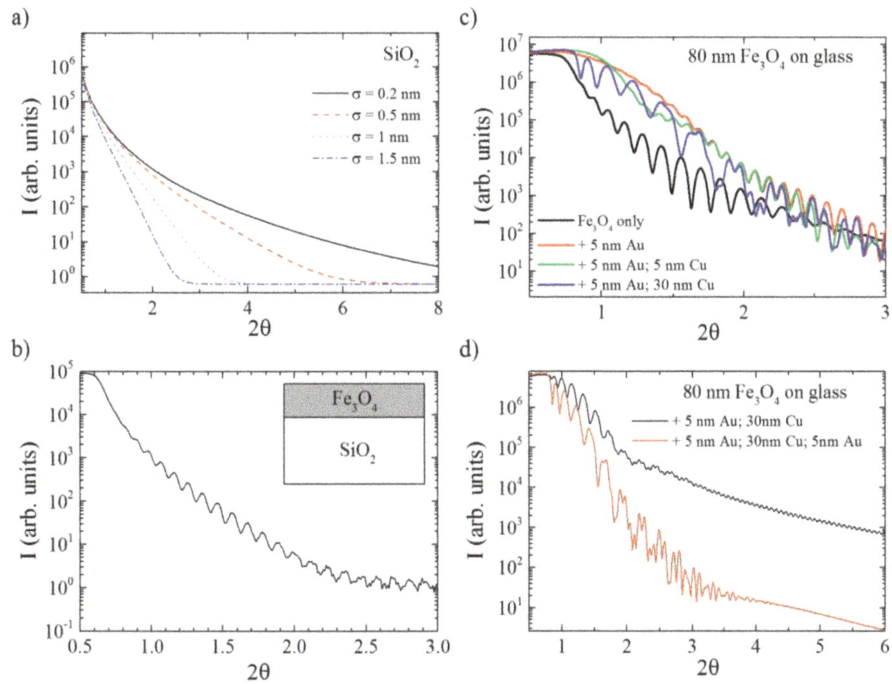

Figure 3.16. Increasing complexity as layers are added to an XRR sample. (a) Substrate only measurement, where changing roughness (σ) has a significant impact on the decrease of I with 2θ. (b) 80 nm Fe_3O_4 on glass where periodic fringes are now observable. (c) 80 nm Fe_3O_4 on glass with various capping layers (Au/Cu). (d) Simulation of two multilayers (80 nm Fe_3O_4:5 nm Au:30 nm Cu) with and without a 5 nm Au capping layer. The general features are similar, but the decay in intensity and structure start to diverge for $2\theta > 2$.

3.5.4 X-rays versus neutrons

The key differences between x-ray reflectometry and neutron reflectometry are the scattering length densities:

- For x-rays β is proportional to Z;
- For neutrons β varies non-monotonically with Z.

An example of this is given in figure 3.17. This can result in stark differences in contrast—hydrogen and deuterium being the best known example. Secondly, as the neutron has spin, it can also interact magnetically with the sample, thus leading to the development of polarised neutron reflectometry and neutron reflectometry with polarisation analysis: used to study magnetism as a function of depth.

3.5.5 Polarised neutron reflectivity (PNR)

There are several beamlines that will carry out polarised neutron reflectometry (PNR), such as POLREF, CRISP [12, 13], and OFFSPEC at the ISIS facility in Oxfordshire, and D17 [14], and ADAM [15] at ILL in Grenoble.

For this measurement technique the neutrons are polarised prior to reaching the sample by use of a magnetic supermirror, [11] then kept stable by a guide field along the beamline. On reflection from the sample surface, if a magnetic interaction has occurred then the neutron may flip spin state.

The combined refractive index for neutrons polarised parallel (+) and antiparallel (−) to the applied field can be described by:

$$n_{\pm} = n_N \pm n_M = 1 - \left(\frac{N\lambda^2}{2\pi}\right)(b_c \pm C\bar{\mu}) \tag{3.66}$$

where N is the density of scattering centres (atoms/cm^{-3}), $\bar{\mu}$ is the average magnetic moment per atom, λ is the neutron wavelength, and $C = 0.2645 \times 10^{-12}$ cm μ_B^{-1}. It can be seen that the result of magnetic contrast is to shift the reflectivity profile with

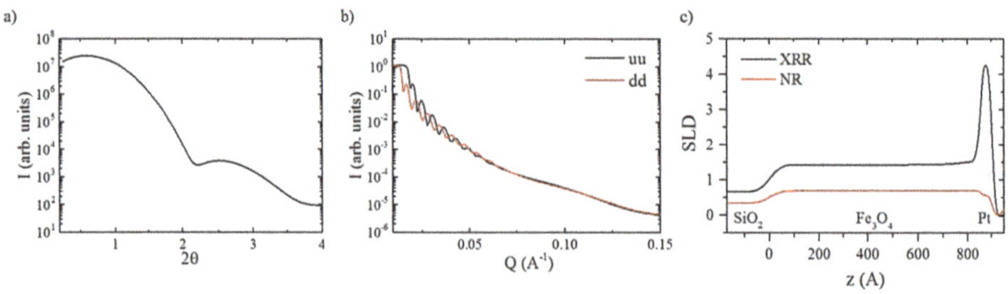

Figure 3.17. Comparison of (a) x-ray and (b) polarised neutron reflectivity measurements of 80 nm Fe_3O_4:5 nm Pt on glass obtained on the Polref beamline, Oxfordshire. The PNR data shows the characteristic splitting of the spin up (uu) and spin down (dd) profiles due to the magnetisation of the thin film (as a function of depth). (c) The corresponding non-magnetic SLD for (a) and (b). The fringes due to the Fe_3O_4 layer are not obvious in (a) due to interface roughness. Contrast matching between the Fe_3O_4 and Pt layers for the NR data results in improved contrast for the Fe_3O_4 layers.

respect to Q (or 2θ). An example of this is seen in figure 3.17, where PNR of a Fe_3O_4: Pt thin film on glass is shown.

The advantage of PNR is that it can give information about magnetisation of a thin film as a function of depth, which is particularly useful for magnetic thin films being developed for spintronics or giant magnetic resistance (GMR) applications [16, 17].

3.6 Examples of diffraction techniques in the literature

In the following sections, several examples of research articles where diffraction techniques were used to characterise or identify a given material will be highlighted. For each article, starter questions are provided, which are aimed at focussing the reader's attention on the key aspects of the article (with respect to this chapter). If you are using this book as part of a course, you could consider framing seminars around discussion of one or more of these papers.

3.6.1 In 'Strain in nanoscale germanium hut clusters on Si(001) studied by x-ray diffraction'

Steinfort *et al* examine the formation of 'islands' of Ge grown on a silicon substrate by molecular beam epitaxy (MBE) [20]. This is a technique that can produce very clean films and interfaces due to ultrahigh vacuum and highly controlled film growth.

Reading through the article, try to answer the following:

1. What is a 'hut cluster'?
2. What experimental methods have been used in this paper?
3. What wavelength of x-rays was used?
4. Define kinematic scattering theory.
5. What do the authors mean by 'completely strained'?
6. Why does inhomogenous strain lead to asymmetric peaks?

The aim of the work was to investigate the impact of strain on the hut clusters observed in these samples using x-ray diffraction. Simple simulations of the expected pattern as the lattice constant $a_y(z)$ was varied enabled the authors to identify the origin of the asymmetric peak in their data. They concluded that the hut clusters exhibited a gradual (continuous) change from fully strained at the Si interface to unstrained Ge atoms at the top of the hut cluster.

3.6.2 In 'Neutron-diffraction study of the Jahn–Teller transition in stoichiometric $LaMnO_3$'

Rodríguez-Carvajal *et al* studied the changing crystal structure of $LaMnO_3$—a perovskite—as a function of temperature [21].

Reading through the article, try to answer the following:

1. What experimental methods were used?
2. What is Jahn–Teller distortion?

3. What is the importance of the Mn^{3+}:Mn^{4+} ratio with regards to the Jahn–Teller distortion?*
4. What information does the DTA/TGA provide?*
5. What are the authors' arguments for a non-cubic primitive cell?
6. What was the wavelength of neutrons used in this study? How does this compare to an x-ray source (Cu, Kα)?
7. What crystal structures do the Pbnm and R3c space groups refer to? How do these differ?

**For questions 3 and 4 you might find it useful to refer to chapters* 1, 6 *and* 8.

Manganites, (La,R)MnO_3, where R is a rare earth element, are a rich family of materials that exhibit varying degrees of competition between magnetic, structural and electric phase transitions. The aim of this work was to obtain high resolution experimental data to test the role of the Jahn–Teller effect with regards to the interplay of structure, magnetism and electric transport in the parent compound $LaMnO_3$. The authors observed evidence of a significant increase in spatial fluctuations of the cubic lattice above a structural transition at 750 K. They attribute these observations to dynamic fluctuations of an underlying orthorhombic distortion.

3.6.3 In 'Neutron-diffraction measurements of magnetic order and a structural transition in the parent $BaFe_2As_2$ compound of FeAs-based high temperature superconductors'

Huang *et al* investigated the magneto-structural phase transition in $BaFe_2As_2$ [22]. Reading through the article, try to answer the following:

1. When were the oxypnictide superconductors discovered (and where)?
2. Are the oxypnictides Type I or Type II superconductors?
3. What crystal structure does this $BaFeAs_2$ compound form in?
4. What is the significance of the (220) rhombohedral reflection splitting to (400) and (040) orthorhombic reflections?
5. What is the significance of the {2a, 4d, 4e} and {4a, 8f, 8i} sites in table 1?
6. What does figure 3 tell us about the magnetic ordering in this compound?

The aim of this work was obtain information on the nature of magnetic state in the $BaFe_2As_2$ parent compound, which exhibits high temperature superconductivity when Ba is substituted by K. The authors observed simultaneous structural (tetragonal to orthorhombic) and magnetic (paramagnetic to antiferromagnetic) phase transitions at ~143 K. They concluded that there is evidence for strong coupling between the magnetic and structural order parameters.

3.6.4 In 'Unconventional magnetic order on the hyperhoneycomb Kitaev lattice in β-Li_2IrO_3: full solution via magnetic resonant x-ray diffraction'

Biffin *et al* used magnetic resonant x-ray diffraction to study the recently synthesised 'hyperhoneycomb' β-Li_2IrO_3 [23].

Reading through the article, try to answer the following:

1. What experimental methods were used?
2. What effect does neutron absorption have on the obtained diffraction pattern?
3. What is MRXD?
4. How is the absolute value of the ordered moment extracted from the neutron data?
5. What are the key features of a Rietveld fit?
6. How are selection rules determined for the magnetic Bragg peaks?

The aim of this work was to determine the magnetic structure of this compound, which exhibits long range order at low temperatures. The authors showed using magnetic resonance x-ray diffraction that this compound exhibits incommensurate magnetic order stabilised by Kitaev interactions—anisotropic bond-dependent interactions ono a 3D honeycomb lattice.

3.7 Questions

3.7.1 Structure factor and reflection rules

The periodicity of the crystal lattice can result in certain *h,k,l* conditions for observable Bragg reflections. The aim of this set of questions is to derive these conditions for some cubic lattices, and then to determine the structure factor for NaCl.

The structure factor allows us to calculate the intensity of diffracted x-rays for a given crystal structure and consists of two main terms—the atomic form factor, f_j (which is dependent on the elements involved as the scattering power of an atom will depend on its charge density), and a translation term, $e^{-i\boldsymbol{G}.\boldsymbol{r}j}$, where $\boldsymbol{r}_j$ is the translation vector for the atoms of the unit cell.

$$S_G = \sum f_j e^{-i.\boldsymbol{G}.\boldsymbol{r}_j} \tag{3.7.1}$$

$$f_j = 4\pi \int \mathrm{d}r n_j(r) r^2 \tag{3.7.2}$$

For certain crystal structures the individual terms of the structure factor may cancel so that only a subset of the possible reflections are observed (otherwise referred to as extinction conditions). This is determined by reducing the crystal lattice to its smallest repeatable unit, determining the translation vector between atoms in this unit and substituting this into equation (3.7.1).

If:

$$\boldsymbol{r} = n_1\boldsymbol{a}_1 + n_2\boldsymbol{a}_2 + n_3\boldsymbol{a}_3 \tag{3.7.3}$$

and:

$$\boldsymbol{G} = h\boldsymbol{b}_1 + k\boldsymbol{b}_2 + l\boldsymbol{b}_3 \tag{3.7.4}$$

Then:

$$\boldsymbol{G} \cdot \boldsymbol{r}_j = 2\pi(hx_j + ky_j + lz_j) \tag{3.7.5}$$

Some useful relationships to remember are:

$$e^0 = 1 \tag{3.7.6}$$

$$e^{-i(2n+1)\pi} = -1 \tag{3.7.7}$$

$$e^{-i(2n)\pi} = 1 \tag{3.7.8}$$

1. *The structure factor of a simple cubic lattice of sodium atoms*
   a) Determine the basis of the SC lattice.
   Hint: A crystal structure is comprised of an infinite repetition of atoms onto a lattice. The basis is the smallest unit (or group) of repetitive atoms.
   b) Determine the translation vectors for the basis with respect to the (0,0,0) coordinates.
   c) By substituting the result of (b) into equation (3.7.1) determine the observed reflections in terms of f_{Na}.
2. *The structure factor of a face centred cubic (FCC) lattice of sodium atoms*
   a) Determine the basis of the FCC lattice.
   b) Determine the translation vectors for the basis with respect to the (0,0,0) coordinates.
   c) By substituting the result of (b) into equation (3.7.1) determine the observed reflections.
3. *The structure factor of a body centred cubic (BCC) lattice of sodium atoms*
   a) Determine the basis of the BCC lattice.
   b) Determine the translation vectors for the basis with respect to the (0,0,0) coordinates.
   c) By substituting the result of (b) into equation (3.7.1) determine the observed reflections.
4. *The structure factor for NaCl*
   Figure Q3.1 shows the NaCl lattice, which is of FCC type, with the Na on the (0,0,0) site and the Cl on the (1/2,0,0) site. How would the calculation of the structure factor differ in this case?
   a) Determine the basis of the FCC lattice.
   b) Determine the translation vectors for the basis with respect to the (0,0,0) coordinates.
   c) By substituting the result of (b) into equation (3.7.1) determine the observed reflections.
   d) What would the average structure factor, $\langle S(hkl) \rangle$, be if the Na and Cl atoms were likely to occupy the (0,0,0) and (1/2,0,0) sites?

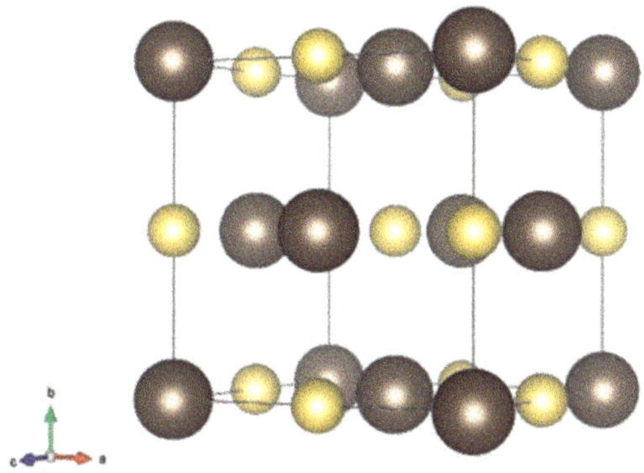

Figure Q3.1. NaCl Lattice. In this representation a Na atom (brown) sits on the (0,0,0) site and a Cl atom (yellow) sits on the (1/2,0,0) site.

5. *Reflection rules*

   How could we use the *h*,*k*,*l* conditions determined in questions 3.7.1–3, to determine which cubic lattice the XRD pattern of figure Q3.2 represents?

   For further related questions, please refer to C Kittel's *Introduction to Solid State Physics* chapters 1 and 2 [18].

3.7.2 Extracting key information from x-ray diffraction data

In the previous set of questions we determined the extinction conditions for some ordered crystal lattices. We will now consider the impact of disorder on the diffraction pattern and the reverse of this process: identifying extinction conditions from their observed reflections. Unless otherwise stated, assume lattice constants are given in Å and $\lambda = 1.54$ Å. You will find table 1.8 (in chapter 1) useful.

1. *Superstructure lines*

   A Cu_3Au alloy exists in its ordered state below 400 °C.

   The basis for this is as follows [Au: (0,0,0); Cu: (0.5,0.5,0),(0.5,0,0.5), (0,0.5,0.5)].

   a) Determine the structure factor for this alloy in terms of f_{Au} and f_{Cu}.
   b) What are the extinction conditions?
   c) If the alloy is heated above 400 °C the site occupancy changes from the above ordered state to a random distribution. Determine the structure factor for this case.
   d) Determine the extinction conditions.
   e) Hence prove that the following are superstructure lines: {100}, {110}, {211}, {221}, {201}.

2. *Heusler alloys*

   a) Given the lattice parameters in figure Q3.3, determine the Miller indices for the marked Bragg reflections.

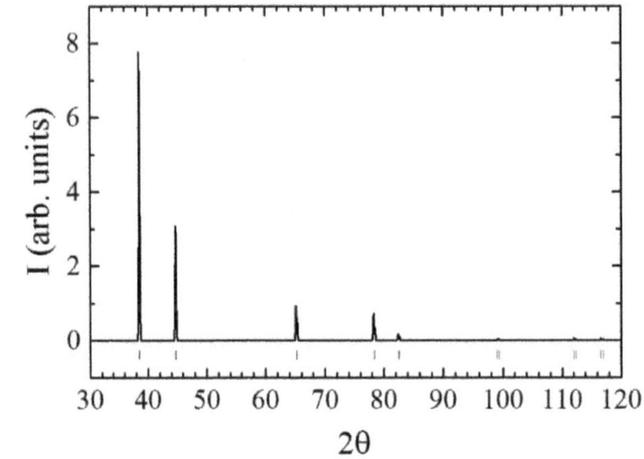

Figure Q3.2. Sample XRD pattern of a cubic structure; λ = 1.54 Å. lattice parameter = 4.0495 Å.

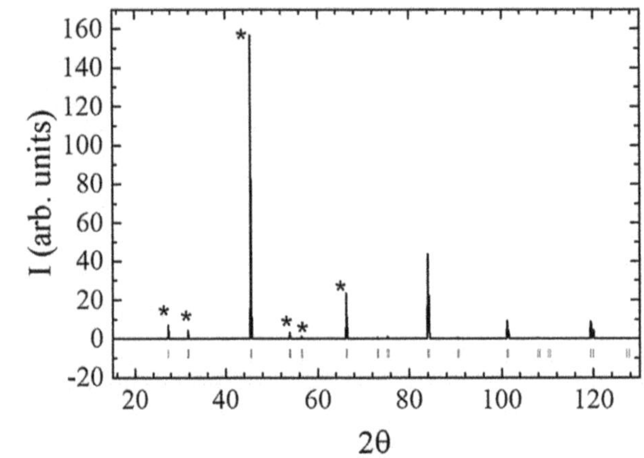

Figure Q3.3. Co_2MnSi XRD pattern in L21 order, where $a = b = c$ = 5.714 23 Å.

b) If you were told that the basis were [Co: (0.25,0.25,0.25), Mn: (0,0,0), Si: (0.5,0.5,0.5)], why might the peak at 2θ = 45° be so pronounced compared to other reflections?

c) What are the extinction conditions for this compound?

3. *Orthorhombic compounds*
   a) Given the lattice parameters in figure Q3.4, determine the Miller indices for the marked reflections.
   b) For the orthorhombic lattice (200) is no longer equivalent to (020). What is the main reason for this?
4. *Rhombohedral/hexagonal compounds*
   a) Given the lattice parameters in figure Q3.5, determine the Miller indices for the marked reflections.
   b) Can you determine any extinction conditions for this pattern?

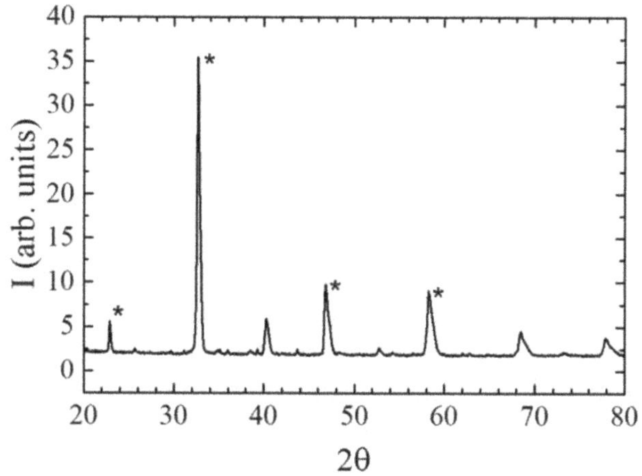

Figure Q3.4. $La_{0.67}Ca_{0.33}MnO_3$ XRD pattern, where a = 5.481 15 Å, b = 5.423 68 Å, c = 7.751 85 Å.

c) The basis for this pattern is [La: (0,0,0.25), Ca: (0,0,0.25), Mn: (0,0,0), O: (0.5,0,0.25)], with the following symmetry operations (x,y,z), $(-x+y,-x,z)$,$(-y,x-y,z)$,$(-y,-x,z+1/2)$,$(-x+y,y,z+1/2)$,$(x,x-y,z+1/2)$. Determine the structure factor and prove/disprove your answer to part (b).

3.7.3 Space groups

1. *Pbnm*

The symmetry operators for this spacegroup are:

$$x,\ y,\ z$$

$$x+1/2,\ -y+1/2,\ -z$$

$$-x+1/2,\ y+1/2,\ -z+1/2$$

$$-x,\ -y,\ z+1/2$$

$$-x,\ -y,\ -z$$

$$-x+1/2,\ y+1/2,\ z$$

$$x+1/2,\ -y+1/2,\ z+1/2$$

$$x,\ y,\ -z+1/2$$

a) Determine the possible atoms in a unit cell attributed to the Wyckoff f site $(x,0,0)$. What is the multiplicity?

b) Determine the possible atoms in a unit cell attributed to the Wyckoff c site (1/4,1/4,1/4). What is the multiplicity?

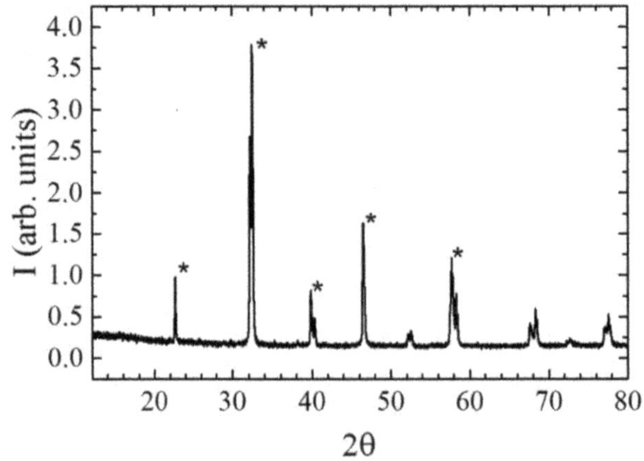

Figure Q3.5. $La_{0.67}Sr_{0.33}MnO_3$ XRD pattern, where $a = b = 5.545\ 44$ Å and $c = 13.428\ 31$ Å.

2. *Pnma*

The symmetry operators for this spacegroup are:

$$x,\ y,\ z$$
$$x+1/2,\ -y+1/2,\ -z+1/2$$
$$-x,\ y+1/2,\ -z$$
$$-x+1/2,\ -y,\ z+1/2$$
$$-x, -y,\ -z$$
$$-x+1/2,\ y+1/2,\ z+1/2$$
$$x,\ -y+1/2,\ z$$
$$x+1/2,\ y,\ -z+1/2$$

a) Determine the possible atoms in a unit cell attributed to the Wyckoff a site (0,0,0). What is the multiplicity?
b) Determine the possible atoms in a unit cell attributed to the Wyckoff c site (x,1/4,z). What is the multiplicity?

3.7.4 Identifying key information in XRR data

1) *Fourier derivation of reflectivity profile for a thin film*

Starting from the equations given in section 3.5.3, show that:

$$R(q) \approx \frac{16\pi^2}{Q^4}\left(\beta_l^2 + (\beta_l - \beta_s)^2 - 2\beta_l(\beta_l - \beta_s)\cos(tQ)\right) \tag{3.7.9}$$

For a single layer, where t is the thickness of the film and $\beta_{l/s}$ is the SLD function for the layer/substrate.

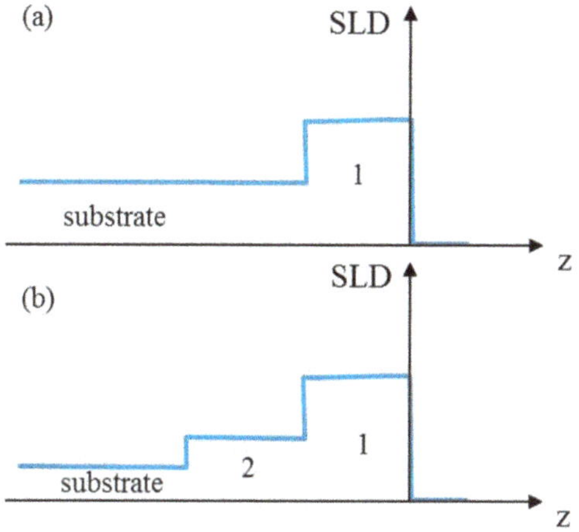

Figure Q3.6. SLD functions for the XRR data given in (a) Q. 3.7.4.2(b)(i) and (b) Q. 3.7.4.2(b)(ii).

2) *Extracting key information from XRR data*

   Using the above relation for $R(Q)$ of a single layer:

   a) Show that the layer thickness, t, can be approximated by $t = 2\pi/\Delta Q$, where ΔQ is the fringe spacing.
   b) The SLD functions for two thin films measured by XRR ($\lambda = 1.54$ Å) are given in figure Q3.6.
      i. For the data in figure Q3.7 estimate the film thickness.
      ii. For the data in figure Q3.8 estimate the thickness of layer 1 and layer 2.
   c) Discuss how the amplitude of the fringes seen in figures Q3.7 and Q3.8 is dependent on the density of the layers.

3) *Extracting key information from XRR data*

   For the examples given in figure Q3.9, a single layer of material was deposited onto a glass substrate using pulsed laser deposition. The resulting XRR plots are shown.

   a) Estimate the thickness of each film.
   b) What would lead to the additional structure seen particularly for Nb and Ta?

3.7.5 Computer aided problems

1) *Simulating x-ray power diffraction (XRD) data*

   Referring to the Fullprof Labscript (appendix F), download and use the Fullprof [8] software to simulate the XRD pattern for the heusler Co_2MnSi, in different levels of disorder.

   In the fully ordered state (L21) the basis can be described by Co at (¼,¼,¼); Mn at (0,0,0), Si at (½,½,½).

   - L21—Co, Mn, and Si each have distinct sites.
   - B2—Mn and Si sites are equally occupied (by Mn and Si).

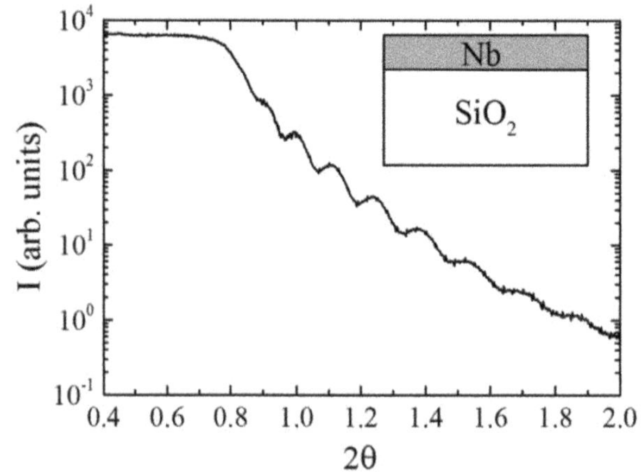

Figure Q3.7. XRR of the single layer film (see Q. 2.7.4.2).

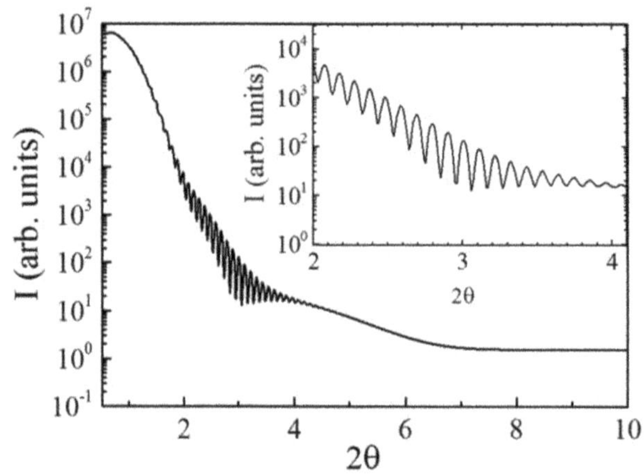

Figure Q3.8. XRR of the bi-layer (see Q 2.7.4.2). The inset shows a magnified section of the main plot.

- DO3—Co and Mn sites are equally occupied (by 2Co and Mn).
- A2—Equal probability of 2Co, Mn, Si at each site.
  a) Identify the peaks that appear or disappear as you change the structure from L21 order to A2 order.
  b) Is there any change in the relative intensity of specific Bragg reflections?
  c) Using what you know of the structure factor, explain the origin in difference between each set of XRD patterns generated.

As a sanity check, you could also determine the structure factor for each of the cases above (as per Q. 3.7.1) to convince yourself of the expected extinction conditions.

2) *Simulating x-ray reflectivity (XRR) patterns*

Referring to the GenX Labscript (appendix G), download and use the GenX [19] software to simulate the following XRR patterns:

- Glass substrate (SiO_2)
- 30 nm Nb on glass
- 30 nm Ta on glass
- 100 nm Ta on glass
- 100 nm Fe_3O_4 on glass
- 100 nm Fe_3O_4 followed by 5 nm Pt on glass

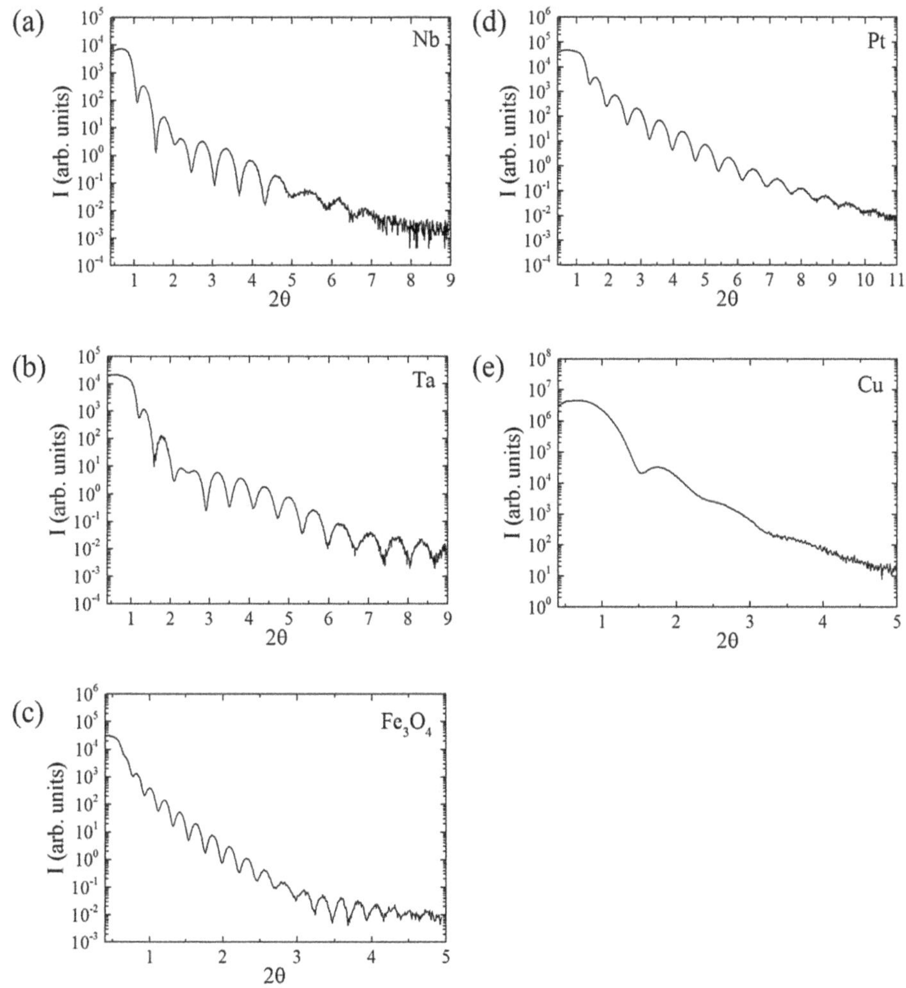

Figure Q3.9. XRR of single layer (a) Nb, (b) Ta, (c) Fe_3O_4, (d) Pt, and (e) Cu films.

Start with a roughness of 10 Å, and use the following values for number density:

Material	Density (g cm^{-3})	Atomic weight	Number density (atoms/A)
SiO_2	2.3	60.084	0.023 05
Cu	8.96	63.5	0.084 97
Nb	8.57	92.906	0.055 55
Ta	16.7	180.95	0.055 58
Fe_3O_4	5.17	231.537	0.013 45
Pt	21.45	195.08	0.066 22

Now, alter the roughness and thickness of each layer.

a) How does the roughness affect the XRR pattern(s)?
b) How does the thickness affect the XRR pattern(s)?
c) How does the density or type of element affect the XRR pattern(s)?
d) Now add extra layers, or build a multilayer structure—how does this affect the XRR pattern?

3.7.6 Calibrating detectors—some basic geometry

1) *1D detector*

Given the instrument layout shown in figure Q3.10, where the specular and off-specular reflection of neutrons off of a thin film surface is monitored by a 2D detector (OSMOND) that has a pixel width of 0.5 mm, and distance from sample to detector, D_1 = 3 m:

a) Determine the angular resolution of the detector when $\theta = 0.5$.
Hint: Consider the difference between the height of a neutron reflected by θ and $\theta + \mathrm{d}\theta$.
b) What would be the angular resolution at $\theta = 0.25$?
c) If you observe a feature that repeats every four pixels (as seen in figure Q3.10), what lengthscale will this correspond to?

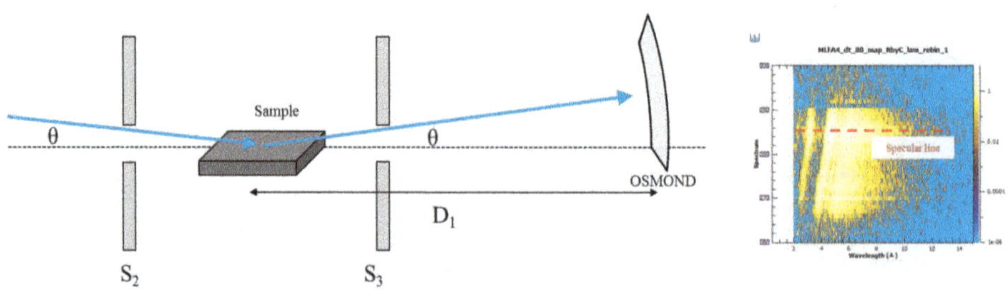

Figure Q3.10. Left: schematic of the 1D detector geometry on PolRef. Right: example dataset where the *y*-axis corresponds to pixel number and the *x*-axis corresponds to wavelength.

References

[1] Dianoux A and Lander G 2002 *Neutron Data Booklet* (Institute Laue-Langevin)
[2] Sivia D S 2011 *Elementary Scattering Theory: For X-ray and Neutron Users* (Oxford: Oxford University Press)
[3] Berreman D W 1955 *Rev. Sci. Instrum.* **26** 1048
[4] Costrell L 1949 *Accurate Determination of the Deadtime and Recovery Characteristics of Geiger-Müller Counters*, U.S. Department of Commerce National Bureau of standards *Research paper RP1965* Volume 42
[5] Cagliotti G *et al* 1958 *Nucl. Instrum.* **3** 223–8
[6] Cagliotti G *et al* 1960 *Nucl. Instrum.* **9** 195–8
[7] McCusker L B *et al* 1999 *J. Appl. Cryst.* **32** 36–50
[8] Fullpr of user manual https://psi.ch/sinq/dmc/ManualsEN/fullprof.pdf
[9] Darwin C G 1922 *Philos. Mag.* **43** 809–29
[10] Ali M *et al* 2011 *J. Appl. Phys.* **110** 054515
[11] Parratt L 1954 *Phys. Rev.* **95** 359
[12] Felici R *et al* 1988 *Appl. Phys.* A **45** 169–74
[13] Penfold J 1991 *Physica* B **173** 1–10
[14] Cubitt R and Fragneto G 2002 *Appl. Phys.* A **74** S329–31
[15] Schreyera A *et al* 1998 *Physica* B **248** 349–54
[16] Uribe-Laverde M A *et al* 2013 *Phys. Rev.* B **87** 115105
[17] Moorsom T *et al* 2014 *Phys. Rev.* B **90** 125311
[18] Kittel C 2004 *Introduction to Solid State Physics* 8th edn (New York: Wiley)
[19] Björck M and Andersson G 2007 *J. Appl. Cryst.* **40** 1174–8
[20] Steinfort A J *et al* 1996 *Phys. Rev. Lett.* **77** 2009
[21] Rodríguez-Carvajal J *et al* 1998 *Phys. Rev. B* **57** R3189
[22] Huang Q *et al* 2008 *Phys. Rev. Lett.* **101** 257003
[23] Biffin A *et al* 2014 *Phys. Rev. B* **90** 205116

IOP Publishing

Characterisation Methods in Solid State and Materials Science

Kelly Morrison

Chapter 4

Microscopy techniques

In this chapter, some common microscopy techniques will be outlined starting with optical microscopy, followed by scanning electron and transmission electron microscopy (TEM) and finally profilometry. The aim will not be to describe these techniques in great detail, but to give a working overview of their capabilities, and the available contrast mechanisms. Finally, some examples of these techniques in practice will be highlighted using examples from the literature.

4.1 Optical microscopy

4.1.1 The optical microscope

In chapter 1 we covered the basics of optics: refraction, reflection, diffraction, resolution and lenses. This included a discussion of the applications of a single convex lens in forming an image, including magnification. The aim of an optical microscope is to aid in high precision work (large working distance; figure 4.1), or to magnify an image so that details can be picked out (imaging; figures 4.2 and 4.3). This is achieved with a combination of lenses, in particular, the eyepiece and objective lenses, with a combined magnification, M, described by:

$$M = M_o \times M_e \tag{4.1}$$

where M_o is the magnification of the objective lens, and M_e is the magnification of the eyepiece lens. For example, an eyepiece with magnification of ×10 coupled with an objective with magnification ×100 would result in an image magnification of ×1000. Annotated examples of typical microscopes, found in the laboratory are given in figures 4.1–4.3.

Common techniques to achieve a uniform light source for imaging are by 'critical' or Köhler illumination [1, 2]. In 'critical' illumination a uniform light source, such as from a large bulb, is focussed onto the condenser (which concentrates light onto the

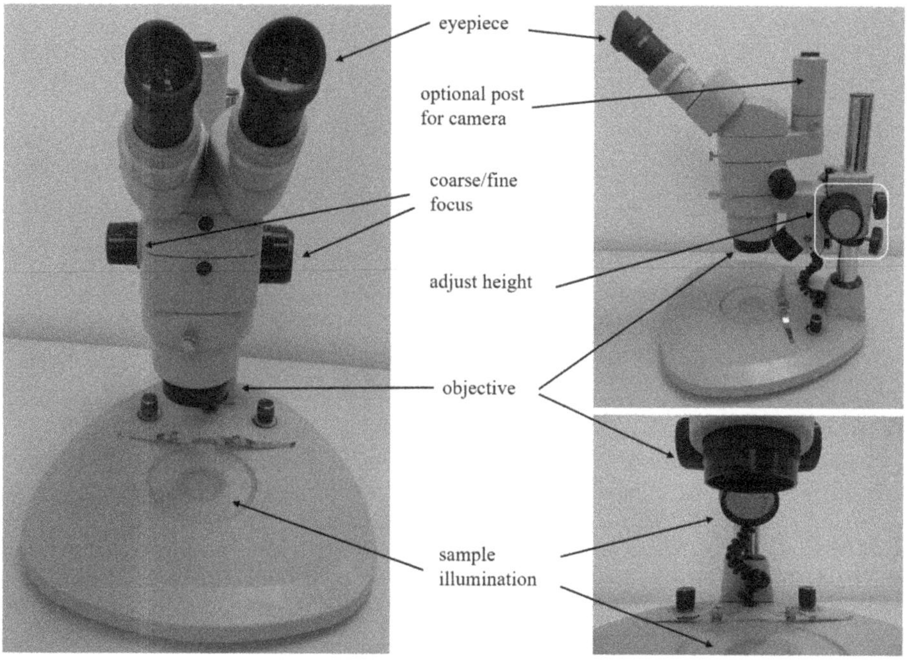

Figure 4.1. Stereo microscope example. More freedom of movement with respect to objective distance from sample stage enables a wide variety of samples to be observed. This type of microscope is commonly used to aid high precision work. Sample illumination can be from above or below, depending on the user's needs.

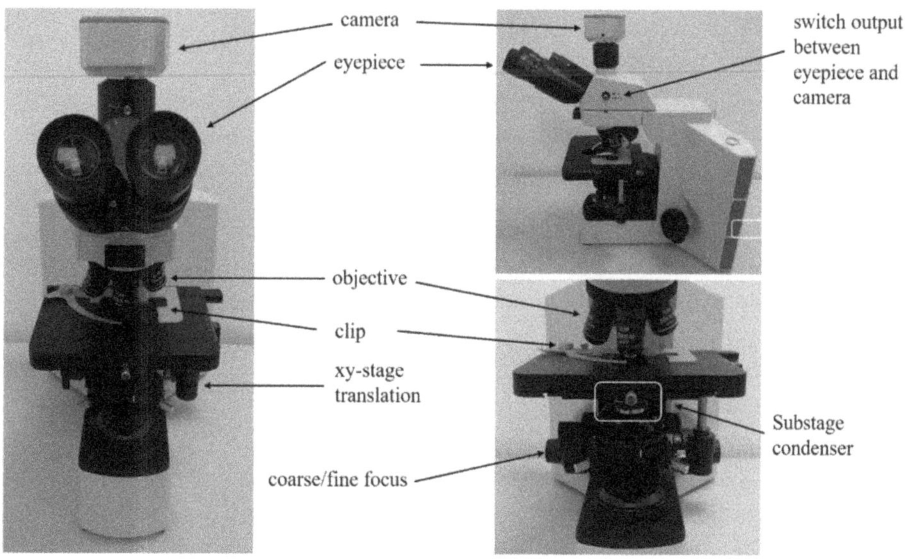

Figure 4.2. Bench microscope example. The clip is used to hold the sample (often a glass slide) in place, with the *xy*-stage translation to enable ease of movement whilst observing under the microscope. A simple prism (moved in/out of the optical path) is used to switch output between the eyepieces and a camera. The substage condenser focusses the light source onto the sample stage; there is usually a clip to enable insertion of filters.

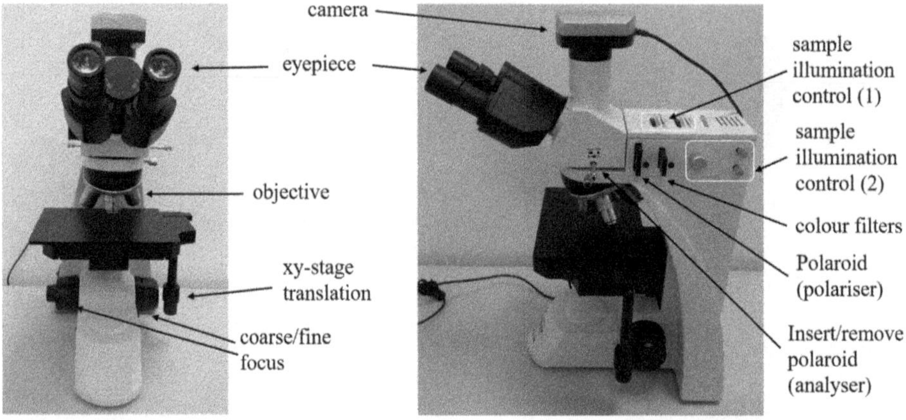

Figure 4.3. Metallurgical microscope example. Sample illumination is achieved by a bulb inside the microscope housing (Köhler illumination). The sample illumination control (1) adjusts the brightness and position of the bulb with respect to the optics, whereas control (2) adjusts the aperture and field diaphragms. Colour filters and polaroids (polariser and analyser) can be used to enhance the contrast of the image. The *xy*-stage translation ensures smooth control of movement (of the sample) whilst observing under the microscope. A prism can be moved in/out of the optical path to switch output between the eyepieces and a camera.

object with uniform intensity). In Köhler illumination, commonly implemented with an integral bulb in the microscope, the image formed by the filament (of the bulb) is focussed onto a collector lens. This results in a defocussed image of the light source at the sample plane and image plane (human eye or camera). Whilst critical illumination requires a large, uniform light source (~5 cm), Köhler illumination requires care to regularly check the alignment.

The objective lens is the primary image-forming lens, whereas the eyepiece is the uppermost lens system that forms a magnified image. It may contain some or all of the following housed in a compact tube: front lens, multi-element lenses, diaphragm, reticule for gauging size and field lens. A reticule, or other standard object is used to calibrate the image magnification so that a lengthscale can be provided on images saved using a camera. The prevalence of open source image processing software, such as ImageJ, makes independent analysis and calibration of images fairly trivial: as long as the image is not resized between saving and calibrating, the number of pixels per reticule spacing can be used to determined the length per pixel.

4.1.2 Use of polarised light

In the example of a metallurgical microscope (figure 4.3) a polaroid (polariser) is identified as one of the optional filters. This can be used in conjunction with a second polaroid (analyser) placed in the beam path. In this example, the polariser can be rotated to change the relative orientation of the polariser and analyser.

4.1.2.1 Sidebar: polarisation

Any electromagnetic wave is comprised of time varying electric and magnetic fields that are at right angles to the direction of propagation of the wave. Normally, light is unpolarised, which means that there is no coherent direction in which the electric field is varying for each wave. By polarising the electromagnetic wave, we are choosing to constrain the direction of the electric field in a particular direction (or more aptly, selecting a specific electric field vector for wave propagation). The simplest example of this is linearly polarised light, where the magnitude of E will vary with time, as shown in figure 4.4. Another common example is circularly polarised light, where the magnitude of E is constant, but its vector will rotate about the direction of propagation in a clockwise (right-handed) or anticlockwise (left-handed) fashion. Techniques to convert from right- to left-handed circularly polarised light include use of a half-wave ($\lambda/2$) plate, or reflection off of a surface (at normal incidence).

To understand how polarised light is produced we need to first discuss optical anisotropy; otherwise referred to as birefringence.

4.1.2.2 Birefringence

Many optical components will make use of birefringence to manipulate the polarisation(s) of an electromagnetic wave. Birefringence will arise as a directional variation in refractive index of light that is largely controlled by the position of atoms within the crystal lattice. As cubic crystals are equivalent by symmetry in the $\{x,y,z\}$ directions they will therefore not exhibit optical anisotropy. Tetragonal, hexagonal and trigonal systems, on the other hand, are uniaxial crystals that have a single optical axis. For example, a hexagonal lattice would be identical in the xy-plane, but not when compared to the z-axis. Examples of uniaxial crystals are ice (H_2O), quartz (SiO_2) or sapphire (Al_2O_3). Orthorhombic, monoclinic, or triclinic crystals are biaxial: there will be two optical axes. An example of this is mica [3]. The different refractive indices in birefringent materials can be used to separate or select

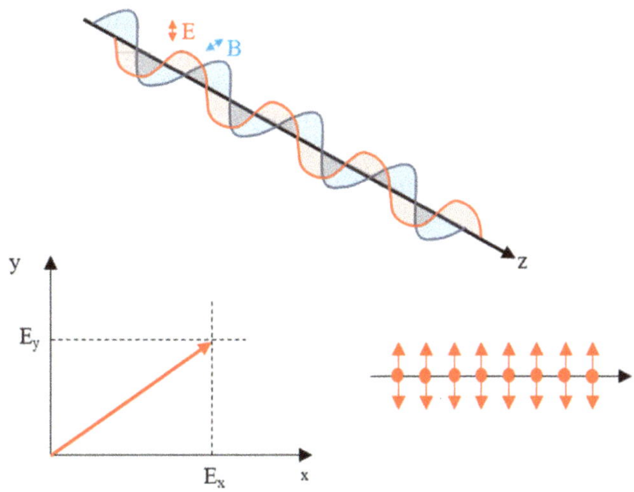

Figure 4.4. Top: Sketch of a polarised electromagnetic wave that is comprised of a constantly varying electric (**E**) and magnetic (**B**) field existing at right angles to one another (x and y) and the direction of propagation (z). Bottom: In an unpolarised beam of light travelling along the z-direction, any direction of **E** in the xy-plane will be just as likely as another. Given that **E** can exist in any direction in the xy-plane, it can also be represented as a combination of vectors E_x and E_y. This is sometimes represented by a combination of in and out-of-plane oscillations of **E**.

out specific polarisation states, or to introduce optical path length differences into an optical system.

Some definitions to note:

- Birefringence—optical anisotropy in a crystalline lattice that results in refractive indices that vary with polarisation. Typically produces extraordinary and ordinary rays.
- Extraordinary ray—created by a birefringent crystal. Does not obey Snell's law of refraction.
- Polaroid/polariser—material which will polarise light that passes through. This is typically a result of birefringence.
- Quarter wave plates—retarding material that slows electromagnetic wave down such that a path length difference of $\lambda/4$ is introduced. Quarter wave plates are often used to convert linearly polarised light to circularly polarised light (and vice versa).
- Half wave plate—as above, except that a path length difference of $\lambda/2$ is introduced. Will also rotate the plane of polarisation of an incoming polarised beam.
- Beamsplitter—optical component that will split an incoming beam into two paths.
- Polarising beamsplitter—optical component that will split an incoming beam into its orthogonal components.
- Phase plate—a spatial filter with a central annulus that introduces a phase shift of $\lambda/4$ relative to the rest of the plate.

The use of polarisers to provide additional contrast is called polarised microscopy. It can be useful to observe grain structure on polished or etched samples [1], such as shown in figure 4.5. In this example, copper samples were polished to 0.5 μm using diamond paste, annealed for 4 h at 900 °C, and then etched in a solution of 25% ammonium hydroxide with a few drops of hydrogen peroxide for a few minutes, followed by a rinse in methylated spirits. The etching process preferentially dissolves copper at the grain boundaries, thus enhancing contrast seen with the microscope.

For transparent samples, polarised light microscopy can be used to provide more information than would be available with bright field imaging. For example, birefringence in the sample would lead to additional contrast when coupled with polarisation analysis. It could also be used to assess stress in a sample as this results in birefringence.

4.1.3 Phase contrast imaging

As discussed in chapter 2, most imaging techniques will measure the intensity of signal and as such, phase information is lost. There are techniques, however, that enable visualisation of changes in phase as a (generally) coherent light source is transmitted or reflected off of a sample. This is the principle of phase contrast imaging, which can be useful, for example, to image biological samples where they

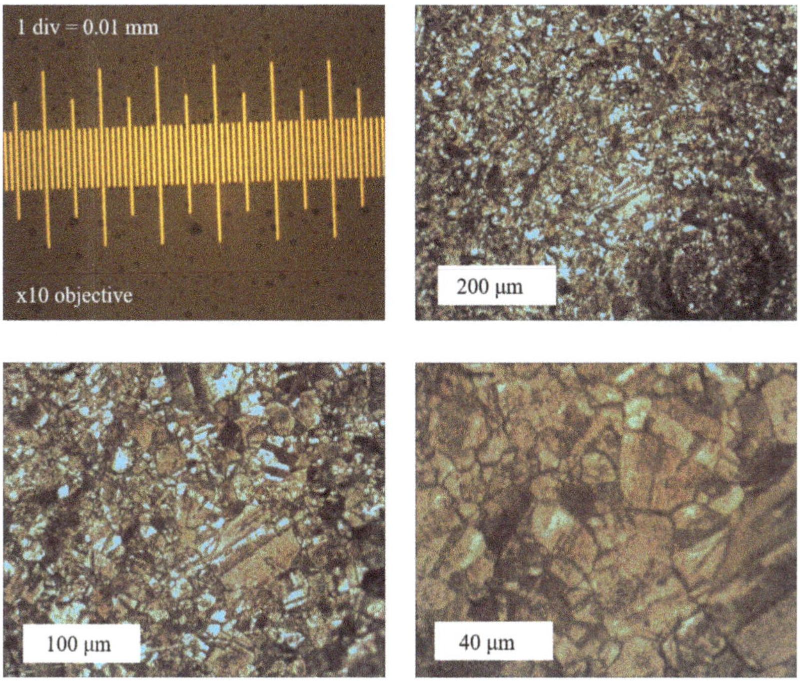

Figure 4.5. Example of polarised microscopy to observe grain structure on an etched copper sample at different magnification. Top left: example reticule image used to calibrate the lengthscale in the images. Magnification is increased from ×100 to ×500 (×10 eyepiece and ×10, ×20, ×50 objective lenses).

may be transparent (and hence have very little contrast), or crystals where contrast at grain boundaries of for different crystallite directions may appear. Such changes in phase arise due to variations in the refractive index of the sample, which would otherwise not be observed.

There are two common techniques employed in phase contrast imaging: the Zernike, and the Schlieren techniques [1, 4]. The Zernike technique requires the use of a phase plate, which introduces a phase shift between light diffracted by the sample, and light that has passed through without interacting, as shown in figure 4.6. The Schlieren technique, on the other hand, provides phase contrast by removing higher order diffraction spots from one side of the objective lens' focal plane with a knife edge. This can, for example, be used to demonstrate subtle variations in air density due to thermal currents as a result of the change in refractive index that would accompany this. An example is given in figure 4.6. Note that the Zernike method can be used to obtain phase contrast in TEM, with appropriate choice of phase plate.

4.1.4 Interference contrast microscopy

Whilst phase contrast imaging can provide qualitative information on changes of phase within a sample, there are instances where quantitative phase mapping is

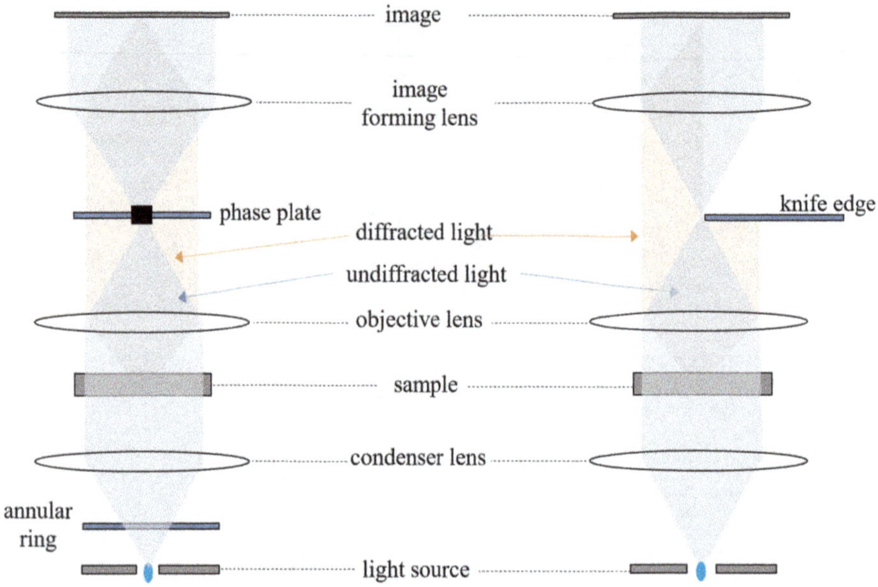

Figure 4.6. Schematics of the Zernike (left) and Schlieren (right) techniques, where phase and amplitude masks are used to produce phase contrast, respectively.

required. This can be achieved using holography or differential interference contrast (DIC).

In holography methods a reference beam interferes with the sample beam, which contains phase variations due to the sample [4]. For this technique to work, both the sample and reference beams must be coherent. It is the use of the reference beam that enables quantitative measurement of phase variations, achieved by monitoring amplitude variations on a camera (CCD). By overlaying the image with an interference pattern, the phase—information that is usually lost—can be recorded onto the CCD [5, 6].

The Nomarksi method, otherwise referred to as differential interference contrast microscopy (DIC) uses the contrast obtained with polarised light, where it is separated into its orthogonal components using a polarising beamsplitter such as the Wollaston prism [7]. This orthogonal split results in a slight spatial separation of the two beams, which will produce an interference pattern when they are recombined. The advantage of using polarised light is that you can use incoherent illumination (such as a tungsten bulb). Finally, the illumination and detection polaroids are cross polarised so that the beam path that does not pass through the sample (and acquire a 90 degree phase shift) is eliminated.

4.2 Electron microscopy

4.2.1 Scanning electron microscopy (SEM)

The scanning electron microscope (SEM) moves past the diffraction limit of an optical microscope (~0.3 μm) by using electrons instead of light. This works due to the wave-particle duality of matter, as discussed in chapter 3 when comparing x-ray and neutron techniques. In the case of an SEM, a column of charged particles (electrons) is used in a similar way to the light microscope, where the electron beam is focussed by a combination of electric and magnetic lenses. The energy of these incident electrons will be dependent on their accelerating voltage: the potential difference between the anode and cathode in the electron gun, with a decrease in the wavelength as the energy is increased. For example, an accelerating voltage of 20 kV would have an associated wavelength of 0.0087 nm and an optimal resolution of 0.44 nm, whereas 100 kV would have an optimal resolution of 0.19 nm [8].

As seen in figure 4.7, there are several key components of an electron microscope:

- The electron source: such as a tungsten filament; solid state crystal (CeB_6 or LaB_6); or field emission gun (FEG).
- Lenses: condenser, scan coils and objective lens.
- Sample stage.
- Detectors.

Depending on the available detectors, different characterisation can be carried out ranging from the straightforward images conveying topological information, to atomic or magnetic contrast [8, 9]. For standard SEM images, an Everhart–Thornley detector is used. This is similar to the scintillator described in chapter 3, but with the addition of a Faraday cage, which enables collection of low energy electrons from a large area by applying a positive voltage bias

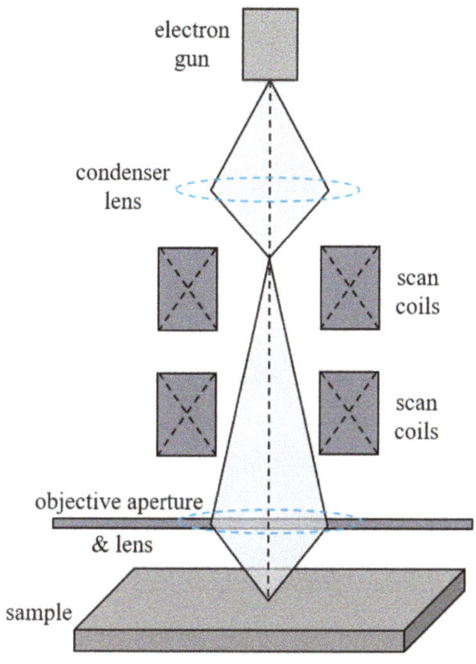

Figure 4.7. Schematic of the lens system for a scanning electron microscope.

The type of information produced and the interaction volume (depth and width) is shown schematically in figure 4.8. The interaction volume describes the area where it is most probable that an interaction will occur. This will generally increase as the accelerating voltage is increased and decrease with increasing atomic number. The escape depth, on the other hand, describes the depth from which electrons produced by various interactions can exit the sample, and this will typically increase with increasing accelerating voltage.

The result of electron interactions with the specimen being measured is the production of different categories of electrons or x-rays, as defined below.

Primary electrons in the case of TEM, are the electrons that pass through the sample undergoing elastic scattering events. As such, their energy is similar to the incident energy of the electron beam. As elastic scattering events are largely due to diffraction these will provide information on crystallographic structure and atomic number.

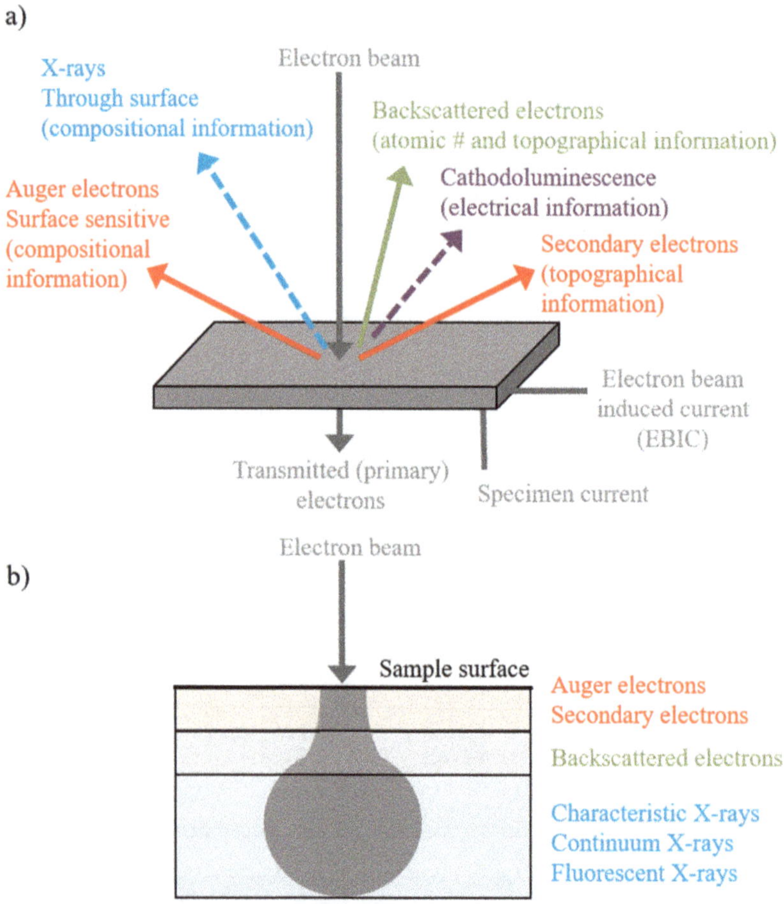

Figure 4.8. (a) Summary of the available information in a scanning electron microscope. (b) Schematic of the electron beam interaction volume (grey area) and the corresponding escape depth for the different electrons and x-rays produced (coloured blocks).

Secondary electrons (SEs) are the primary source for imaging in an SEM. These are the electrons that are produced by inelastic scattering, which have an average energy of 3–5 eV (compared to ~keV of incident beam). The low energy makes collection easy by applying a voltage bias to the Faraday cage of the Everhart–Thornley detector. Only SEs produced in the top ~5 nm for metals or ~50 nm for insulators will escape. As such, they largely provide topographical information, with the number of SEs escaping increasing for rough features. This results in images that are relatively easy to interpret, as they resemble optical images, with the added advantage of a larger depth of field.

Secondary electrons are categorised into four types: SE1 to SE4. SE1 are the electrons produced by interaction near the sample surface, hence they provide high resolution information. SE2 are produced from backscattered electrons that interact with the sample on exiting, and as such will have a higher escape depth (and thus a lower resolution). SE3 are produced when backscattered electrons exit the sample and interact with the objective lens or other parts of specimen chamber. Finally, SE4 are produced by the primary beam as it passes through final aperture. As the detector cannot distinguish between the different types of secondary electrons, SE2 to SE4 all add a low resolution background to the overall detected signal. As the magnification is increased, the contrast seen from SE2–4 disappears and becomes a continuous background over the detail conveyed by SE1.

Backscattered electrons are produced largely by elastic scattering (retaining 60%–80% of their incident energy), where the electrons are scattered back out of the sample. They present an alternative imaging opportunity to SE detection, but with a larger source depth, presenting both atomic and topographical information. Where SE images may provide compositional contrast for highly polished samples, in practice they are usually dominated by topography. BSE images, on the other hand will be dominated by compositional contrast, as demonstrated by the sketch of electron yield shown in figure 4.9. The challenge with BSE imaging lies in their detection, as the higher energy electrons would be difficult to capture with the Everhart–Thornley detector. Instead, a detector is mounted directly over the sample, to capture the electrons that escape at angles close to normal incidence. With regards to the source of information, the escape depth and width (~μm) of BSE varies inversely with the atomic number, Z, whilst the production of BSE will increase with increasing Z.

Auger electrons, characteristic x-rays, continuum and fluorescent x-rays, as discussed in more detail in chapter 5, are the result of electron excitation processes and can be used to extract compositional information.

The **electron beam induced current** (EBIC) is another probe that is used to characterise semiconducting samples. The basic principle relies on the depletion region present at a pn junction or Schottky barrier (as discussed in more detail in chapter 7). Put simply, there exists an area of the sample—the depletion region—where no charge carriers exist. If an electron beam is incident on this region then it will generate electron–hole pairs which move away producing a measurable current. Applications of EBIC detection include mapping defects in semiconductor crystals or backwards engineering integrated circuits [9].

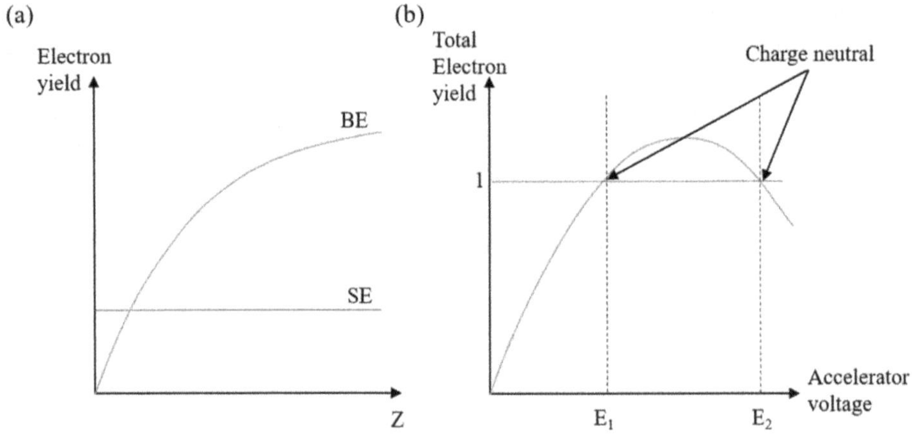

Figure 4.9. (a) Sketch of the electron yield for secondary and backscattered electrons (SE and BSE, respectively) as a function of atomic number, Z. (b) Variation of the total electron yield as a function of accelerator voltage. The point at which the total electron yield is equal to one, E_1 and E_2, will be sample dependent.

Cathodoluminesence is the light emitted from a sample, such as the recombination of electrons and holes in a direct or indirect semiconductor.

4.2.2 Sources

Electron sources can be categorised as either thermionic or field emission. Thermionic emission is the production of electrons from a heated metal as a result of the available thermal energy exceeding the work function of the cathode. This is the working principle behind the cathode ray tube described in chapter 3. In the case of electron sources, the beam of electrons produced by thermionic emission are accelerated towards an anode with an aperture, which is held at a respective potential difference categorised as the accelerating voltage. In an SEM, the electrons that pass through the aperture are then focussed by a set of lenses onto the sample surface.

The brightness of the electron source describes the beam current density per unit area per solid angle (measure of the number of electrons per second that can be directed at a given area). This is influenced by the type of electron gun used, but will generally increase linearly with the accelerating voltage. The spot size at the sample surface (controlled by the condenser lens) will largely determine the available resolution for a given category of electrons or resulting radiation detected. The electron beam energy spread will determine the extent of aberrations. For example, low accelerating voltage will lead to chromatic aberration if beam spread is too large.

Examples of thermionic sources include the tungsten electron filament, and LaB_6 or CeB_6 crystals; a schematic of which is given in figure 4.10. In the case of a tungsten filament, an inverted V-shaped wire of tungsten, about 50–100 μm long is heated to about 2800 K. Whilst this can produce a stable electron source, with a brightness of the order of 10^6 A cm^{-2} sr^{-1}, the lifetime is typically of the order of

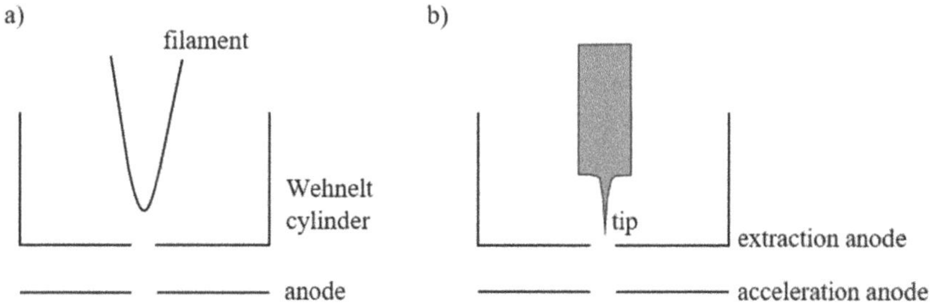

Figure 4.10. Sketch of the thermionic (a) and field emission (b) electron gun geometries. In thermionic emission, the current passing through the filament determines the rate of production of electrons. For field emission, the potential difference on the extraction anode determines the electron production.

100 h due to melting of the filament. The high temperature of the filament can also result in a large energy spread (and thus more aberrations) compared to other electron sources. Nevertheless, it is the cheapest available electron source and can work at relatively low vacuum, which makes operation and replacement less costly.

Single crystal lanthanum hexaboride (LaB_6) or cerium hexaboride (CeB_6), on the other hand, have lifetimes of the order of 1500 h and an order of magnitude increase in brightness (~10^7 A cm^{-2} sr^{-1}) due to a lower work function (operates at 1800 K). In this case, the crystal is held between two wires, which heat it by passing a current through. The advantages of this source are a higher signal-to-noise ratio (due to the increased brightness), smaller spot size, and much longer lifetime. In addition, the quality of the source degrades gradually with time, so it is possible to predict when it may fail (unlike with W).

In the case of field emission, electrons are stripped off of a material by applying a large enough electric field (>10^9 V m^{-1}). This is made possible by fabricating an emission tip with a very small radius (~100 nm), kept under ultra high vacuum to preserve its structure. As shown in figure 4.10, electrons are pulled off of the tungsten tip by applying an electric field to the extraction anode, and then accelerated by applying a larger potential difference to the acceleration anode. The small tip radius improves emission (brightnesses of 10^9 A cm^{-2} sr^{-1} are possible) and focusing ability. Another advantage of the field emission gun is that it has well defined energies (energy spread 0.5 eV compared to 1–2 eV for thermionic emission).

4.2.3 SEM resolution and contrast

The different contrast available for secondary and backscattered electron imaging has already been discussed. In short, atomic contrast is stronger for BSE imaging, and available for SE imaging only for well-polished samples (where topography does not dominate). With regards to resolution, the spot size—that is the size of the focussed electron beam at the sample—will largely determine the resolution of the image.

The spot size itself is controlled by the condenser and objective lenses. One option would be to decrease the spot size by increasing the current to the condenser lens, however, decreasing the spot size in this way could decrease the brightness of the beam after the objective due to a beam spread larger than the available aperture. This reduction in intensity will have a knock-on effect on the signal picked up by the detector. As the image is controlled by statistics such that more counts produce a better signal, this means that smallest spot size may not always yield the highest resolution (for same image acquisition time and aperture size) [10].

It should also be remembered that the depth of information may also affect the resolution of the image (similar to the differences seen for SE and BSE imaging). If the accelerating voltage is increased then the interaction volume will increase. For dense samples, a higher acceleration voltage leads to lower wavelengths and thus better resolution. For less dense (e.g. biological) samples, however, lower accelerating voltages are preferable due to the increase in interaction volume that would decrease the effective resolution.

Lastly, the depth of field, which determines the length over which the sample will be in focus, will be determined by the aperture size and working distance, as demonstrated in figure 4.11. As the depth of field is controlled largely by how tight the cone of focussed electron beam is (i.e. the area of the electron beam equivalent to the practical resolution), a larger aperture, which increases the spot size will also result in a decrease of the depth of field. Similarly, a smaller working distance will decrease the depth of field [8].

4.2.4 Sample preparation and beam damage

With regards to sample preparation, some consideration will need to be made with regards to the suitability of the sample for imaging. For example, due to the low vacuum required, samples should not contain an aspect that could be easily vaporised (such as water or a solvent). In addition, the sample should be firmly fixed to the sample holder; usually an aluminium SEM stub. This is particularly important in the case of nanoparticles, where again, the vacuum could pull the particles off of the stub and contaminate the chamber. While biological samples will need more preparation to make them suitable for SEM imaging (due to their high

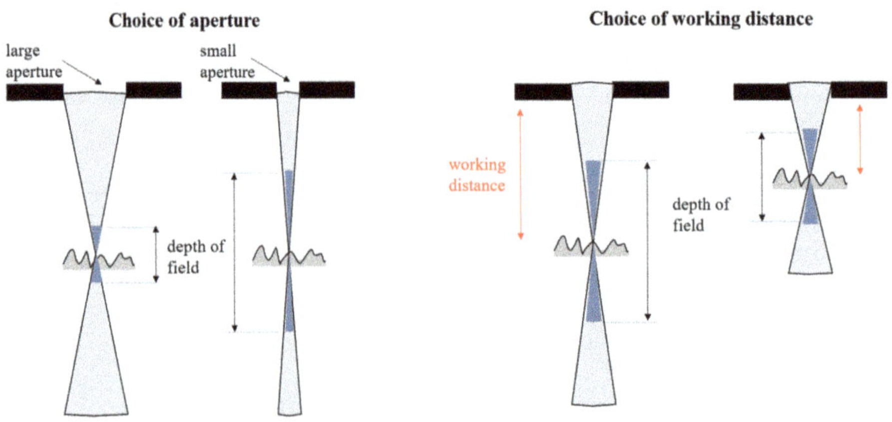

Figure 4.11. Sketch of the change in depth of field as the aperture size and working distance is varied. The darker shaded area shows the equivalent depth of field, for each example.

water content in particular), this is not the focus of this book. Instead the reader is referred to [8].

With regards to mounting samples, a commonly employed technique is to use double-sided carbon sticky tabs, which are designed to be the size of the standard SEM subs. These have the advantage of being conductive, as well as the ease of use (such as positioning of the sample or removing by soaking the stub in acetone). For example, nanoparticles can be easily affixed to such carbon tabs with a brush. Alternatives include various glues and epoxies, where conductive epoxy is preferred. It should be noted that the glue/epoxy must not outgas much and must be left to dry fully before use.

Another requirement of samples is that they are conductive. If this is not an intrinsic property of the sample, this can be achieved by coating the sample in gold or carbon. For chemical analysis, carbon is preferred as it has a lower atomic number and a low absorption factor for x-rays (i.e. will not heavily impact on the escape depth or contrast). If a coating is not used on non-conductive samples, then a build up of negative charge can occur, which will distort the image. Common image artefacts that arise due to this include: lines on the screen; abnormal contrast; image drift; or breaks in the image. An example of this is shown in figure 4.12. (Note that as secondary electrons are lower energy than back scattered electrons, SE imaging will be more susceptible to charging effects.)

In the case where it is not possible to coat the sample (due to further measurements being required, for example), knowledge of the electron emission coefficient can be used to select a suitable accelerating voltage. In this case, choice of an accelerating voltage that falls between E_1 and E_2 defined in figure 4.9 would result in a build up of positive charge. This has less of an effect on the image (only preventing some electrons from escaping), thus making SE imaging possible without a conductive coating. Typical accelerating voltages for this regime are in the range of 100 V–1 kV [8].

Lastly, care should be taken to ensure the sample is kept clean, as demonstrated by the difference in images shown in figures 4.12(a) and (c). Both show metallic thin films, but in the case of (c), the sample has either been dropped (picking up carbon or dust from another surface) or handled by the user without using gloves. The presence of carbon on the surface of the sample results in areas of very dark contrast as it burns off in reaction to the electron beam. A similar effect—although to a lesser extent—is seen in a corner of (a) where the electron beam was focussed on a small area for a prolonged period of time to correct for astigmatism.

4.2.5 Transmission electron microscopy (TEM)

The scanning transmission electron microscope (STEM) and TEM are similar to scanning electron microscopy except that the electron beam passes through the sample before detection. As shown, in figure 4.13, for STEM, the electron beam is focussed to a point and transmitted through. As a result, it can image sample thicknesses of the order of 100 nm to 1 μ, with the added advantage of not requiring post sample optics. Given that the electron beam is focussed onto a specific point, it

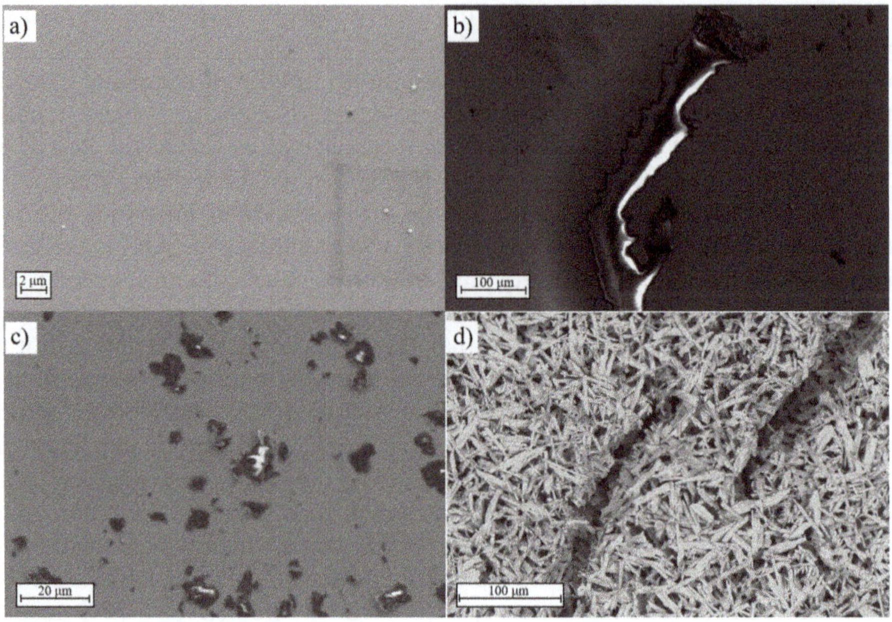

Figure 4.12. Example scanning electron microscope images. (a) Sample damage on a metallic thin film. The slightly darker contrast in the bottom right hand corner is the result of beam exposure during SEM alignment and astigmatism correction. (b) Sample charging effects on a metallic thin film on a glass substrate. Damage to the film has exposed part of the substrate, which is insulating. As a result, this area in the centre of the image charges up, producing a bright edge and shadow effect on the left hand side. Excessive charging may also cause the image to drift over a single scan. (c) Carbon residue on a metallic thin film. The dark flakes are dirt (carbon) that have been burnt by the electron beam during imaging (hence the dark contrast). (d) Example of the depth of field available with the SEM. Image is of a $YBa_2Cu_3O_7$ pellet prepared by solid state synthesis. Details of lower lying layers can be seen in the trenches in this sample.

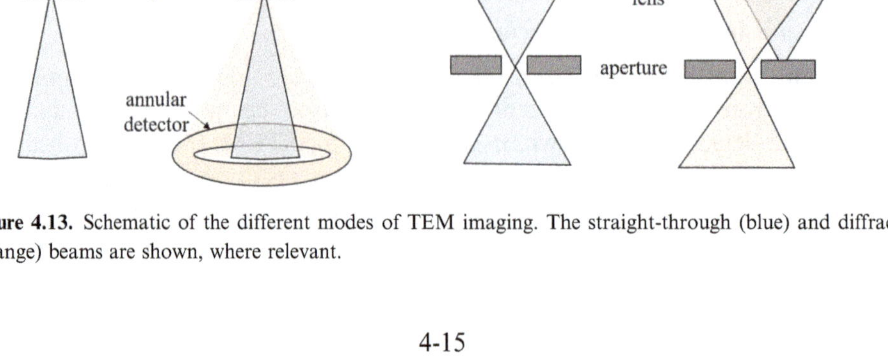

Figure 4.13. Schematic of the different modes of TEM imaging. The straight-through (blue) and diffracted (orange) beams are shown, where relevant.

also offers the possibility of high resolution. For conventional TEM, on the other hand, the electron beam is run in flood mode, where the whole sample area is irradiated and a set of image forming lenses after the sample are used to produce the required image (figure 4.13). There are two main modes available with both STEM and TEM (albeit obtained in different ways): dark field and bright field.

In bright field (BF) microscopy both the diffracted and undiffracted electrons are collected to form an image of the specimen on a bright background. Practically, this is realised by detecting the central ring of diffracted electrons passing through the sample (the blue shaded area in figure 4.13). In dark field (DF) microscopy, on the other hand, only the diffracted radiation is used to form an image. This results in the specimen appearing bright against a dark background and is achieved with an annular detector in the case of STEM (ADF), or by displacing the aperture or tilting the beam in the case of conventional TEM.

If the angle of DF imaging is high enough, then the contrast achieved from diffraction of the electron off of the atoms in the sample is lost. This usually occurs for angles exceeding 20 mrad, where it is argued that thermal effects (incoherent scattering due to vibrations of the atoms) blur the Bragg diffraction spots to the point where they no longer dominate the information available. In addition, if information is collected and integrated over a large area, such as the annular ring of an ADF detector, the phase information that might usually contain atomic position is also lost. This is the underlying principle of HADDF—where a large enough angle is selected such that the majority of contrast will arise from Rutherford scattering (i.e. elastic scattering off of an atom) [10–12]. The intensity of the HAADF STEM images typically varies with the atomic number, $Z^{1.7}$ [11, 13].

Sample preparation for TEM will vary depending on the sample, and what type of cross-section is required. To summarise, the difficulty arises from preparing the ultrathin specimens required without introducing artefacts or damage from the thinning process. Some of the potential methods for doing this are outlined in detail in [10].

As an example, STEM/HAADF, STEM/BF, and high resolution TEM (HRTEM) images of Fe_3O_4:Pt deposited onto glass are shown in figure 4.14. In this example, the sample was prepared by lifting out a lamella from the thin film and thinning to a thickness of ~100 nm using a focussed ion beam (FIB) with Ga source. To preserve and mark the original surface a ~50 nm layer of Au was sputtered onto the sample before removing a slice with FIB, in addition to a ~1 nm Pt layer laid *in situ*. This technique was chosen so as to prepare a slice perpendicular to the film surface to study the interfaces of the bilayer. In the STEM/BF image you can see contrast due to separate polycrystal grains, whereas in HAADF this is dominated by Z contrast (the grain boundaries are no longer visible). This indicates that the Fe_3O_4 is continuous phase. The HRTEM also shows 2–3 nm crystallised Pt stacked on top of the Fe_3O_4. As the sample studied was measured prior to being prepared for TEM, there are traces of a thermal grease on the top surface—indicated by the bright line above the first Pt layer in the BF image. Energy dispersive x-ray spectroscopy has also been used to determine the chemical composition along the sample confirming the identification of each layer.

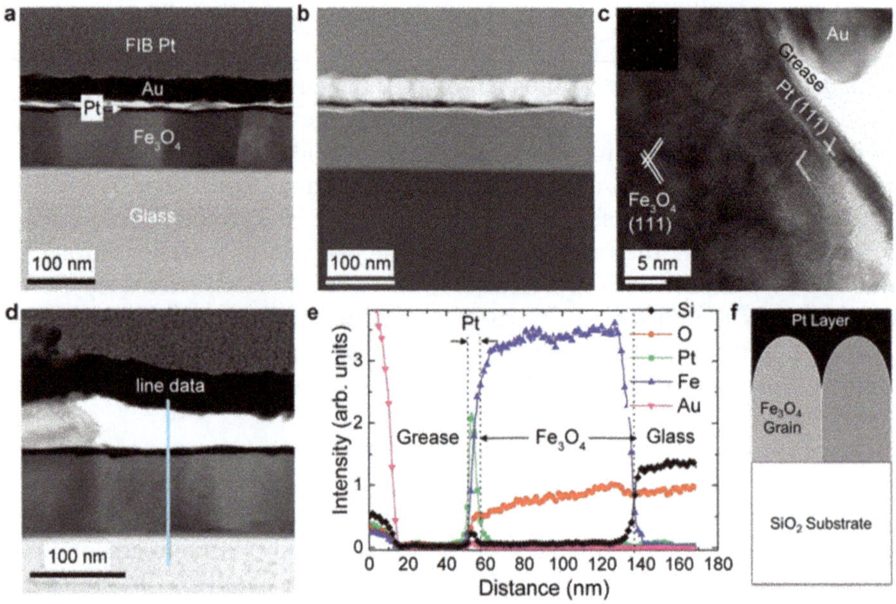

Figure 4.14. TEM analysis of a Fe_3O_4:Pt bilayer deposited onto glass. (a) STEM/BF image showing polycrystalline grains. (b) HAADF images showing continuous phase for the Fe_3O_4 layer. (c) Conventional HRTEM of the bilayer, showing stacking of the Pt layer on the Fe_3O_4 film. (d) and (e) STEM/BF image of the thin film stack and corresponding EDX line-scan performed perpendicular to the interfaces of the layers, respectively. (f) Schematic of the grain growth in this film, determined from a combination of XRR and TEM. Figure reproduced from [14]. © 2016 Caruana *et al.*

Additional spectroscopic information often available in an SEM or TEM include energy dispersive spectroscopy (EDS or EDX), electron backscatter diffraction (EBSD), and x-ray photospectroscopy (XPS), are discussed in chapter 5.

4.3 Profilometry

Profilometry is a non-contact method for determining the surface profile of an object, including step height, curvature, flatness and thickness of thin films or coatings. Two commonly employed profilometry methods are optical and scanning microscopy, both of which are outlined here.

4.3.1 Optical profilometry

Optical profilometry is an extension of optical microscopy that uses either the interference of light off of regular features, or reactive focussing (confocal profilometry) to extract topographical information. In the case of an interference based optical profilometer, either a monochromatic source is used, where the resulting image will be a combination of the surface and the interference pattern produced, or a white light source will be used, with colour separation in a CCD and extensive analysis.

The interference patterns produced by monochromatic light sources incident on different apertures was demonstrated in chapter 1. In the Young's double slit experiment, for example (figure 1.8(d)), a plane wave incident on two slits produced a sequence of bright and dark fringes on a screen some distance, D, away. In chapter 2, this was described mathematically using the idea of Fourier diffraction (section 2.6.1). Alternatively, the same behaviour could be derived by considering the path difference between two rays incident on each slit that are diffracted to the same point on the screen. In this case, a plane wave incident on both slits could be described in three dimensions by:

$$E(\mathbf{r}, t) = E_0 \cos(\mathbf{k} \cdot \mathbf{r} - \omega t + \varphi) \tag{4.2}$$

where E_0 is the amplitude, ω is the angular frequency of the light source ($= 2\pi f$), $\mathbf{k}$ is the wavevector ($k = 2\pi/\lambda$), and φ is a phase shift. If we set up a measurement where the incident light source is monochromatic (single wavelength, λ and frequency, f), then we can show that the path difference introduced by travelling the two possible optical paths can result in the observed maxima and minima.

We can, for simplicity, set $t = 0$, as we are only concerned in the spatially varying components of this wave. We can also choose to set the phase shift equal to zero at the slits. This leaves:

$$E = E_0 \cos(\mathbf{k} \cdot \mathbf{r}) \tag{4.3}$$

Now consider the phase shift introduced between two waves that travel different paths, $\mathbf{r}_1$ and $\mathbf{r}_2$.

$$E_1 = E_0 \cos(\mathbf{k} \cdot \mathbf{r}_1) \tag{4.4}$$

$$E_2 = E_0 \cos(\mathbf{k} \cdot \mathbf{r}_2) \tag{4.5}$$

where

$$E_{tot} = E_1 + E_2 \tag{4.6}$$

If $\mathbf{k} \cdot \mathbf{r}_1 = 0$ and $\mathbf{k} \cdot \mathbf{r}_2 = \pi$ (or similarly $\mathbf{k} \cdot \mathbf{r}_1 = n\pi$ and $\mathbf{k} \cdot \mathbf{r}_2 = (2n + 1)\pi$) then the intensity of the combined waves will equal zero. This is the main premise of interference measurements such as:

- Diffraction grating and double slit experiment;
- Michelson, Sagnac, Mach–Zender interferometers;
- The Fabry–Pérot interferometer (discussed in more detail in chapter 5);
- The Air wedge;
- Newton's ring.

For Young's double slit example, the path difference arises from the separation of the slits, d. The fact that a different path length can lead to interference in this way lends itself readily to profilometry—determining the topological landscape of a material or device being investigated. This typically means the measurement of surface roughness, step heights, or 3D objects, and is useful, for example, in determining thin film thickness, identifying defects (voids or particles) and thus the defect density—all of which are complementary data for more complex characterisation.

The basic components of an optical profilometer will be:

- Coherent light source (i.e. single λ);
- Collimating, focussing, and collection optics;
- Camera;
- Computer software.

A schematic of such a simple setup is given in figure 4.15. Here you have a reference beam (in this case a mirror kept a known distance from the beam splitter) and a second path that is altered by the sample being investigated. Whilst a standard image of the surface will be obtained, overlaid on top of this will be an interference pattern arising from the changes in height across the sample. This can be deconvolved using software available with the profilometer, usually based on using fast Fourier transforms.

In the case of confocal profilometry, the light source is focussed at a point on the sample surface, and the focus changed to maximise the intensity measured by the CCD, where the image will be in focus. This spot is then rastered across the sample surface and the focus adjusted accordingly. The change in focus across the sample surface will provide information on the topography. This technique works well because of the limited depth of field available with an optical microscope (not the case for electron microscopy). To aid in detection of the focal point, a pattern can be projected onto the sample surface, thus providing a reference for analysis of the image formed.

4.3.2 Scanning tunnelling microscope (STM)

Scanning tunnelling microscopy (STM) is a high resolution microscopy technique used to image individual atoms (0.1 nm lateral and 0.01 nm depth resolution). It was developed by Binning and Rohrer in 1982 [15], for which they won the Nobel prize in 1986 [16]. The basic principle makes use of quantum tunnelling, whereby electrons

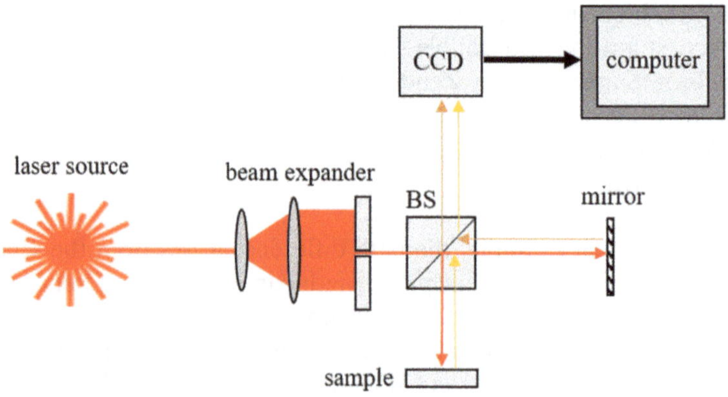

Figure 4.15. Schematic of a basic optical profilometer, where BS is a beam splitter.

tunnel across the vacuum between the sample probe and the sample surface. It consists of an atomically sharp tip that is scanned across a sample surface and whilst it is most often employed in ultra high vacuum it can also be used in air or liquid [17, 18]. Challenges of this measurement usually involve: (1) minimising vibrations, as this will limit the resolution, and (2) preparation of the tip.

For low bias voltage, V_T, the tunnelling current, I, can be approximated by [19]:

$$I \approx 18\frac{V_T}{10^4\Omega}\frac{k}{d}A_{\text{eff}}e^{-2kd} \tag{4.7}$$

where d is the distance between tip and sample, A_{eff} is the effective area of the tip, and k is defined by $2k = 1.025\Phi^{1/2}$, where Φ is the average work function. As such, there is an order of magnitude increase in the current for each Angstrom the tip moves closer to the surface. This makes it a particularly sensitive probe (a 2% tolerance on current would result in a 0.1 nm tolerance in the gap).

There are two common imaging modes:

1) Constant current mode: where the height between the sample surface and the tip is kept constant by using a feedback loop measuring the tunnelling current. The stylus is raised or lowered to keep the distance between it and the sample surface constant (inferred by the current that flows through the stylus–sample 'circuit' shown in figure 4.16) and a contour map is generated by recording the vertical distance moved by the stylus whilst scanning above the sample surface.
2) Constant height mode: where the feedback loop is switched off and the tip is scanned at a constant height. Changes in the tunnelling current can be used to infer topography. The advantage of this mode is faster imaging (due to removal of a feedback mechanism), which would make it more suitable for studying time dependent effects, however, it would require relatively flat surfaces to begin with.

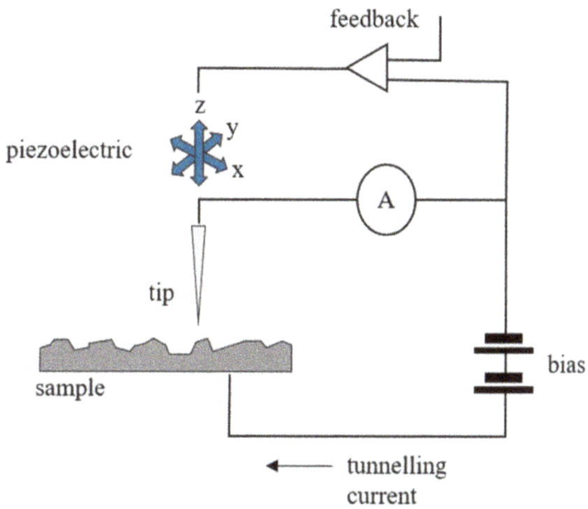

Figure 4.16. Schematic of an STM measurement. An atomically sharp tip is rastered across the sample surface, with feedback control to maintain constant current (without need for constant height).

STM can also provide information on local electronic properties, however, some knowledge of the size, and chemical nature of the tip is required to sufficiently interpret such data. This is complicated by difficulties of routinely preparing atomically sharp tips.

4.3.3 Atomic force microscope (AFM)

Atomic force microscopy (AFM) is another high resolution profilometry technique used to probe the sample surface to nanometer resolution (i.e. to determine surface roughness). Unlike STM, it does not require a conductive sample, and can also image surfaces in air or liquid. It works by scanning the sharp tip (nm) of a cantilever across a sample surface, as shown in figure 4.17. Image areas are typically <150 × 150 μm with a maximum roughness of the order of 10–20 μm. Deflections of the cantilever due to forces experienced by the tip will vary according to Hooke's law and are often measured by tracking the deflection of a laser spot reflected from the top of the cantilever. This is projected onto a four quadrant photodiode, as shown in figure 4.17, where deflection is calculated from the difference in A + B and C + D; comparison of A + C and B + D gives lateral or torsional bending of the cantilever. Depending on the type of probe, forces that the tip will experience include: van der Waals, chemical bonding (interatomic and intermolecular), electrostatic, capillary, magnetic, or adhesion.

There are three imaging modes [9, 20]:

1) Contact mode: where the probe is pressed against the sample surface. This is the simplest imaging mode, which can be run in either constant or variable force mode. In constant force mode, a feedback mechanism is employed to keep the height constant. The change in height as the sample is scanned can then be plotted to directly show the topography. In variable force mode this

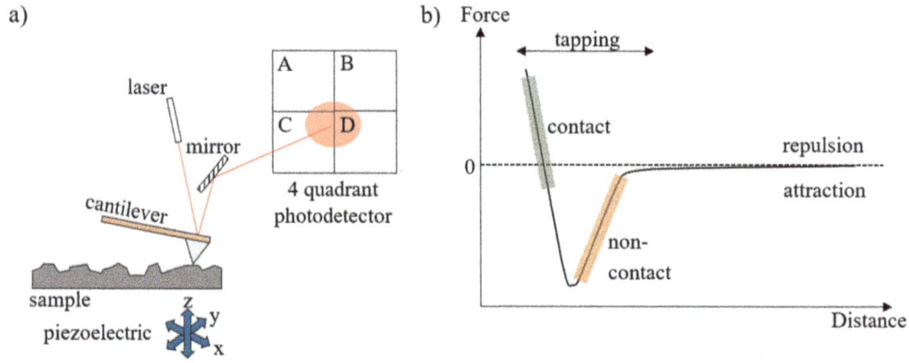

Figure 4.17. Basics of AFM. (a) A cantilever is brought into contact with the sample surface and scanned in x-/y- to map out the sample height. Deflections of the sample height are measured using a laser reflected off of the cantilever onto a four quadrant photodetector. (b) Schematic of the force experienced by the cantilever as a function of distance from the surface. As the tip moves into contact, there is a shift from attractive to repulsive forces. The three imaging regimes are defined by which area of this curve the measurement is operating in.

feedback is switched off and the measured deflection of the tip can be used to plot the sample topography. (Note that this requires a smooth surface or the tip may be damaged.) The advantage of variable force mode is that it can have higher resolution. Disadvantages of contact mode include damage to the sample, and 'stick-slip' movement, where adhesive or frictional forces between the tip and the sample result in image artefacts.

2) Non-contact mode: where the probe is held away from sample surface and long-range forces between the surface and tip result in deflection. In this mode the tip is oscillated about a resonant frequency such that small deflections will cause an observable shift in amplitude. As the tip is now kept a distance away from the sample surface the area of interaction will have decreased, resulting in a potential increase of resolution.
3) Intermittent contact or 'tapping' mode: where the probe height is oscillated at an amplitude greater than in non-contact mode. When the tip comes into contact with the sample surface it will engage, reducing the amplitude of oscillation and shifting the resonant frequency. A feedback loop can be used to maintain constant height and thus amplitude. One concern with this imaging mode is the impact of capillary forces due to adsorbed water, which can 'trap/stick' the tip on the surface unless the spring constant (of the tip) is high enough. For this reason, tips for tapping mode usually have much higher spring constants than tips for non-contact mode.

Extensions of AFM include: magnetic force microscopy (MFM), as discussed in chapter 8; electric force microscopy (EFM), which probes ferroelectric regions of the surface with a conductive tip [21]; scanning Kelvin probe microscopy (SKPM), which maps the surface potential and capacitance gradient using a conductive tip in non-contact mode [22]; scanning capacitance microscopy (SCM) [23]; scanning ion conductance microscopy (SICM), which scans a nanopipette in an ionic liquid [24, 25]; force-distance spectroscopy, where a force-distance curve is obtained, similar to that seen in figure 4.17 [26]; nanoindentation, where the hardness of the sample is measured by indenting the surface with the AFM tip, and imaging the resulting deformation [27]; nanolithography, where the cantilever is used to modify the surface of the sample [28]; lateral force microscopy (LFM), which measures torsional forces experienced by the cantilever determined by frictional and adhesive forces on the surface [29]; near field scanning optical microscopy (NSOM) [30], where a nanoaperture in the cantilever is used to optically excite the surface as the topography is mapped; tip enhanced Raman spectroscopy (TERS), as mentioned in chapter 5; and scanning thermal microscopy (SThM) [31], where the resistance of the tip is monitored as it is scanned in contact mode, providing information on local temperature.

4.4 Examples of microscopy techniques in the literature

In the following section, I will highlight several examples of research articles where microscopy techniques were used to characterise or identify a given material. For each article, starter questions are provided, which are aimed at focussing the reader's attention on the key aspects of the article (with respect to this chapter).

4.4.1 In 'Atomic-scale structural and electronic properties of $SrTiO_3$/GaAs interfaces: a combined STEM-EELS and first principles study'

Hong *et al* investigated the electronic and structural properties of GaAs grown on $SrTiO_3$ with a focus on the interface between the oxide and semiconductor [32].

Reading through the article, try to answer the following:

1. What is the motivation for this work?
2. Identify the key features of the STEM and EELS techniques.
3. How was the sample prepared prior to STEM (include the initial thin film deposition)?
4. What does HAADF stand for? What type of contrast will it provide?
5. Are the structures shown in figure 5 distinctly different? Or different projections of the same unit cell?

The aim of this work was to investigate the energetics of Ga and As termination at the $SrTiO_3$:GaAs interface. The authors combined experiment and theory to show that the SrO/Ga and SrO/As interfaces can be promoted under oxygen rich or poor conditions, respectively. In addition, using atomically resolved information they calculated the density of states for each interface thus determining that they were metallic in nature, with a Fermi level pinned to the conduction band minimum.

4.4.2 In 'Intermixing and periodic self-assembly of borophene line defects'

Liu *et al* investigated line defects in the 2D synthetic material—borophene [33].

Reading through the article, try to answer the following:

1. List the similarities and differences between borophene, boron and graphene.
2. How were the borophene samples prepared?
3. What experimental technique was used to study the line defects in this system, and why was this suitable?
4. What do the authors mean by 'hollow hexagons' and how does this correspond to the $\nu_{1/6}$ and $\nu_{1/5}$ structures?
5. How does scanning tunnelling spectroscopy differ from STM?
6. What are the main conclusions of this paper?

The aim of this work was to study the evolution of line defects and structural phases present in borophene in order to develop an atomic level understanding of the material. The authors observed two borophene phases that act as building blocks for different borophene crystalline phases, which can be described using a simple mixing ratio (of the two).

4.4.3 In 'Resolving strain in carbon nanotubes at the atomic level'

Warner *et al* investigated the atomic displacements of single-walled carbon nanotubes using high resolution transmission electron microscopy (HR-TEM) [34].

Reading through the article, try to answer the following:

1. Why is aberration corrected TEM necessary for imaging carbon nanostructures with atomic resolution?
2. What is a Moiré pattern?
3. What is the difference between chiral and non-chiral SWNT? How do the authors identify this?
4. What is the mechanism for introducing strain into these nanotubes?
5. How was the distortion in the SWNT determined?
6. What are the key differences between the models for different strain types shown in figure 4?

The aim of this work was to identify the strain present in single walled carbon nanotubes using high resolution TEM. By selecting a non-chiral nanotube the authors were able to identify atomic positions, and use this to build a map of their relative placement with respect to an unstrained nanotube model. This was then used to determine the type of strain present to be non-uniform shear strain.

4.4.4 In 'Atomic structure and dynamics of single platinum atom interactions with monolayer MoS_2'

Li *et al* investigated the dynamics of Pt atoms on a MoS_2 surface [35].

Reading through the article, try to answer the following:

1. How were the samples produced?
2. What type of imaging was used?
3. How did the authors distinguish between Mo, S and Pt?
4. At what site was Pt most stable for (a) clean MoS_2, and (b) MoS_2 with carbon impurities?
5. How was Pt hopping controlled in this experiment?
6. How was DFT used to explain the stable site for Pt on clean MoS_2?

The aim of this work was to investigate the dynamics of Pt atoms on a MoS_2 surface. The authors demonstrated that there are clear differences between surfaces with and without carbon impurities. They also identified a stable site for clean MoS_2, where the energy barrier for escape to another site (as calculated by DFT) was of the order of 3 eV.

References

[1] Haynes R 1984 *Optical Microscopy of Materials* (New York: Springer)

[2] Bradbury S and Bracegirdle B 1998 *Introduction to Light Microscopy* (Abingdon: BIOS Scientific)

[3] Fox M 2010 *Optical Properties of Solids, Oxford Master Series in Condensed Matter Physics* (Oxford: Oxford University Press)

[4] Mertz J 2010 *Introduction to Optical Microscopy* (Englewood, CO: Roberts and Company Publishers)

[5] Indebetouw G, Tada Y and Leacock J 2006 *Biomed Eng. Online* **5** 63

[6] Depeursinge C D *et al* 2002 Digital holography applied to microscopy *Proc. SPIE 4659, Practical Holography XVI and Holographic Materials VIII*
[7] Preza C, Snyder D L and Conchello J-A 1999 *J. Opt. Soc. Am.* A **16** 2185–99
[8] Flegler S L, Heckman J W Jr and Klomparens K L 1993 *Scanning and Transmission Electron Microscopy: An Introduction* (Oxford: Oxford university Press)
[9] Amelinckx S, van Dyck D, van Landuyt J and van Tendeloo G (ed) 1997 *Handbook of Microscopy: Applications in Materials Science, Solid-State Physics and Chemistry, Methods II* (Weinheim: VCH Publishers)
[10] Goodhew P J, Humphreys J and Beanland R 2001 *Electron Microscopy and Analysis* 3rd edn (New York: Taylor & Francis)
[11] Brydson R (ed) 2011 *Aberration-Corrected Analytical Transmisison Electron Microscopy* (New York: Wiley)
[12] Williams D B and Carter C B 2009 *Transmission Electron Microscopy: A Textbook for Materials Science* 2nd edn (New York: Springer) chs 3 and 22
[13] Hartel P, Rose H and Dinges C 1996 *Ultramicroscopy* **63** 93–114
[14] Caruana A J *et al* 2016 *Phys. Status Solidi RRL* **10** 613–17
[15] Binning G, Rohrer H, Gerber C and Weibel E 1982 *Phys. Rev. Lett.* **49** 57–61
[16] http://nobelprize.org/educational/physics/microscopes/scanning/
[17] De Feyter S and Schryver F C D 2005 *J. Phys. Chem.* **109** 4290–302
[18] Song J P *et al* 1993 *Surf. Sci.* **296** 229–309
[19] Bai C 1992 *Scanning Tunneling Microscopy and its Application, Springer Series in Surface Sciences* 32 (Berlin: Springer)
[20] Bowen W R and Hilal N 2009 *Atomic Force Microscopy in Process Engineering, Introduction to AFM for Improved Processes and Products* (Amsterdam: Elsevier)
[21] Girard P 2001 *Nanotechnology* **12** 485
[22] Melitz W, Shen J, Kummel A C and Lee S 2011 *Surf. Sci. Rep.* **66** 1–27
[23] Matey J R and Blanc J 1985 *J. Appl. Phys.* **57** 1437–44
[24] Hansma P K *et al* 1989 *Science* **243** 641–3
[25] Chen C, Zhou Y and Baker L A 2012 *Ann. Rev. Anal. Chem.* **5** 207–28
[26] Cappella B and Dietler G 1999 *Surf. Sci. Rep.* **34** 1–104
[27] Schuh C A 2006 *Mater. Today* **9** 32–40
[28] Wendel M *et al* 1996 *Superlattices Microstruct.* **20** 349–56
[29] Carpick R W, Ogletree D F and Salmeron M 1997 *Appl. Phys. Lett.* **70** 1548
[30] Dunn R C 1999 *Chem. Rev.* **99** 2891–928
[31] Gomès S, Assy A and Chapuis P 2015 *Phys. Stat. Solidi* A **212** 477–94
[32] Hong L *et al* 2017 *Phys. Rev. B* **96** 035311
[33] Liu X *et al* 2018 *Nature Materials* **17** 783–8
[34] Warner J H *et al* 2011 *Nature Materials* **10** 958–62
[35] Li H *et al* 2017 *ACS Nano* **11** 3392–403

IOP Publishing

Characterisation Methods in Solid State and Materials Science

Kelly Morrison

Chapter 5

Spectroscopy techniques

In this chapter, the various commonly available spectroscopy techniques will be outlined. They are separated into five main themes: inelastic scattering, electron spectroscopy, fluorescence, absorption and resonance techniques. The aim will be to provide the reader with a broad understanding of the available techniques so that they can make an informed decision on which one would be suitable for a given application. Finally, some examples will be highlighted by selected publications.

5.1 What is spectroscopy?

Spectroscopy is defined as the measurement and characterisation of interaction of light (electromagnetic radiation) and matter. For example, when a material is illuminated by an electromagnetic wave (photons) there are three general outcomes:

1) No interaction.
2) A photon will scatter inelastically off of an atomic orbital electron (Compton scattering for high energies).
3) A photon will be absorbed, resulting in electron emission, generation of an Auger electron, or fluorescence.

These interactions can be utilised to extract information such as the chemical composition, density, thickness, oxidation state, electronic band structure, or magnetic and lattice (phonon) excitations of a material. Indeed, the wealth of available information has led to several variations of similar techniques, such as the myriad manifestations of photoelectron or absorption spectroscopy. Thus, it is useful to simultaneously discuss similar techniques and in this chapter it is achieved by categorising them into the following broad themes:

- Energy loss (inelastic scattering);
- Electron spectroscopy (primarily photoemission techniques);

doi:10.1088/2053-2563/ab2df5ch5

- Fluorescence;
- Absorption;
- Resonance.

5.2 Energy loss spectroscopy

In chapter 3, the concept of inelastic scattering was briefly mentioned in the context of x-ray diffraction as a 'nuisance' to the spectrographer. Whilst this may be true for diffraction measurements, there is useful information to be obtained if one knows how to look for it. As was previously stated, the majority of scattering events, when an electromagnetic wave interacts with a material, will be elastic. The difficulty with any inelastic spectroscopy is then to detect the remaining 1%. In addition, the wavelength (or energy) of the wave must now be measured in order to determine how much energy was lost. As the inelastic scattering event will cause a change in energy of the wave, this manifests as a change (or shift) in the wavelength. This energy loss is therefore sometimes referred to in terms of this wavelength shift (quantified by the wavenumber, with associated units of cm^{-1}) requiring a wavelength dispersive measurement.

Much like diffraction, inelastic scattering can be characterised by the initial and final wavevectors, $\boldsymbol{k_i}$ and $\boldsymbol{k_f}$, as shown in figure 5.1, where $|\boldsymbol{k}| = 2\pi/\lambda$. In this case, however, as energy (and momentum) is not conserved, these two wavevectors are no longer equal ($\boldsymbol{k_i} \neq \boldsymbol{k_f}$) and the momentum wavevector, $\boldsymbol{Q}$ is now determined by:

$$\boldsymbol{Q} = \boldsymbol{k}_i - \boldsymbol{k}_f \tag{5.1}$$

If we define the incident wavevector, $\boldsymbol{k_i}$, as:

$$\boldsymbol{k}_i = \left(0,\, 0,\, \frac{2\pi}{\lambda_i}\right) \tag{5.2}$$

Then the momentum transfer, $\boldsymbol{Q}$, can be determined by:

$$\boldsymbol{Q} = \frac{2\pi}{\lambda_f}\left(-\sin 2\theta \cos \varphi,\, -\sin 2\theta \sin \varphi,\, \frac{\lambda_f}{\lambda_i}\cos 2\theta\right) \tag{5.3}$$

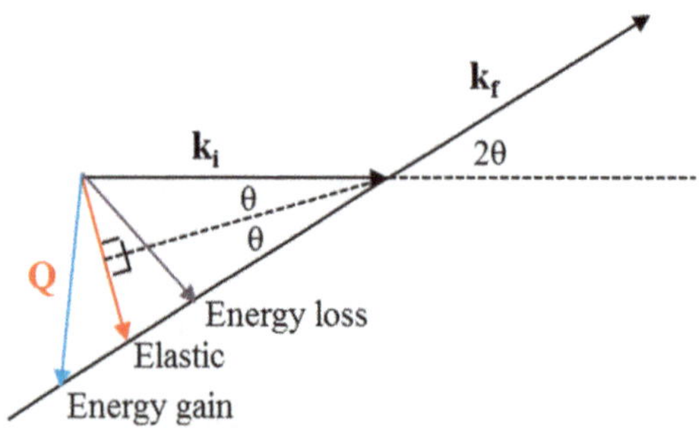

Figure 5.1. Schematic of an inelastic scattering event, where $\boldsymbol{k_i}$ and $\boldsymbol{k_f}$ are the initial and final wavevectors, respectively, and $\boldsymbol{Q}$ is the momentum transfer.

Therefore, we can calculate $\boldsymbol{Q}$ if the incident wavelength is known, and the scattering angles, θ and ϕ, can be determined alongside the final wavelength λ_f. Alternatively, the magnitude of $\boldsymbol{Q}$ can be determined using the vector sum:

$$Q^2 = k_i^2 + k_f^2 - 2k_i k_f \cos 2\theta \tag{5.4}$$

In addition, if the energy transfer, ΔE, is defined by $\Delta E = \hbar\omega$ then the energy loss or gain can also be determined from knowledge of the initial and final wavelengths as follows.

For x-rays (or other electromagnetic waves):

$$\Delta E = \hbar\omega = 2\pi c\left(\frac{1}{\lambda_i} - \frac{1}{\lambda_f}\right) = 2\pi\hbar(f_i - f_f) \tag{5.5}$$

For particles, such as neutrons:

$$\Delta E = \hbar\omega = \frac{h^2}{2m_\text{n}}\left(\frac{1}{\lambda_i^2} - \frac{1}{\lambda_f^2}\right) \tag{5.6}$$

where m_n is the mass of the neutron (or other particle used).

Common types of inelastic scattering include:

- Phonons—lattice vibrations. Inelastic scattering with phonons will result in a slight increase in temperature. Energy loss is typically less than 1 eV.
- Plasmons—collective electron waves. 5–30 eV energy loss, with a short mean free path. A common feature of inelastic spectroscopy, but does not provide much useful data.
- Spin waves—magnetic vibrations, where a perturbation of the orientation of one spin (precession) will excite a neighbouring spin, resulting in a collective excitation (much like a magnetic 'mexican wave').
- Inner shell excitation—inelastic scattering of an electron with an inner shell electron may result in ionisation of the atom as the inner shell electron is excited enough to overcome the binding energy of the atom. This would result in characteristic emission of x-rays as a higher energy electron decayed to fill the space that this process will have created.

Lattice excitations, or phonons, can be thought of as collective excitations of the atoms within a crystal lattice (or molecule). They can be described as symmetric, asymmetric, longitudinal, transverse or occupying acoustic or optical branches of the phonon modes. Examples of longitudinal and transverse lattice excitations are given in figure 5.2, where it can be seen that longitudinal phonons arise when atom movement (within a plane) is parallel to the direction of propagation of the wave, and transverse motion is perpendicular to the direction of propagation (indicated by the wavevector, k).

In chapter 1 the concept of crystal planes was presented in order to describe the collective position of atoms within a crystal lattice. In the context of phonon excitations, these planes can be used to simplify the description of collective atomic

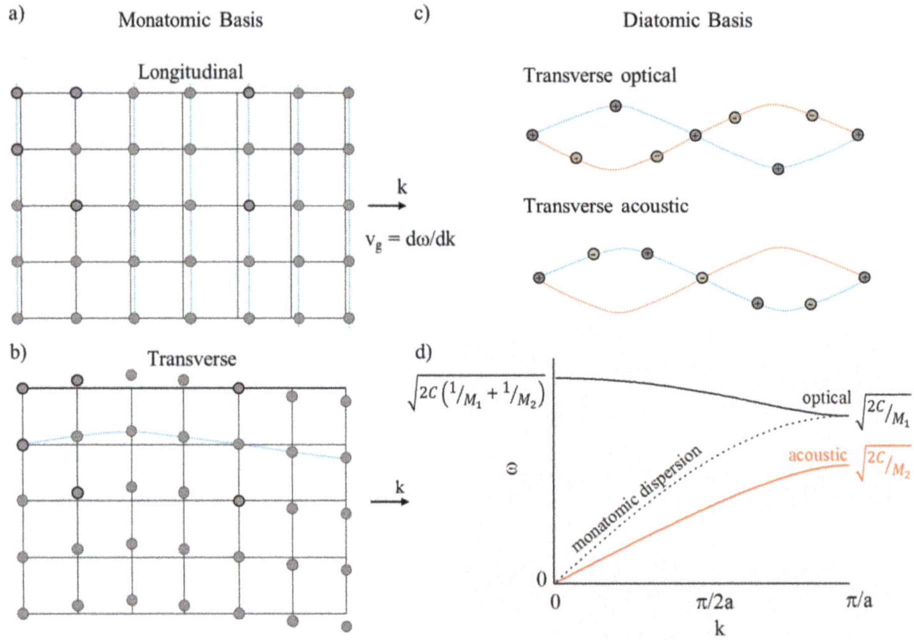

Figure 5.2. Examples of collective lattice vibrations (phonons) in a monatomic and diatomic basis. (a) and (b) longitudinal, and transverse lattice waves, respectively, where k indicates the wavevector (i.e. direction of propagation) and the dashed blue line indicates the displacement of individual atoms from their equilibrium positions. (c) Transverse optical and acoustic waves for a diatomic primitive basis. (d) Example phonon dispersion curves for (a)–(c).

displacements, as is seen in figure 5.2. For example, for a simple cubic system there are three key planes to consider: the [100] (edge), [111] (face diagonal), and [111] (body diagonal).

We consider first a simple system, where the primitive basis used to describe the unit cell is a single atom (monatomic basis). For any collective excitation there will be a force, F, associated with the displacement of each atom, and this could be described collectively by defining a force constant between nearest neighbour planes, C. By treating the movement of atoms within a plane in a similar way as springs, where the force experienced by each atom is a linear function of the displacement and force constant C:

$$F = C(u_{s+1} - u_s) - C(u_{s-1} - u_s) \tag{5.7}$$

we can start to describe the force acting on each atom, and therefore the equation of motion ($F = ma$) associated with it. A solution to these equations (derived in full here [1]) yields the following dispersion relation:

$$\omega^2 = \left(\frac{2C}{M}\right)(1 - \cos ka) \tag{5.8}$$

where C is the force constant for one atom of a plane, M is the mass of the atom, k is the wavevector, and a is the lattice constant. At the first Brillouin zone $k = \pi/a$ and the slope of this dispersion relation will be:

$$\frac{d\omega^2}{dk} = \left(\frac{2Ca}{M}\right)\sin ka = 0 \tag{5.9}$$

Therefore:

$$\omega^2 = \frac{4C}{M}\sin^2\frac{ka}{2} \tag{5.10}$$

And:

$$\omega = \sqrt{\frac{4C}{M}}\left|\sin\frac{ka}{2}\right| \tag{5.11}$$

(Outside the first Brillouin zone for a given direction we expect a reproduction of lattice motions within the first Brillouin zone.) As with any wave, the group velocity, which describes the speed of propagation of the wave, will be equal to:

$$v_g = \mathrm{d}\omega/\mathrm{d}k \tag{5.12}$$

For $ka \ll 1$ for the equations above this reduces to $v = \omega/k$ (i.e. linear). This can be seen in figure 5.2(d) where there is a linear dispersion at low values of k for the monatomic basis.

If we extend this example to a primitive basis with two atoms, then additional phonon modes start to appear due to the increase in degeneracy of the system. In this case the phonon dispersion develops two distinct branches—acoustic and optical—as shown in figure 5.2. These branches describe the relative motion of the different atoms with respect to one another for a given mode, where for acoustic phonons, both atoms move in phase/sequence, and for optical phonons, the atoms move $\pi/2$ out of phase. In the general case, where there are p atoms in the primitive basis, there will be theee acoustic branches (one longitudinal and two transverse) and $3p - 3$ optical branches. For the diatomic basis this results in three acoustic and three optical (one longitudinal and two transverse) branches.

A similar treatment for the diatomic model leads to a dispersion relation of the form:

$$M_1M_2\omega^4 - 2C(M_1 + M_2)\omega^2 + 2C^2(1 - \cos ka) = 0 \tag{5.13}$$

which has two distinct solutions that relate to the optical and acoustic branches, as shown in figure 5.2(d).

Spin wave excitations (figure 5.3) will be observed in any magnetic material that is excited by a microwave field. They will also be present in most magnetic materials as a result of thermal perturbations. The exact nature of the energy loss due to inelastic scattering with a spin wave will be driven by the exchange interaction. For example,

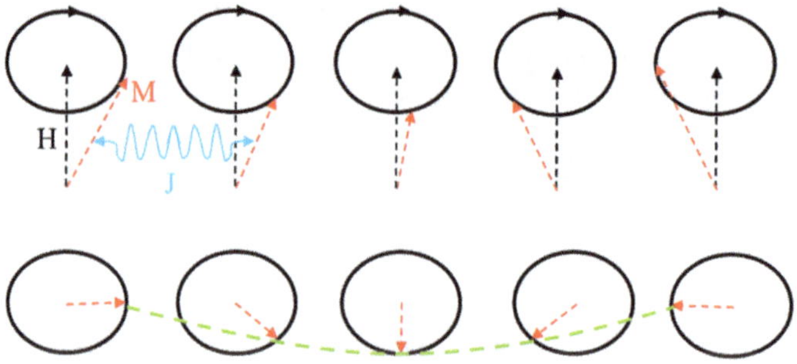

Figure 5.3. Example of a spin wave excitation. Top: side on view, where precession of the spin moment (M) about the applied magnetic field, H, is evident. The collective excitation will be driven by the interaction, J, between neighbouring spins. Bottom: Top down perspective, where the precession of the spin can be traced as a wave (green dashed line).

for ferromagnets, the exchange interaction can be estimated by the Heisenberg model where the ground state energy is:

$$U = -\sum_{ij} J\hat{S}_i \cdot \hat{S}_j \tag{5.14}$$

where J is the exchange interaction between the spin, S, on electron i and j. For a 1D chain with two nearest neighbours this relation simplifies to:

$$U = -2J\sum_{i} \hat{S}_i \cdot \hat{S}_{i+1} \tag{5.15}$$

which can be solved in a similar way to the phonon dispersion relations to obtain:

$$\hbar\omega = 4JS(1 - \cos ka) \tag{5.16}$$

where at small k:

$$\hbar\omega \approx 2JSk^2a^2 = Dk^2 \tag{5.17}$$

5.2.1 Inelastic neutron spectroscopy (INS)

In inelastic neutron spectroscopy, neutrons are scattered off of the sample to be measured, and measured by a detector setup optimised for determination of ΔE. For time of flight measurements, for example, the bunch of neutrons with varying energy that was useful for diffraction (where wavelength does not change) now pose a problem with regards to determining both Q and ΔE. To work around this, the neutrons are monochromated using a set of choppers so that λ_i is now well known (direct geometry). The final wavelength can then be determined by the usual time of flight method:

$$\lambda \sim \frac{ht}{m_n l} \tag{5.18}$$

where t is the time delay and l is the length of the beamline.

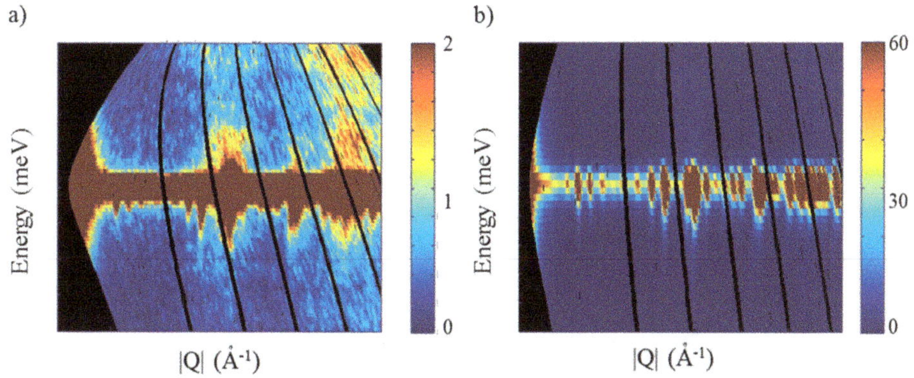

Figure 5.4. Example inelastic neutron scattering data for $La(Fe,Si)_{13}$ polycrystal, obtained on the MARI beamline, at ISIS Oxfordshire. Both images show the same data, but with different intensity scales. (a) Inelastic scattering from phonons can be observed at an energy transfer of 10–20 meV. (b) Low resolution Bragg diffraction peaks (the elastic line at $E = 0$). The black areas indicate 'blind spots' where the detectors are absent, or cannot measure neutrons due to the geometry of the sample area with respect to the detector bank.

An example of inelastic neutron scattering data for a magnetic sample—$La(Fe,Si)_{13}$—is given in figure 5.4. At $E = 0$ a broad line is observed that can be attributed to elastic scattering, as evident when the intensity scale is redefined, as seen in figure 5.4(b). At $\boldsymbol{Q} = 0$ the principal diffraction maxima can be seen (i.e. the first diffraction spot, which contains all information). At higher energies, such as 10–20 meV between $|\boldsymbol{Q}| = 3$–6 Å^{-1}, inelastic scattering from lattice vibrations can be observed. Close to the elastic line, at $|\boldsymbol{Q}| \sim 3.5$ Å^{-1}, the indications of a spin wave (magnetic excitations) can be seen.

Another method for measurement of inelastic neutron scattering is the triple-axis spectrometer. In this configuration the incident wavelength is selected by a crystal monochromator: using Bragg diffraction to separate different wavelength and selecting by the diffraction angle, θ. This wave is then scattered off of the sample, and the scattered waves are selected by wavelength using a second crystal monochromator. The addition of the analyser crystal is what defines this as a 'triple' axis spectrometer (figure 5.5), and it differs from time of flight techniques by the breadth of information it can collect. For example, now that the incident and final wavelengths are being selected, this measurement is limited to scans of constant $|\boldsymbol{Q}|$ (where $\boldsymbol{Q} = \boldsymbol{k}_i - \boldsymbol{k}_f$ and λ_i, λ_f are scanned in a 'locked coupled' mode) or ΔE (where λ_i could be kept constant and λ_f is scanned). The advantage is, however, higher resolution (figure 5.5).

5.2.2 Raman spectroscopy

Raman spectroscopy is an optical technique based on the Raman inelastic scattering process. In this case, the origin of inelastic scattering is the interaction of the photon (or electromagnetic wave) with the vibrational excitations of the sample to be studied (phonons). This can be thought of classically, where the interaction of the electromagnetic wave with a molecule can induce a vibration (or absorb one), or on

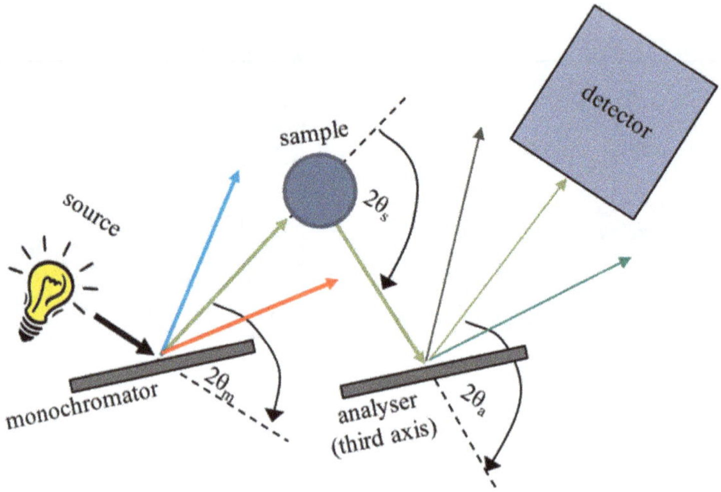

Figure 5.5. Schematic of a triple axis spectrometer. The analyser is simply a second monochromating crystal. Choice of angles θ_m, θ_s, and θ_a will select the incident wavelength at the sample, reflected beam, and wavelength of the beam at the detector.

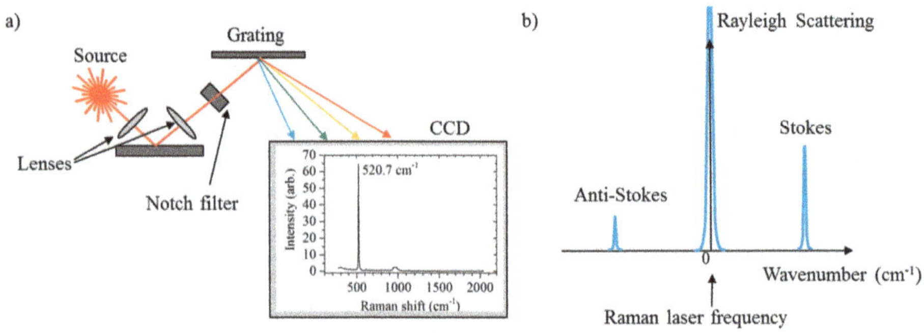

Figure 5.6. (a) Schematic of Raman spectroscopy and example calibration spectra for Si. (b) Schematic of the types of scattering of an electromagnetic wave off of a surface. Elastic Rayleigh scattering is the most probable event, followed by Stokes scattering, and then anti-Stokes.

a quantum level, where changes in polarisability in and around an atom results in splitting of the electronic energy levels referred to as virtual states. For the frequency range that is studied (up to THz) the information that can be accessed by this technique includes optical phonons and molecule vibrations. As every molecule has its own unique vibrational fingerprint, Raman spectra can be routinely used as an identification technique [2].

A schematic representation of a typical Raman spectrometer is shown in figure 5.6. Such a system consists of a laser source, lenses for focussing and collecting the laser spot on the surface of the sample, a notch filter for removing the majority of elastically scattered light, a grating to split the Raman spectra according to wavelength, and a CCD for detection,. Most commercial Raman systems, such as those from Renishaw or Jobin Yvon, are confocal Raman microscopes where the excitation and scattered light are delivered and collected through the same microscope objective in what is referred to as the 'backscattered' geometry.

By convention, inelastically scattered light that loses energy is referred to as Stokes scattering, and light that gains energy is referred to as anti-Stokes scattering. As can be seen from figure 5.6, Stokes scattering is more likely to occur than anti-Stokes; this is related to the population density of the vibrational states that would lead to anti-Stokes scattering (i.e. it is easier to excite a vibrational state than for the photon to interact with a low occurrence of that same vibrational state). As the energy of the vibrational state and the temperature will determine its population, the relative intensity of the Stokes and anti-Stokes peaks can also be used to measure temperature [3]. The convention is for Raman spectra to be plotted relative to the frequency used during the measurement (i.e. the laser frequency at zero) with the Stokes scattered light intensities plotted on the positive x-axis in wavenumber (cm^{-1}). This means that the peak positions are in general the same regardless of which laser is used.

The main drawback of Raman, much like most inelastic spectroscopy techniques, is the weakness of the signal that is produced, with a differential cross section, $d\sigma_R/d\Omega$ of the order of 10^{-32}–10^{-27} $cm^2\ sr^{-1}$ molecule [4]. For some materials, referred to as Raman inactive, you will not be able to get a signal at all. Whilst the Raman cross section, σ_R, is generally dependent on the wavelength of light used:

$$\sigma_R \propto \frac{1}{\lambda^4} \tag{5.19}$$

resonance effects that occur close to the frequency of light corresponding to an electronic transition can lead to enhanced scattering, which can be useful for increasing the signal of the measurement. If this is not the case, however, the user might be tempted to increase the signal by increasing the fluence (power per unit area) of the incident laser. This would be achieved by increasing the power of the laser itself, or by increasing the magnification of the optical system, so that the same power is focussed on a smaller spot. If not careful, increasing the fluence in this way can lead to photobleaching (i.e. burning) of organic molecules, or local changes in temperature of inorganic materials with poor thermal conductivity. In addition, depending on the material being studied, this local increase in temperature could itself lead to a phase change.

Raman peaks will be typically well defined, particularly so when compared to fluorescence features. There are many factors that can influence the shape of Raman features, however, the quality of the crystal structure is particularly important. As a result, materials with well defined crystalline structures such as semiconductors will normally have the best defined peaks. This is why it has become standard practice within the Raman community for Raman systems to be calibrated using the main mode of silicon located at 520.7 cm^{-1} (as seen in figure 5.6), before any measurement. The scan range will usually be set for 200–2000 cm^{-1} as this is the active range for most molecules studied, however, for some lighter elements (such a H atom attached to a C atom) stretching lines can be found at a wavenumber as high as 3000–3500 cm^{-1}. As the position of these peaks will also be dependent on strain (for a given material) they can be used, for example, to scan across a quantum well to image the strain between epitaxially grown layers [5], or mapping of strain across a graphene flake [6].

A popular application of Raman over the last decade has been for characterisation of carbon structures such as carbon nanotubes, graphene, graphite and amorphous carbon. Examples of this have been given in figure 5.7. In graphene, for example, there are three peaks that are principally used to characterise the quality:

1. The defect 'D' peak at ~1350 cm^{-1}
2. The graphitic 'G' peak at ~1580 cm^{-1}.
3. The second order defect '2D' peak at ~2600 cm^{-1}. This peak is present whether the d peak is or not, due to the presence of the Dirac cone in graphene (which allows for double resonance to occur) [7].

The ratio of these peaks can convey a lot of information. Where the D and G peaks can indicate the quality of the sample, the ratio of the G and 2D peaks can indicate how many layers there are. For example, for very amorphous carbon, there will be a

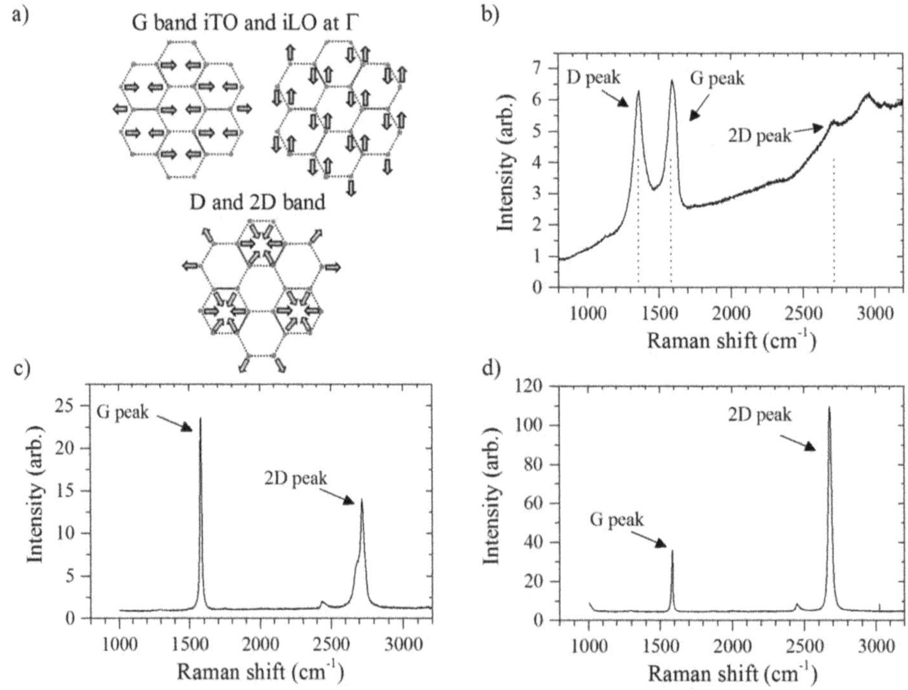

Figure 5.7. Raman spectra is various form of carbon/graphite. (a) Schematic of the phonon modes that lead to the D, G and 2D peaks. (b) Amorphous carbon. (c) Few layers of mechanically exfoliated graphene. (d) Single layer of mechanically exfoliated graphene.

broad D peak in the Raman spectra, whereas in graphite, the G peak will be very high, with evidence of a small D peak and some 2D peak, as seen in figure 5.7. The ratio of the D and G peak can also be used as an indicator of thickness for up to three layers. If the 2D peak is more than 1.5 times higher than the G peak then it indicates one layer, if the intensities are equal then it indicates two layers, and if the 2D peak is approximately half the intensity of the G peak it indicates three layers or more.

In summary, the position, intensity and FWHM of a Raman peak can be affected by:

- Doping;
- Temperature;
- Local environment;
- Strain;
- Phase;
- Particle size;
- Crystal quality.

Thus it can be a very powerful tool when investigating phase changes in a material, or for non-destructive assessment of sample quality.

Extensions of Raman include surface enhanced Raman scattering (SERS) and tip enhanced Raman scattering (TERS). In SERS plasmonic interactions driven by gold nanoparticles dispersed over the sample to be measured, are used to enhance the signal by a factor of 10^6–10^9. This measurement is difficult to reproduce exactly, however, due to the difficulty with uniform distribution of the Au particles as well as their size distribution (and the effects that this will have on the enhancement itself). In TERS, the plasmonic interaction is limited to the end of a metallised AFM tip (or sharpened optical fibre), which is held in non-contact mode above the sample, whilst a laser is incident on the surface. Whilst this is more controlled than SERS, the enhancement produced by individual tips will need to be calibrated against a standard such as silicon.

5.2.3 Brillouin light scattering spectroscopy (BLS)

Brillouin light spectroscopy is an inelastic spectroscopy technique that measures scattering of phonons in the GHz range. Whereas in Raman spectroscopy the frequency shift was of the order of 10^{13} Hz for a 10^{14} Hz incident light source, the focus for BLS is much smaller frequency shifts (10^8–10^{11} Hz range). Whilst Raman will be used to observe inelastic scattering of light from vibrational and rotational modes of lattice vibrations, BLS is generally used to measure the elastic properties of a material by directly probing the propagation of sound waves within the sample. Alternatively, it has also been used as a sensitive measurement of magnetic phenomena such as spin waves if coupled with suitable polarisers.

In order to measure such subtle differences, a highly stable, monochromatic source will be required in addition to a Fabry–Pérot interferometer in place of the diffraction grating used for Raman, as shown in figure 5.8. The Fabry–Pérot

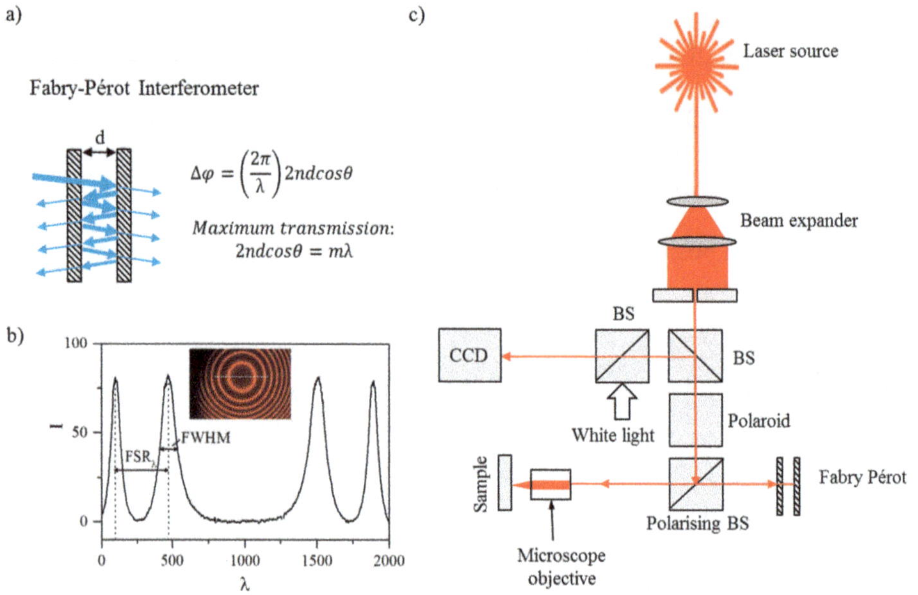

Figure 5.8. (a) Schematic of the Fabry–Pérot etalon. (b) Definition of the free spectral range, FSR, and full width half maximum, FWHM, for a Fabry–Pérot etalon. The inset is the corresponding interference pattern, where the dashed white line indicates the linescan shown. (c) Schematic of the micro-BLS setup for imaging of magnons.

interferometer works by interference of light that is reflected multiple times from two parallel mirrors, a schematic of which is given in figure 5.8(a). It consists of two plane parallel mirror surfaces, with reflection coefficient, R, and separated by distance, d. As the incident light passes into the interferometer it is subsequently reflected off of the internal surfaces a number of times, thus introducing a phase difference, $\Delta\varphi$:

$$\Delta\varphi = \left(\frac{2\pi}{\lambda}\right)2n_{\mathrm{FP}}\, d\cos\theta \tag{5.20}$$

where n_{FP} is the refractive index of the etalon, d the mirror separation, θ the angle of incident light with respect to the normal, and λ is the wavelength of light used. Due to the interference of light that eventually exits the interferometer, and will thus have some integer multiple phase difference (compared to the incident light); this will result in an interference pattern with maxima that occur when:

$$2n_{\mathrm{FP}}\, d\cos\theta = m\lambda \tag{5.21}$$

Therefore, the mirror spacing, d controls the sensitivity to a particular wavelength. In practice, this separation will often be controlled by a piezoelectric motor. The Fabry–Pérot can be used on its own or in tandem with a second interferometer in order to increase the sensitivity, as defined by the finesse, F, of the interferometer:

$$F = \frac{FSR_\lambda}{FWHM_\lambda} = \frac{\pi\sqrt{R}}{1 - R} \tag{5.22}$$

where R is the reflectivity of the mirrors, $FSR_\lambda = \Delta\lambda$ is the free spectral range, and $FWHM_\lambda$ describes the full width half maximum of the maxima peaks, as shown in figure 5.8(b).

Examples of this technique in practice include the study of surface acoustic waves in an optical fibre [8], measurement of confined spin waves [9], and micro-focussed BLS [10], as shown in figure 5.8(c), where a polarising beam splitter is used to separate the Rayleigh and Stokes scattering.

5.2.4 Electron energy loss spectroscopy (EELS)

Electron energy loss spectroscopy involves the measurement of electrons that are scattered inelastically when passing through a material; commonly employed with TEM [11]. Whilst this technique does involve characterisation of inelastic scattering, it borrows heavily from concepts in absorption spectroscopy, such as absorption edges (due to inner electron ionisation) and fine structure oscillations. It is particularly useful for characterisation of low Z elements such as He, Li and Be, where fluorescence techniques such as XRF and EDS struggle.

In EELS a magnetic spectrometer is used to measure the energy of electrons which have scattered off of a material surface. This works by passing the electrons—charged particles—through a magnetic field, where the electrons will be deflected by the Lorentz force:

$$\mathbf{F} = q\mathbf{E} + q\mathbf{v} \times \mathbf{B} = q\mathbf{v} \times \mathbf{B} \tag{5.23}$$

where the electric field, $\mathbf{E} = 0$, q is the charge on the electrons, $\mathbf{v}$ is their velocity, and $\mathbf{B}$ is the magnetic field. The level of deflection can therefore be controlled by the magnetic field within the spectrometer, which will either detect single energies (serial measurement) or multiple energies simultaneously by employing the use of a strip detector (and the angular spread of the electrons as they exit the magnetic spectrometer). Note that this detector is the magnetic analogue of the hemispherical detector discussed in the next section (electron spectroscopy).

A typical EELS spectrum will have an intense peak at $E = 0$, where the electrons have scattered elastically or transmitted through the sample, with low loss regions close to 20 eV that are due to plasmons. If we zoom in on the scan far enough, then characteristic edges should be observable, similar to x-ray absorption spectroscopy. It is these edges which can be used to determine which elements are present, and quantify their relative fraction(s) [12]. Due to the large background close to $E = 0$ (i.e. the elastic line) it is commonly the higher energy absorption edges (110–2000 eV) that are used for analysis. Extracting quantitative information from this measurement, however, requires careful analysis and sample preparation, as the measurement is convoluted with additional energy loss effects (such as plasmons) as well as multiple scattering events through thicker samples.

5.3 Electron spectroscopy

The generation of electrons when a material is illuminated by an electromagnetic wave, such as x-rays, is the focus of electron spectroscopy. It includes Auger electron spectroscopy (AES), x-ray photoemission spectroscopy (XPS), and angular resolved photoemission Spectroscopy (ARPES). As the concepts and techniques employed are similar, they will be first outlined here, followed by a more detailed description of each method.

5.3.1 Overview

5.3.1.1 Photoemission spectroscopy

In chapter 1 we discussed the photoelectric effect: the emission of electrons when a material is illuminated with an electromagnetic wave (i.e. photons) with energy that exceeds the work function of the material, Φ, (the energy required to release the outermost electron). In photoemission spectroscopy (PES) this principle is used to probe the energy levels of a material in order to determine the chemical composition as well as information about the types of bonds that have been formed.

Similar to the development of x-ray diffraction, the groundwork of PES was laid by Manne Siegbahn and his son. For example, Manne won the Nobel prize in 1925 for 'his discoveries and research in the field of x-ray spectroscopy'. This was for the development of apparatus and methods for improved accuracy when mapping x-ray spectra. His son, Kai M Siegbahn, later won the Nobel Prize in 1981 for 'his contribution to the development of high-resolution electron spectroscopy'.

In a typical PES measurement, a schematic of which is given in figure 5.10, a monochromatic beam of x-rays is focussed onto the sample of interest. Electrons emitted by the photoelectric effect are then collected by an energy analyser, which measures the number of electrons per unit time at a kinetic energy, E_K. The binding energy, E_B, can then be determined using:

$$E_B = E_i - E_K - \Phi \tag{5.24}$$

where Φ is the work function of the spectrometer, and E_i is the energy of the incident x-rays.

Knowledge of the binding energy is useful when mapping the chemical specificity (i.e. the valence of individual elements, such as Fe^0, Fe^{2+}, Fe^{3+}). This can be achieved by applying Koopman's theorem where *the binding energy of an electron is*

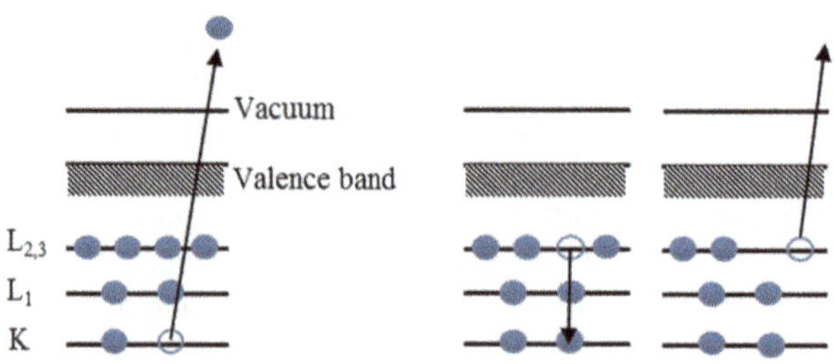

Figure 5.9. Generation of an Auger electron.

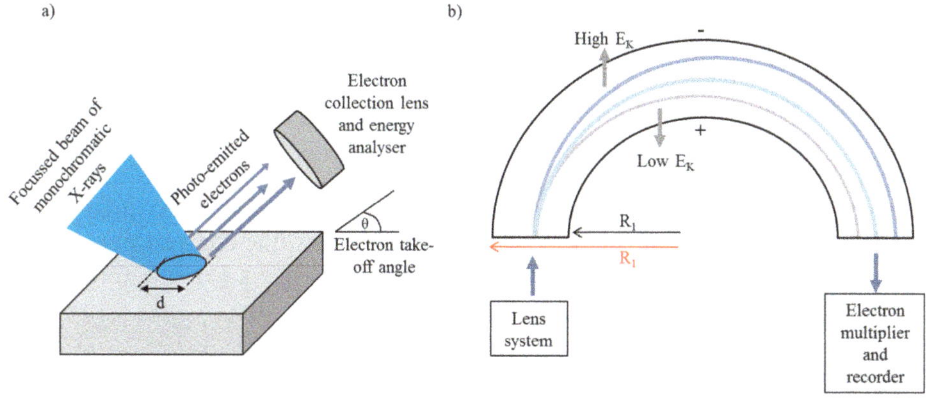

Figure 5.10. Schematic of a standard XPS measurement. (a) Geometry of the XPS measurement, where d is the spot size of the x-ray beam on the sample, and θ is the angle between the sample surface and detector cross section. (b) Schematic of the hemispherical energy analyser.

the difference in energy between the initial state, where an atom has n electrons, and the final state, where the atom is ionised (n − 1 electrons) and the electron has escaped. This is summarised by:

$$E_{\mathrm{B}} = E_{\mathrm{final}}(n - 1) - E_{\mathrm{initial}}(n) = \varepsilon_{\mathrm{k}} \tag{5.25}$$

where ε_{k} is the orbital energy of the ejected electron. It should be noted that there are relaxation processes (rearrangement of the electrons within the energy shells), relativistic and correlation effects that result in small changes to this value, such that:

$$E_{\mathrm{B}} = -\varepsilon_{\mathrm{k}} - \delta\varepsilon_{\mathrm{relax}} + \delta\varepsilon_{\mathrm{relativistic}} + \delta\varepsilon_{\mathrm{correlation}} \tag{5.26}$$

If these perturbations ($\delta\varepsilon_{\mathrm{relax}}$, $\delta\varepsilon_{\mathrm{relativistic}}$, $\delta\varepsilon_{\mathrm{correlation}}$) can be considered small with respect to ε_{k}, however, then Koopman's theorem holds.

5.3.1.2 Generation of Auger electrons

Photoemission results in the liberation of an electron, typically from a core–shell. This results in an unoccupied lower energy level, which is soon filled by a higher energy electron decaying to fill that space. Sometimes this results in the emission of a photon that has energy equal to the difference in energy levels (of the higher and lower energy levels), but it is also possible for it to result in fluorescence (where the energy of the emitted photon is not equal to the absorbed photon), or the emission of another electron—known as an Auger electron.

Generation of Auger electrons, as shown in figure 5.9 is a three electron process:

1) The first electron is emitted from the sample via photoemission or electron collision.
2) An electron from a higher energy level relaxes to fill the vacated lower energy state.
3) The transition energy from this relaxing electron is imparted to another electron, which then escapes.

Auger electrons have characteristic kinetic energy irrespective of the incident energy. This is different to PES, where the kinetic energy is the difference between the ejected electron and the incident photon energy.

5.3.1.3 Hemispherical detector

The energy of the ejected electrons, E_k, once collected, is typically measured using a hemispherical energy analyser. In a detector such as this, the charged particles are directed between two curved, charged plates and experience a Lorentz force:

$$\mathbf{F} = q\mathbf{E} + q\mathbf{v} \times \mathbf{B} = qV \tag{5.27}$$

where in contrast to the detectors used in EELS, the magnetic field, $\mathbf{B} = 0$, $\mathbf{F}$ and $\mathbf{E}$ are the force and electric field (both vector quantities), q is the charge of the electrons, $\mathbf{v}$ is the velocity of the electrons, and V is the electric potential, which is perpendicular to the magnitude of the force, $|F|$. This results in bending of the electron beam, which is proportional to their speed and the potential difference between the two plates, ΔV. As such, high energy particles will not change direction fast enough to avoid hitting the side of the analyser (negatively charged plate). Similarly, low energy particles will not have enough energy to make it through the analyser before they hit the positively charge plate. For the rest of the electrons, the divergence of the electron beam will be proportional to their kinetic energy and can be detected by a multi-channel detector. There is therefore, an energy range between which the electrons could reach the detector and a path that follows the curvature of the hemispherical detector for electrons with the following energy [13]:

$$E_k = q\Delta V\left(\frac{R_1 R_2}{R_2^2 - R_1^2}\right) \tag{5.28}$$

where R_1 and R_2 are the radii of the outer and inner charged plates, respectively.

There are two measurement modes that the analyser can be used in: constant analyser energy (CAE) and constant retard ratio (CRR). In CAE (also known as fixed analyser transmission, FAT) the electrons are decelerated (retarded) to a user defined 'pass energy' before they enter the analyser, which has a potential difference matched to this energy (see equation 5.28). In this measurement mode, the choice of energy will determine the resolution and quality of the measurement. For example, low pass energies would have the advantage of high resolution, but fewer electrons will be detected and thus the signal will be low. For high energies the reverse is true—lower resolution, but larger signal. Note that as the pass energy is kept constant throughout the measurement, the resolution is also constant.

In CRR (also known as fixed retard ratio, FRR), the electrons are decelerated to a ratio of their original kinetic energy. In this measurement mode, the lens system will scan through the energies to be measured, and the voltage on each plate will be simultaneously scanned such that ΔV (and the pass energy) matches the ratio:

$$E_{pass} = \frac{E_k}{\text{retard ratio}} \tag{5.29}$$

The resolution will also be proportional to the retard ratio:

$$\text{Resolution} = \frac{k}{\text{retard ratio}} \tag{5.30}$$

where k is a parameter that depends on the quality of the instrument (0.002%–2%). Note that as this is a percentage resolution of the energy, E_k, this implies that the resolution of CRR measurements will decrease for higher kinetic energies. For CAE, as the pass energy is kept constant, the resolution will be constant throughout the measurement.

In principle, the intensity of the peaks measured by the analyser will be proportional to the number of atoms per unit volume, the incident flux (of x-rays), and the ionisation cross section, σ_i. This is complicated by variation in the detector's efficiency with energy and sensitivity factor, which can be calibrated for by measuring a known sample such as Au. (Incidentally, this is also how the work function of the spectrometer is determined.)

If the detector can be considered calibrated then, the concentration of element A in atomic % could be determined from:

$$C_A = \frac{\frac{I_A}{F_A}}{\sum_n \left(\frac{I_n}{F_n}\right)} \times 100\% \tag{5.31}$$

where I_A and I_n are the peak areas for element A and n, and F_A and F_n are the corresponding sensitivity factors. The sensitivity factor, F, will include: the cross section of emission for that particular element and orbital transition; instrumental factors such as the transmission function, efficiency of the detector, and corrections for stray magnetic field effects; and the electron attenuation length.

5.3.1.4 Information depth

PES is often a surface sensitive measurement due to inelastic scattering of the photoemitted electrons with atoms inside the material. For each additional inelastic scattering event after the electron has been ejected, it will either gain/lose energy, or be reabsorbed. This behaviour is characterised by the Beer–Lambert law:

$$I_d = I_0 e^{-d/\lambda_a \sin\alpha} \tag{5.32}$$

where α is the electron take off angle (see figure 5.10), d is the depth, and λ_a is the attenuation length: the average distance an electron with energy, E_k, can travel. As you can see from figure 5.11, the key result of this behaviour is that approximately 66% of the electrons ejected will originate from a depth of λ_a, and 95% from a depth of $3\lambda_a$. This leads to a characteristic lengthscale beyond which information cannot be usefully retrieved.

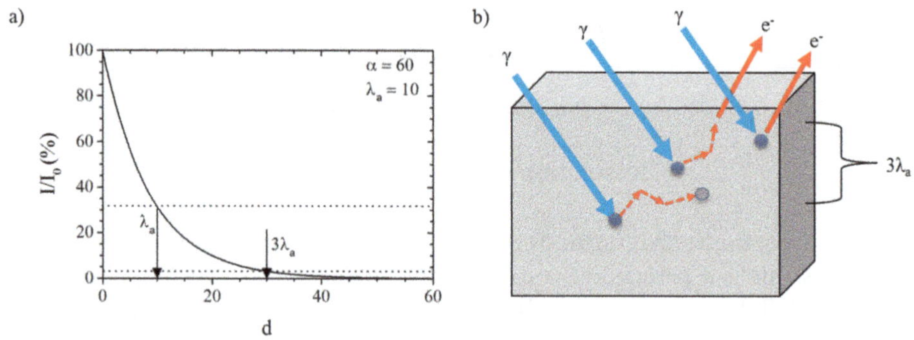

Figure 5.11. Inelastic scattering events in XPS (a) Decrease in intensity of electrons emitted at a depth, *d*, with respect to electrons escaping at the surface, I_0. (b) Schematic of the path(s) an electron may take (red arrows) when excited with a photon (blue lines). Either the electron escapes (solid red arrow) or is reabsorbed.

Another quantity that is often used to describe the lengthscale over which electrons can travel within a material is the inelastic mean free path (IMFP). As a general rule of thumb, the attenuation length is about 90% of the IMFP and both will vary with the energy of the incident electrons, E_A, as follows:

$$\lambda_a = \frac{538a_A}{E_A^2} + 0.41a_A(a_A E_A)^{0.5} \tag{5.33}$$

where E_A is in units of eV, $a_A{}^3$ is the volume of the atom in nm^3 that the electron is interacting with, and λ_a is in units of nm [14].

5.3.2 X-ray photoelectron spectroscopy (XPS)

XPS is a surface sensitive technique that measures the energy of photons emitted when a sample is illuminated with x-rays. Unlike x-ray diffraction, the source here is typically Mg or Al due to the lower energy characteristic lines (1253.6 and 1486.6 eV, respectively). The advantage of this is a larger ionisation cross section, σ_i, ($\sigma_i \propto 1/(E_i)^3$), whilst still being high enough to generate photoexcitation in most elements. The linewidth of the characteristic K_α peaks used in typical measurements are also particularly low (0.9 eV for Al, 0.7 eV for Mg), thus aiding energy resolution. In addition, the generation of x-rays relies on the anode (Al or Mg, for example) to be at high potential, rather than grounded, as was the case with CRTs in chapter 3. This enables the construction of 'twin anode' assemblies, where a measurement can be completed with one, or both of the available anodes. Switching between these two anodes has the advantage of selectively choosing the information depth of the measurement.

In the measurement, the sample and the spectrometer are grounded together so that the Fermi levels are matched and the work function of the measurement, Φ, (i.e. the spectrometer) can be reliably determined. The binding energy or kinetic energy of the photoemitted electrons can then be calculated using equation (5.24). For reference, it is not unusual to obtain 0.1% atomic resolution; whilst spatial resolution might be limited to microns.

As the binding energy, E_B, measured by XPS will be sensitive to the local chemical and physical environment as well as the oxidation state of the atom, it can be used to distinguish between similar chemical states. For example, XPS data given in figure 5.12 demonstrates that for different oxidation states of iron oxide (Fe_xO_y) subtle shifts in the binding energy for the Fe ^2p_j states is observed.

Variations of XPS include angular resolved XPS, which has the advantage of mapping out the elemental concentration as a function of depth, hard x-ray photoelectron spectroscopy (HAXPEX), which has the advantage of larger penetration depths (and thus is no longer surface sensitive), and small area XPS (SAXPS), which is optimised for higher resolution.

This variation of the information depth (3λ) with energy, E_A, and take-off angle, α, is utilised in both HAXPEX and angular resolved XPS. For HAXPEX, higher energy x-rays (e.g. 15 keV rather than usual 1.5 keV) will result in an increase of the attenuation length, and thus the volume of sample that can be sampled. This means that the techniques need no longer be surface sensitive, requiring UHV. The compromise, however, is the reduced ionising cross section, σ_i, which is compensated by using a higher incident flux of radiation, as is typically available at a synchrotron source.

For angular resolved XPS, the angle of the detector can be varied with respect to the sample in order to collect ejected electrons at varying take-off angles. By analysing the information obtained in this way, in conjunction with the Beer–Lambert Law defined by equation (5.32), we can obtain depth dependent information. For example, the intensity of ejected electrons will drop off faster at normal incidence (i.e. for $|\alpha < 90|$). This means that by decreasing the angle between the sample surface and the detector, (as seen in figure 5.13) the data obtained will be averaged over a smaller thickness. As such, a set of measurements for varying angles, can be compared to determine which elemental states exist further below the surface. In a similar fashion, two sets of measurements could be obtained with the Al or Mg anode sources, with characteristic energies of 1486.6 and 1253.6 eV, respectively. In this case the change in information depth will be $6\sin\alpha$ and $7\sin\alpha$ (equation (5.32)), respectively, due to the different energies of the source, whereas by varying the angle between 15° and 90° we could obtain changes in the information depth of 0.8λ to 3λ, as shown in figure 5.13.

Small area XPS requires some adaption to the standard measurement geometry in order to enable high resolution spatial mapping of the sample surface. This is achieved by focussing the monochromated x-ray source onto the sample surface [15], and in some cases, rastering this across the surface in order to build up a spatial map of the elemental composition. In such a way, the spatial resolution will only be limited by the spot size of the x-ray probe, where focussing of the x-ray source can be achieved by a combination of electrostatic or electromagnetic lenses and a monochromator.

Finally, destructive XPS could also be employed—where the surface of the sample is milled (sputtered) away as the spectra is collected. Whilst this can, in principle, provide information about the elemental composition below the surface, it is also complicated by implantation effects of the ion beam itself.

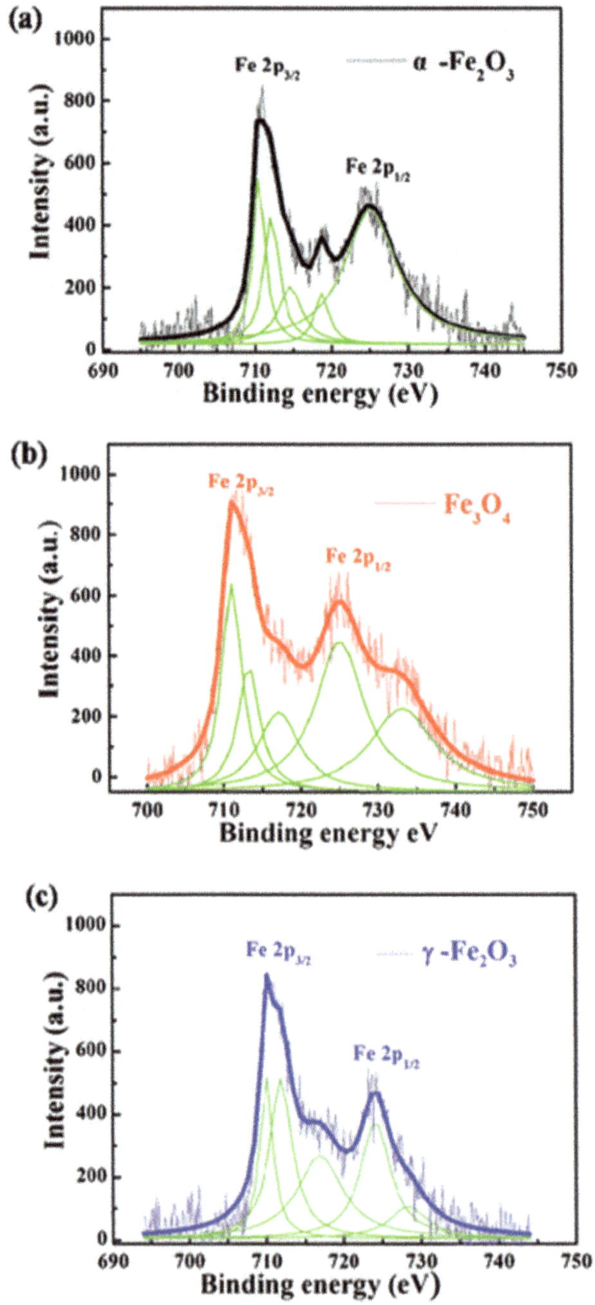

Figure 5.12. XPS for the Fe ^2p_j peaks of FeO_x nanoparticles. Subtle changes in the Fe environment (such as bonding, density or p-/n-type dopants) alters the position of the Fermi level, resulting in observable shifts of the characteristic peaks. Reproduced from [15] with permission from The Royal Society of Chemistry.

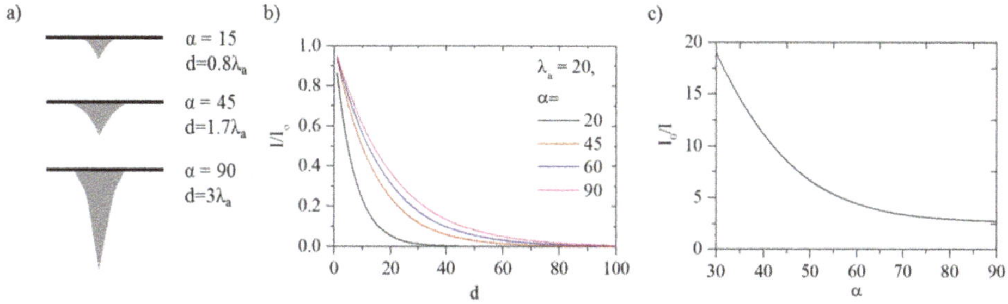

Figure 5.13. Available depth information with ARXPS. (a) Schematic of the change in the probe volume as the angle, α, is varied. (b) Variation in the normalised intensity calculated from the Beer–Lambert law for several incident angles. (c) Variation of the normalised intensity originating from a depth of $1\lambda_a$ as a function of angle, α.

5.3.3 Ultraviolet photoemission spectroscopy (UPS)

UPS is another variant of PES, where the excitation radiation is in the ultraviolet (UV) range ($\lambda = 10$–400 nm, $f = 10^{14}$–10^{16} Hz) rather than x-rays ($\lambda = 0.01$–10 nm, $f = 3 \times 10^{16}$–3×10^{19} Hz). The lower frequency of the UV source results in a lower incident energy, $E_i < 100$ eV, and thus higher energy and momentum resolution. For example, a commonly used ultraviolet source is the helium discharge lamp, which has characteristic emission lines for He I or II at 21.2 eV and 40.8 eV, respectively. As can be seen from equation (5.33), this lower E_i also results in a much lower information depth, which makes the technique extremely surface sensitive (5 Å at $E_i = 20$–100 eV). Thus, for UPS measurements ultrahigh vacuum and often *in situ* growth of films to be studied is required.

In addition, the photoemission driven by this measurement will involve the valence electrons (i.e. the outermost electrons) of the sample to be measured, rather than the core states excited by XPS [16]. This makes this technique particularly useful for studying the electronic states of a material, such as is the case with ARPES.

5.3.4 Angle resolved photoemission spectroscopy (ARPES)

ARPES is another variant of PES, also known as angle resolved ultraviolet photoelectron spectroscopy (ARUPS), where the excitation radiation is ultraviolet rather than x-ray. The difference between UPS and ARPES is that the detector is now scanned so that momentum and energy can be detected, as shown in figure 5.14. In this way, the band structure, $E(\mathbf{k})$, of the sample can be determined, and it is for this reason that this technique is often referred to as band mapping. The advantage of using UV incident radiation for this technique is that the lower incident energy, E_i, results in higher energy and momentum resolution.

As we are mapping reciprocal space, however, to obtain any useful information requires the sample to be a well aligned single crystal. In addition, the extreme surface sensitivity of the UPS technique (5 Å at 20–100 eV) often requires the crystals to be cleaved *in situ* (i.e. under vacuum). The resolution function in this

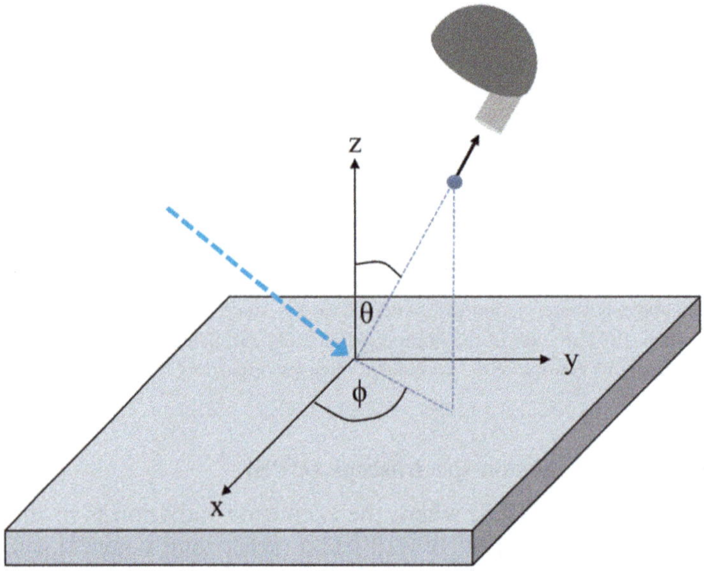

Figure 5.14. Measurement geometry for ARPES. The energy analyser is scanned across the sample at angles ϕ and θ to the sample surface.

measurement is also fairly complex given that we are now mapping in 3D. Whilst energy resolution will largely be determined by the analyser, momentum resolution will be limited by the analyser and E_k, and spatial resolution by the spot size of the UV radiation as well as the stability of the source.

For an in depth treatment of the theory of ARPES, and the various assumptions made, I refer the reader to [18] and the references therein. In its simplest sense (i.e. assuming non-interacting particles, and that the photon momentum is negligible), the momentum can be determined by first determining the photoelectron momentum, p:

$$p = \sqrt{2mE_k} \tag{5.34}$$

where m_e is the electron mass, and E_k is the measured kinetic energy. If the measurement of E_k is angle resolved, then the momentum, p, can be regarded as a vector ($\boldsymbol{p}$), which is related to the wavevector, $\boldsymbol{k}$, by:

$$\boldsymbol{p} = \hbar \boldsymbol{k} \tag{5.35}$$

This is useful as it provides a way for us to interpret the information in reciprocal space, as is standard for band mapping. In this case, the magnitude of the total wavevector, $|\boldsymbol{k}|$, can be described by:

$$|\boldsymbol{k}| = \frac{1}{\hbar}\sqrt{2mE_k} \tag{5.36}$$

where its x-, y-, z-components are described by:

$$k_x = |\boldsymbol{k}| \sin \theta \cos \varphi \tag{5.37}$$

$$k_y = |\boldsymbol{k}| \sin \theta \sin \varphi \tag{5.38}$$

$$k_z = |\boldsymbol{k}| \cos \theta \tag{5.39}$$

Therefore, whilst the energy analyser can determine E_K (as is standard in PES), by scanning the analyser through θ and ϕ, with respect to a well-defined crystal surface, $E(\boldsymbol{k})$ can be determined.

Some of the information available with this technique include:

- Elasticity;
- Electrical/magnetic properties;
- Optical properties;
- Energy band dispersion $E(k)$;
- Energy eigenvalues for delocalised electrons.

For example, mapping the bandstructure of superconductors [17, 18], topological insulators such as Bi_2Se_3 [19], as discussed in more detail in chapter 7, or measurement of quasiparticle dispersion relations and lifetimes [18].

An extension of this technique is spin resolved ARPES (Spin-ARPES), which can provide information on spin degeneracy of the bandstructure. By passing the collected electrons to a spin detector, which relies on the asymmetry of scattering (of spin polarised electrons) off of a crystal, the difference in spin polarised up/down electrons can be measured. In this way, a distinction can also be made between spin polarised and non-spin polarised states [20].

5.3.5 Auger electron spectroscopy (AES)

Auger electron spectroscopy is also a surface sensitive technique that involves measurement of an emitted electron. In this case, however, the electron studied—the Auger electron—is not the direct result of photoemission, but an alternative relaxation process, as described earlier. As such, the initial excitation process can be driven by incident x-rays or an electron beam. The convention with AES, unlike XPS, is to use x-ray notation to identify the Auger peaks. As Auger relaxation is a three electron process, the x-ray notation for the ejected electron, relaxed electron and Auger (emitted) electron will be listed as subscripts for each peak (in that order). For the example Auger generation given in figure 5.9 the kinetic energy of the Auger electron could therefore be described by:

$$E_{\mathrm{KL_2L_3}} \sim E_{\mathrm{K}} - E_{\mathrm{L_2L_3}} - E_{\mathrm{L_2L_3}} \sim E_{\mathrm{K}} - 2E_{\mathrm{L_2L_3}} \tag{5.40}$$

where K, L_2, L_3 correspond to the $^1s_{1/2}$, $^2p_{1/2}$, and $^2p_{3/2}$ energy levels and the bold in the subscript here has been used to distinguish from kinetic energy, E_K, used earlier. (In some notation this might be simplified to KLL.) Whilst the calculation of the energy of the Auger electron can be quite complex, an empirical relationship between the energy of the Auger electrons with respect to the next element in the periodic table can be used to simplify analysis [13]:

$$E_{KL_2L_3}(Z) = E_K(Z) - [E_{L_2L_3}(Z) + E_{L_2L_3}(Z + 1)] \quad (5.41)$$

where $E_K(Z)$ is the energy of the K level, and $E_{L2,3}$ is the energy of the $L_{2,3}$ level. As the kinetic energy of the Auger electron is the characterised quantity here, AES data is always plotted as a function of the kinetic energy, E_K, rather than the binding energy, E_B. Data is also usually collected in constant retard ration (CRR) mode and displayed as a direct intensity or differential (dN/dE) versus E_K. Note that Auger electrons are observed in standard XPS spectra and can be identified by the absence of a satellite peak with large intensity that corresponds to a characteristic emission.

Information available from Auger and photoelectron peaks can be combined to reliably measure the chemical shift with sufficient accuracy (i.e. free from artefacts arising from electrostatic charging of the sample and small errors in the spectrometer calibration). This is known as the Auger parameter, α, and involves the determination of the separation of the photoelectron and Auger lines as well as their chemical shift. Or simply, the binding energy of the photoelectron, E_B, and the kinetic energy of the Auger transition, E_K, such that:

$$\alpha = E_B + E_k \quad (5.42)$$

Examples of this in practice can be found in [21].

5.4 Fluorescence techniques

Whilst photoemission spectroscopy was concerned with measuring the energy of electrons emitted from a material (and thus determine the binding energy), fluorescence techniques are concerned with measuring the x-rays emitted. Fluorescence occurs when a material absorbs electromagnetic radiation, so that an electron can be excited or ejected, only to later emit electromagnetic radiation, usually at a longer wavelength (i.e. energy is lost). When an electron is photoemitted, this leaves a hole in the core–shell of the atom, which a higher energy electron will fill by relaxing and emitting a photon.

5.4.1 X-ray fluorescence (XRF)

X-ray fluorescence is a non-destructive characterisation technique that can be used to identify what chemical elements are present in a material. This is achieved by identifying the position of characteristic peaks that are emitted when the excited material relaxes and emits an x-ray with energy equal to the difference between different energy levels. In this measurement a beam of x-rays is focussed onto the sample, and the emitted x-rays are collected into either an energy dispersive or a wavelength dispersive detector.

With an energy dispersive detector, the energy difference is measured directly using a semiconductor detector (such as a Si p-i-n junction). As semiconductors have a Fermi energy that occurs between the conduction and valence bands, they will not usually conduct electricity. The exceptions to this are when the semiconductor is excited thermally (as you will see in chapter 7, the resistance will decrease as temperature increases, unlike metal systems), or through interaction with an

Table 5.1. Advantages and disadvantages of energy and wavelength dispersive detectors.

Property	Energy dispersive	Wavelength dispersive
Energy resolution	Typically of the order of 100 eV	+ Better energy resolution (~10 eV)
Artefacts and background	– Need to be aware of measurement artefacts such as escape electrons, (typically 1.74 keV below the true position), internal fluorescence, and signal processing errors. – Due to detector composition, need to be wary of low levels of Si (from the p-i-n) or Be (from some x-ray windows).	+ No artefacts + Lower background enables detection of smaller amount of elements
Elemental resolution	– Due to requirements of the detector in an SEM (namely inclusion of a robust 'x-ray window'), measurement is detector dependent, but typically $Z > 4$.	+ Able to resolve light elements such as Be ($Z = 4$), with the right crystal
Detector geometry	+ Collects all x-rays (no specific geometry).	– The detection geometry requires detector to move to a specific angle, thus losing some of the intensity.
Measurement time	+ Measurement relatively fast. Multiple energies can be detected during a scan, without needing to move the detector.	– Measurement is time consuming due to selective measurement of each wavelength.

electromagnetic wave that generates an electron–hole pair (which can be conducted). This property is utilised by the p-i-n semiconductor detector: it is typically kept cold with liquid nitrogen to limit thermal excitation of electrons to the conduction band, and a voltage bias is applied to encourage excitation by an incoming electromagnetic wave. The electrical signal that is produced (charge pulse) will be proportional to the energy of the x-ray that generated it, and this can be measured as a voltage (which is typically amplified before detection). It should be noted that small artefacts could be present in the measured spectrum due to fluorescence in the x-ray window (often Be), of the detector itself (Si). This might be of the order of 0.5%–1% of the total signal.

Wavelength dispersive detectors, on the other hand, will use Bragg diffraction from a crystal with known lattice spacing, d, in order to separate the incident x-rays by wavelength. As this does not rely on carrier excitation, it is less prone to thermal instabilities, therefore does not require active cooling. A summary of the advantages and disadvantages of each type of detector is given in table 5.1.

Whilst XRF can be used to determine if an element is present in a sample, it is also possible to determine the relative fraction of each element present. This can be done by comparing the intensity of the observed peaks to a known standard, as outlined by Castaing [22]:

$$\frac{C_i}{C_{(i)}} = K\frac{I_i}{I_{(i)}} \tag{5.43}$$

where C_i is the concentration of element, i, K is a sensitivity factor (determined by the instrument, Z and absorption and fluorescence effects), I_i is the measured intensity for a characteristic peak of element, i, and $C_{(i)}$ and $I_{(i)}$ are the concentration and intensity for a standard sample. Alternatively, the height of the K_α peaks for two elements A and B could be compared using the Cliff Lorimer ratio technique [23]:

$$\frac{C_A}{C_B} = K_{AB}\frac{I_A}{I_B} \tag{5.44}$$

$$C_A + C_B = 100\% \tag{5.45}$$

where K_{AB} is the Cliff–Lorimer factor for elements A and B and $C_{A/B}$ and $I_{A/B}$ are the concentration and intensity of element A/B, respectively. In the first approximation, comparing the relative height(s) of the K or L peaks to one another will work quite well for determination of the relative concentrations; however, this relies on negligible absorption and fluorescence effects. For this assumption to remain valid requires a thin sample (the thin film approximation), which can also limit the signal.

In the lab, XRF is a useful tool for confirmation of a sample's stoichiometry, but it is also commonly used in industry and construction sectors to identify materials on site using a handheld unit (which matches to characteristic lines for elements, such as those given in table 5.2).

5.4.2 Energy dispersive x-ray spectroscopy (EDS, EDX, XEDS)

Energy dispersive x-ray spectroscopy (known as EDS, EDX or XEDS) is a variant of XRF used commonly in scanning electron microscopes, where the sample is 'excited' by charged particles rather than x-rays. For example, in a scanning electron

Table 5.2. Some energies of the strongest characteristic lines for the elements 3–11 and 20–28 in the periodic table.

Element	$K_{\alpha 1}$ (keV)	Element	$K_{\alpha 1}$ (keV)	$L_{\alpha 1}$ (keV)
(3) Lithium	0.05	(20) Calcium	3.69	0.34
(4) Berylium	0.11	(21) Scandium	4.09	0.39
(5) Boron	0.18	(22) Titanium	4.51	0.45
(6) Carbon	0.28	(23) Vanadium	4.95	0.51
(7) Nitrogen	0.39	(24) Chromium	5.41	0.57
(8) Oxygen	0.52	(25) Manganese	5.90	0.64
(9) Fluorine	0.68	(26) Iron	6.40	0.70
(10) Neon	0.85	(27) Cobalt	6.93	0.77
(11) Sodium	1.04	(28) Nickel	7.48	0.85

microscope, a steady stream of electrons is reflected off of the sample surface, sometimes resulting in emission and thus fluorescence.

As EDS can be integrated into most common SEMs or TEMs, it can be an invaluable tool for actively identifying features in an SEM image. Thus, it is often used to map elemental concentrations across sample surfaces (if the user is interested in phase segregation), or to collect complementary information. Due to size restrictions within the SEM column, it is more common to find energy dispersive detectors in use for this technique.

5.5 Absorption spectroscopy

We saw earlier when discussing PES that the number of electrons that escape a material follows an exponential relationship, characterised by the Beer–Lambert law and an attenuation length, λ_a. For absorption spectroscopy, this relationship is used to define 'absorbance' such that it varies linearly as the concentration of material is increased (thus simplifying measurement of the concentration).

In a general absorption spectroscopy measurement, as shown in figure 5.15, the transmittance, I/I_0, can be used to quantify what fraction of incident radiation passes through the sample cell. It can be characterised by:

$$T = \frac{I}{I_0} = 10^{-\varepsilon lc} \tag{5.46}$$

where ε is the molar absorptivity of the sample at wavelength, λ, l is the length of sample, and c is the molar concentration. By taking the negative log of I/I_0, we obtain the definition of absorbance, A:

$$A = -\log10\left(\frac{I}{I_0}\right) = \varepsilon lc \tag{5.47}$$

which now varies linearly with length and concentration. In absorption spectroscopy either the transmittance (I/I_0) or the absorbance will be plotted.

5.5.1 UV-visible absorption spectroscopy

UV-visible absorption spectroscopy is the study of the change in absorbance across the UV-visible range (as summarised in table 5.3), and is used primarily to measure

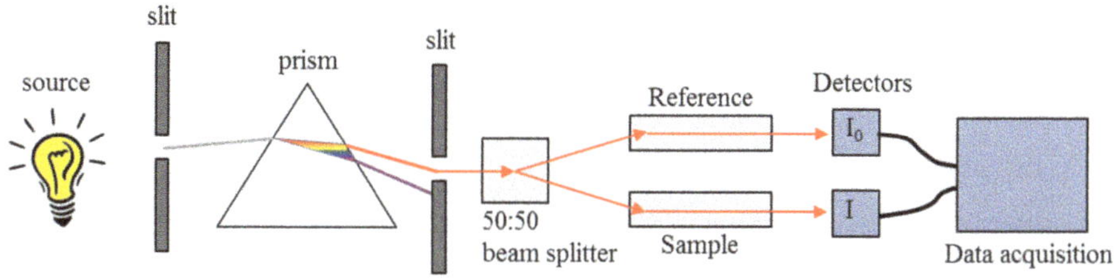

Figure 5.15. General schematic of an absorption spectroscopy measurement, where the wavelength is selected by a prism.

Table 5.3. Wavelength and frequency range for UV to IR.

Colour	Wavelength (nm)	Frequency (10^{14} Hz)
Ultraviolet	10–400	3–120
Violet	380–435	114–130
Blue	435–500	130–150
Cyan	500–520	150–156
Green	520–565	156–169.5
Yellow	565–590	169.5–177
Orange	590–625	177–187.5
Red	625–740	187.5–222
Infrared	>750	>225

the concentration of a sample loaded into the measurement with respect to a reference cell. Examples of its application include the study of reaction rates, enzyme kinetics, or quality control of paints or inks. It can also be used to measure the transmission of thin films, such as indium tin oxide, ITO, a transparent conducting oxide [24]. For example, transparent conducting oxides are useful as the building blocks for most touchscreen displays, where it is necessary for the electrical connections to not interfere with the visible light display of the screen. As indium is a declining resource there have been recent efforts to replace ITO with another oxide, thus requiring UV-visible absorption (or transmission) spectroscopy to confirm the degree of transparency of alternative materials.

5.5.2 Infrared absorption spectroscopy

Infrared (IR) absorption spectroscopy, much like Raman, is used to probe the vibrational properties of a sample and can act as a 'fingerprinting' tool for more complex materials (i.e. identify the structure rather than just the chemical elements present). Whilst symmetric molecules are often Raman active, they will typically be IR inactive and vice versa. Powdered samples are commonly pressed into a pellet for measurement, with the IR inactive KBr powder used as a binding agent. A diffraction grating is then used to separate wavelengths in order to scan energy.

Variations of IR spectroscopy include Fourier transform IR, where an interferometer consisting of two parallel mirrors is used to measure the IR absorption spectra, which is returned by performing a Fourier transform on the output data. Unlike Raman, in IAS, missing intensities (rather than peaks) with respect to wavenumber (i.e. absorbance) correspond to vibrational properties.

5.5.3 X-ray absorption spectroscopy (XAS)

X-ray absorption spectroscopy (XAS) is defined as the measurement of the absorption coefficient, μ, as a function of the x-ray energy, E_i. In general, the absorption coefficient, μ, will be proportional to the density, ρ, atomic mass, A, atomic number, Z, and energy of excitation, E:

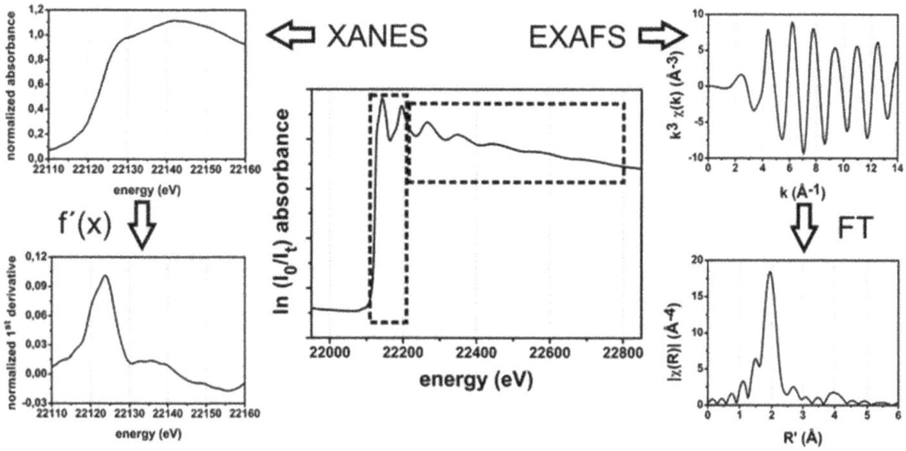

Figure 5.16. Example XANES and EXAFS spectra and analysis of a ruthenium compound collected in transmission mode. Left: normalised XANES spectrum and the first derivative, $f'(x)$. Right: EXAFS spectrum and its Fourier transform. Figure reproduced from [26] with permission from The Royal Society of Chemistry.

$$\mu \approx \frac{\rho Z^4}{AE^3} \tag{5.48}$$

The measurement of absorption can either be achieved in transmission mode, where the intensity of the x-ray source is measured before and after the sample (i.e. a direct measurement), or indirectly by measuring the x-rays (fluorescence mode) or Auger electrons (electron yield mode) produced by photoemission.

Variable energies can be achieved with a synchrotron source, where the incident wavelength can be tuned to the absorption edge of core electrons (K-, L-, M-) for the element of interest. There are three key features of the absorption spectra as a function of energy: a general decrease in μ with increasing energy; sharp, step-like, rises at 'absorption edges'; and oscillations just above the edges, as seen in figure 5.16.

The absorption edges are due to the ionisation of core electrons, and are thus categorised by the shell that they correspond to. For example, absorption of a photon by a K shell electron would be labelled a K absorption edge—as seen in figure 5.17. Note that electrons closer to the core of the atom become accessible as the energy is increased, thus the K edge (1s) in figure 5.17 occurs at higher energy than the L_1, L_2, and L_3 edges (which correspond to 2s, $2p_{1/2}$, $2p_{3/2}$, respectively). As the absorption edges are determined by the energy levels in the atom, they are element specific. In addition, a chemical shift will be observed as the valence of the atom increases (the edge will shift to higher energies). Finally, the oscillations seen just above the absorption edge are due to interference effects as the photoelectron undergoes multiple scattering events.

As seen in figure 5.16, features close to the absorption edge (within 30 eV) are classified as x-ray absorption near edge structure (XANES or NEXAFS), whereas

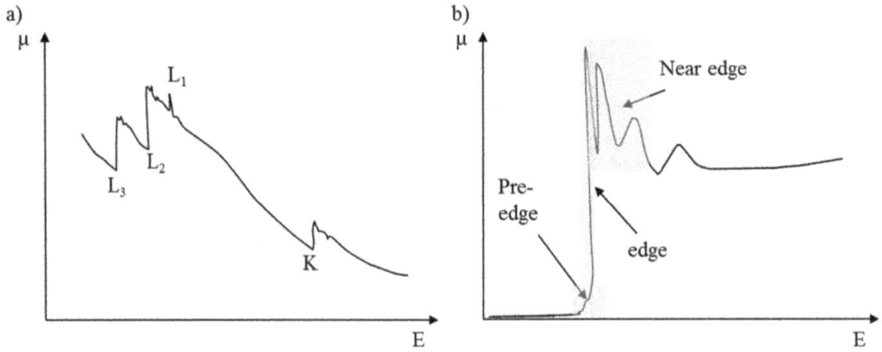

Figure 5.17. Features in an absorption measurement. (a) Schematic of multiple absorption edges. (b) Terminology used when discussing edge features.

the oscillations after the edge are classified as extended x-ray absorption fine structure (EXAFS). The study of both these features is generally referred to as x-ray absorption fine structure (XAFS). In XANES, using single or multiple scattering models, valence shell information, chemical species, and local geometry can be extracted. In EXAFS the oscillations can provide information on the local structure and interatomic distance [25].

Variations of XAS include x-ray magnetic circular dichroism (XMCD) first demonstrated by G van der Laan in 1986 [27] and G Schütz in 1987 [28], and x-ray magnetic linear dichroism (XMLD). In XMCD and XMLD, two sets of absorption spectra are obtained with opposite polarisations of circularly or linearly polarised x-rays. The difference between the absorption spectra can then be used to extract magnetic information such as spin–orbit coupling by integrating the area of this difference spectra and applying sum rules. Further details on the measurement and data analysis can be found in [29, 30].

5.5.4 Mössbauer spectroscopy

The Mössbauer effect describes resonant absorption and emission of gamma rays, where there is no energy loss. In its simplest form, it involves the use of an isotope of the element to be studied (such as $^{57}Fe_{5/2}$, which has a linewidth of 5×10^{-9} eV), which is moved towards sample so that the Doppler shift results in a shift of frequency:

$$f = \left(\frac{c \pm v_r}{c \pm v_s}\right) f_0 \tag{5.49}$$

where c is the speed of light, v_r is the velocity of the detector/sample with respect to the source, v_s is the velocity of the source with respect to the detector, and f_0 is the original frequency. As the energy, E is equal to hf, this can be written as:

$$E = h\left(\frac{c \pm v_r}{c \pm v_s}\right) E_0 \tag{5.50}$$

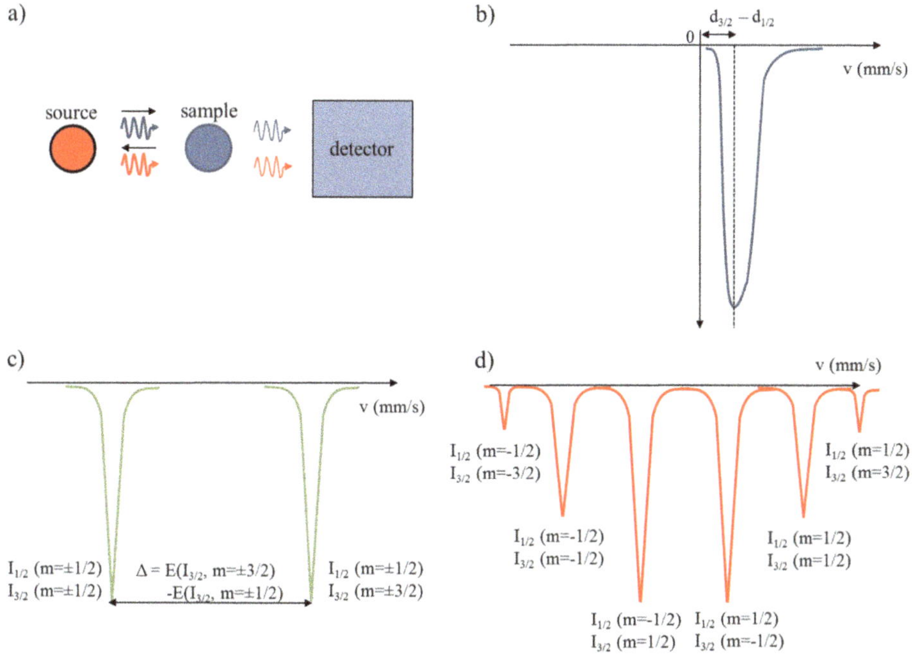

Figure 5.18. Schematic of Mössbauer spectroscopy and some example spectra types. (a) Schematic of the measurement, where movement of the source with respect to the sample and detector, results in a slight red or blue shift of the emitted gamma rays. (b) Example of the impact of isomer shift on the absorption spectra. (c) Example of quadrupole splitting. (d) Example of magnetic (Zeeman splitting), where the intensities of the absorption peaks is related to the angle between the gamma ray and the angular spin moment.

where E_0 is the energy of the emitted gamma ray in the source frame. A schematic of this measurement is given in figure 5.18. Due to the movement of the source, spectra obtained during this measurement are often plotted as absorption versus velocity, mm/s.

Information available from this type of spectroscopy includes isomer (chemical) shifts, quadrupole splitting, and hyperfine splitting, examples of which are given in figure 5.18. Isomer shifts result from small differences in the electron charge density closest to the nucleus and are typically measured relative to a known absorber. It can be used to determine various types of chemical information such as valency states. Quadrupole splitting arises from the nuclear quadrupole moment (m_I = 3/2, 1/2), which can be split in the presence of an asymmetric electric field, producing a doublet. Magnetic, or hyperfine splitting, occurs in the presence of a magnetic field (whether internal or externally applied) such that the nuclear levels for a state with spin of l split into $(2I + 1)$ sublevels As transitions between levels can only occur when m_I changes by 0 or 1, this results in multiple allowed transitions, such as those shown in figure 5.19. The magnitude of absorption will depend on the orientation of the nuclear spin moment with respect to the Mössbauer gamma ray. Further information can be found in [31].

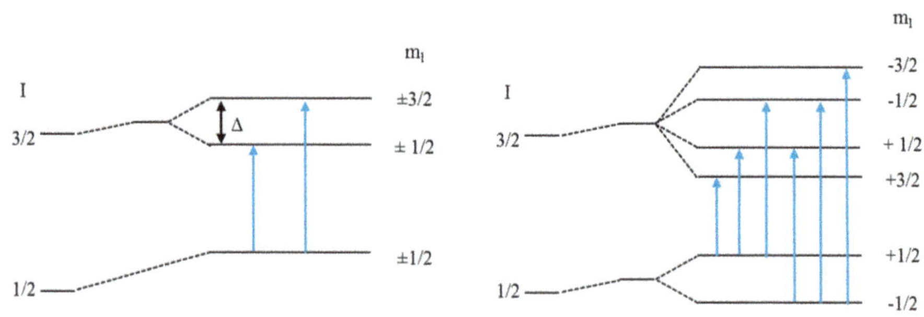

Figure 5.19. Examples of the energy level splitting for figure 5.18(c) and (d).

Table 5.4. Examples of nuclear magnetic moment from periodic table. For a full list, see [35], p 365.

Element *(most abundant isotope with non zero spin)*	Nuclear spin, I	Abundance	Nuclear magnetic moment, μ (μ_n)	Gyromagnetic ratio, γ (MHz/T)
H^1	1/2	99.98	2.792	42.6
C^{13}	1/2	1.108	0.702	10.7
N^{14}	1	99.64	0.404	3.1
O^{17}	5/2	0.04	−1.893	−5.8
Mn^{55}	5/2	100	3.461	10.5
Fe^{57}	1/2	2.245	0.09	1.4
Co^{59}	7/2	100	4.639	10.1
Ni^{61}	3/2	1.25	0.746	3.8

5.6 Resonance techniques

The resonance techniques discussed here largely comprise magnetic resonance measurements where nuclear or spin magnetic moments result in resonant absorption frequencies, which can be analysed to obtain information such as:

- Electronic structure: energy level splitting (emergence of fine structure), such as that seen in figure 5.19.
- Structure of complex molecules: contributes to chemical shifts.
- Precession of an electron spin: contributes to the line width of the resonance absorption peak.
- Internal magnetic field: contributes to a shift in the absorption peak.
- Collective spin excitations (spin waves).

To understand the mechanism behind magnetic resonance techniques, we need to first define magnetic moment, the gyromagnetic ratio, and the resonant frequency condition. A broad overview of magnetism and the associated units is given in chapter 8, which the reader is encouraged to review if required. For reference, the nuclear spin, magnetic moment, and gyromagnetic ratio for a selection of elements is given in table 5.4.

Unlike the spin angular (S), orbital (L) and total (J) momentum defined for the electron in chapter 1 (and which forms the basis of spectroscopic notation), for nuclei with multiple proton and neutrons, the total angular momentum, I, is used to characterise the nuclear magnetic moment, μ. These two quantities are related by:

$$\mu = \gamma \hbar I \tag{5.51}$$

where I is the total angular momentum of the nucleus, otherwise referred to as the 'nuclear spin' (with values of 1/2, 3/2), and γ is the gyromagnetic ratio. The nuclear spin, is determined by the combination of neutron and proton spins in the nucleus, where:

1. If the number of neutrons and the number of protons is even, $I = 0$.
2. If the total number of neutrons and protons is odd, then I will have half integer spin (1/2, 3/2, 5/2, ...).
3. If the number of neutrons is odd and the number of protons is odd then I will be an integer.

For a given value of I, there will be $2I + 1$ possible configurations that become apparent due to magnetic splitting (hyperfine structure). Each level will have a quantum number, m, associated with it (e.g. for $I = 5/2$, $m = -5/2, -3/2, -1/2, 0, 1/2, 3/2, 5/2$).

Thus, the gyromagnetic ratio for nuclei can be written as:

$$\gamma = \frac{\mu \mu_{\mathrm{n}}}{\hbar I}\left[\frac{\mathrm{rad}}{\mathrm{s}}\right] = \frac{\mu \mu_{\mathrm{n}}}{hI}\,[\mathrm{Hz}] \tag{5.52}$$

where the nuclear magnetic moment is expressed in terms of the nuclear magneton, μ_{n}. Like the Bohr magneton, which was a useful unit for expressing the electron spin, the nuclear magneton is defined by:

$$\mu_{\mathrm{n}} = \frac{e\hbar}{2m_{\mathrm{p}}} \tag{5.53}$$

where e is the charge on the electron, m_{p} is the proton rest mass, and $\mu_{\mathrm{n}} = 5.0509 \times 10^{-27}$ J T^{-1}.

Now that we have defined a magnetic moment for the nucleus, we can derive the energy, E, associated with the interaction of this nuclei with an applied magnetic field:

$$E = -\boldsymbol{\mu} \cdot \boldsymbol{B}_{\mathbf{a}} \tag{5.54}$$

where $\boldsymbol{B}$ is the applied magnetic field. If $\boldsymbol{B}$ is applied along the z-axis (as shown in figure 5.20) then the interval energy, U:

$$U = -\mu_z \cdot B_0 = \gamma \hbar I_z B_0 \tag{5.55}$$

where B_0 is the magnitude of the magnetic field along the z-direction, and I_z is the total nuclear magnetic moment along the z-axis. The energy difference between two

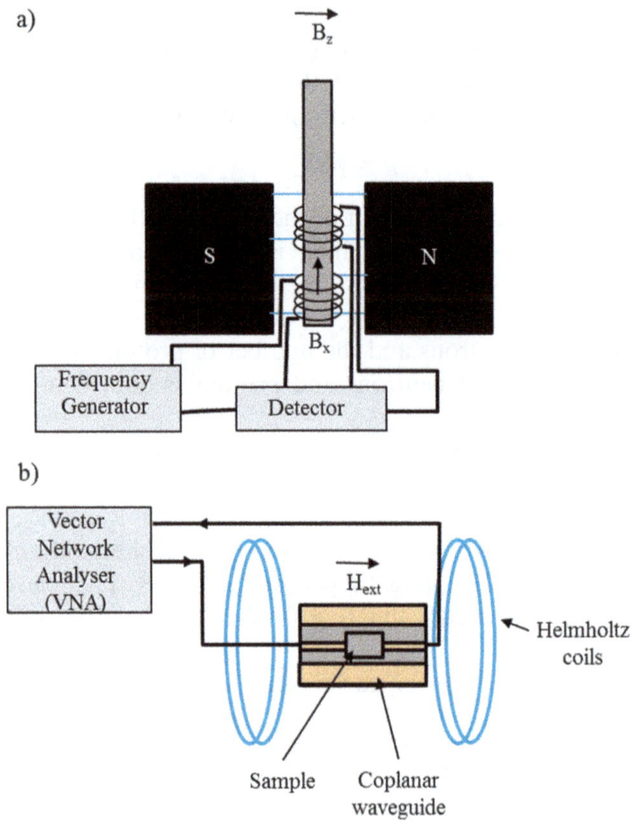

Figure 5.20. Schematic of magnetic resonance measurements. (a) NMR spectroscopy, where the RF coils surround a test tube to be studied. (b) Ferromagnetic resonance, where a coplanar waveguide is used to drive the RF magnetic field. In both cases a DC bias magnetic field is applied perpendicular to the RF field.

levels can therefore be determined as $\Delta E = \hbar\omega = U_1 - U_2 = \gamma\hbar B_0$ for $\Delta I = 1$. So the frequency required to excite a nucleus by one energy level is:

$$\omega_0 = 2\pi f_0 = \gamma B_0 \tag{5.56}$$

This is known as the operating or Larmor frequency, which can also be thought of in terms of the rate of precession of a magnetic moment around an external magnetic field. For protons $\gamma = 2.675 \times 10^8$ s^{-1} T^{-1}, therefore the resonance condition will be 42.58 Tesla per MHz, whereas for electrons ($\gamma = 1.76 \times 10^{11}$ s^{-1} T^{-1}) it will be 28 Tesla per GHz.

For general resonance measurements a static magnetic field is applied along one axis, and the sample is perturbed by a perpendicular RF field. The time evolution of the magnetic moment is characterised by the following Bloch equations:

$$\frac{\mathrm{d}M_x}{\mathrm{d}t} = \gamma(\boldsymbol{M} \times \boldsymbol{B})_x - \frac{M_x}{T_2} \tag{5.57}$$

$$\frac{dM_y}{dt} = \gamma(\boldsymbol{M} \times \boldsymbol{B})_y - \frac{M_y}{T_2} \tag{5.58}$$

$$\frac{dM_z}{dt} = \gamma(\boldsymbol{M} \times \boldsymbol{B})_z - \frac{M_z(t) - M_0}{T_1} \tag{5.59}$$

where T_1 and T_2 are characteristic timescales associated with relaxation of the magnetic moment (spin-lattice and spin-spin, respectively), and M_0 is the steady state (nuclear) magnetisation.

Some of the characteristic features that are analysed in resonance measurements are the position of the peaks (chemical shifts) relative to a reference sample, where the shift is characterised by σ:

$$\sigma = \frac{f_{\text{sample}} - f_{\text{ref}}}{f_{\text{ref}}} \times 10^6 \tag{5.60}$$

As these shifts are typically of the order of Hz for a Larmor frequency of MHz, σ is measured in units of parts per million (ppm).

5.6.1 Nuclear magnetic resonance (NMR)

Nuclear magnetic resonance is a commonly employed characterisation technique for the analysis of complex molecules. It is also the framework of magnetic resonance imaging (MRI). As outlined already, magnetic resonance techniques measure the absorption of RF energy by a magnetic nucleus in a magnetic field. This occurs through the precession of the angular momentum of a spinning nucleus around an applied magnetic field, with a characteristic (resonant) frequency that is proportional to the magnetic field. The measurement itself will either scan the RF frequency at fixed magnetic field, or vice versa (scan the magnetic field at fixed frequency), whilst measuring the absorption.

The Larmor frequency, f_0, will be affected by electronic screening, such as is common with larger nuclei, where f_0 decreases as the electronic screening increases. As a result of this, a shift in the absorption peak (with respect to a reference) will be observed, which makes the technique useful for determining the local electronic structure of complex molecules. Quantitatively, as the fractional shift in the frequency will be the same, it is often easier to obtain high quality data at higher magnetic fields (i.e. the dispersion may be better).

In addition to chemical shifts, spin–spin interactions can result in a shift of the resonant frequency; known as Knight shifts. This is a phenomena specific to metals, where the local magnetic field produced by the metallic atom in response to an applied magnetic field enhances the effective field that the atom experiences. The result is an effective increase in B_0, which then increases f_0.

In general, with the magnetic field applied along z, the NMR frequency is given by:

$$\omega = \gamma H_0(1 + K_{zz} - \sigma_{zz}) \tag{5.61}$$

where K_{zz} is the Knight shift, and σ_{zz} is the chemical shift. A detailed review of NMR applied to C_{60} and fulleride superconductors can be found in [32].

5.6.2 Electron spin resonance (ESR)

Electron spin resonance is similar to NMR, except that now electron spins are excited rather than nuclei. In this case, the measurement requires the sample to have unpaired electrons so that the total angular momentum is non-zero and the *g*-factor can be assumed to be equal to 2. In this case the frequency is kept fixed and the magnetic field is scanned. The resonance condition is:

$$f_0 = \frac{g\mu_B B}{h}$$

where g is the *g*-factor, which is equal to 2 for a free electron, and μ_B is the Bohr magneton. For a detailed review of ESR, see *Theoretical Foundations of Electron Spin Resonance* [33, 34].

5.6.3 Ferromagnetic resonance (FMR)

In ferromagnetic resonance, the sample to be measured is probed by exciting with a microwave signal carried along a microstrip or cavity and measuring the resultant transmission profile. Typically, either the frequency of excitation will be scanned at constant field, or the DC magnetic field will be scanned at constant frequency. This technique is commonly employed to measure the dynamic properties of magnetic materials, in particular thin films for spintronics applications [35, 36]. As with other resonance techniques, the total moment of the sample is aligned along a DC magnetic field, and a perpendicular RF field is applied in order to cause a small precession. This is achieved by placing the sample on a waveguide that generates an electric field transverse to the direction of propagation and a magnetic field in-plane and also perpendicular to the propagation of the EM signal (TEM mode of transmission). A vector network analyser is typically used to collect the resonance spectra, resulting in data such as that given in figure 5.22.

There are two common choices for waveguide with FMR measurements—the microstrip, or the grounded coplanar waveguide, as shown in figure 5.21, and summarised in table 5.5. The microstrip can be made using standard PCB technology as it is simply a strip of conductor a few microns wide, separated from a ground strip. The coplanar waveguide, on the other hand, consists of three strips, two of which are shorted to the ground plate for easier control of the electric field and RF losses. The quality of the FMR signal will largely be determined by the choice and design of the waveguide (its Q factor), where radiation losses and alternative transmission modes might need to be taken into account.

For isotropic thin films with the field applied in plane, the resonance condition is described by [37]:

$$f_0 = \frac{\gamma}{2\pi}\sqrt{(\mu_0 M_s + B)B} \qquad (5.62)$$

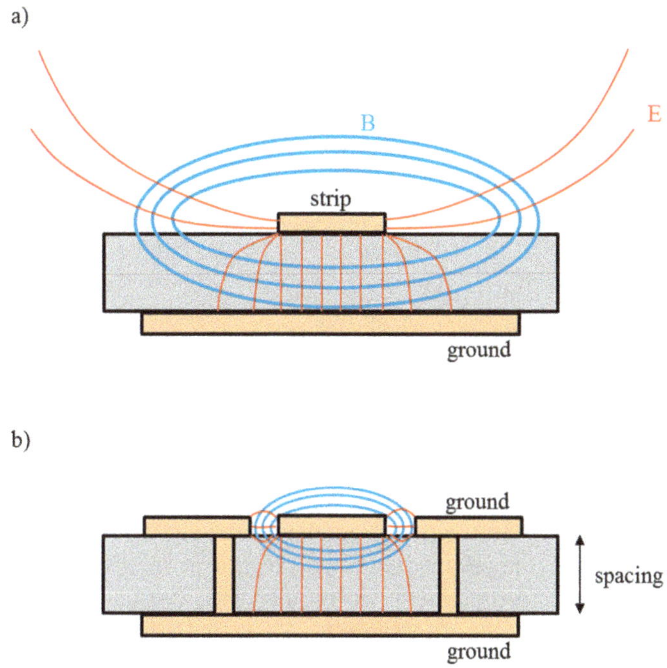

Figure 5.21. Schematic of the two common types of waveguides for FMR (a) the microstrip and (b) the coplanar waveguide. In both cases the mode of EM propagation is chosen to be TEM—transverse electric field to propagation. Energy losses in the coplanar waveguide are easier to control.

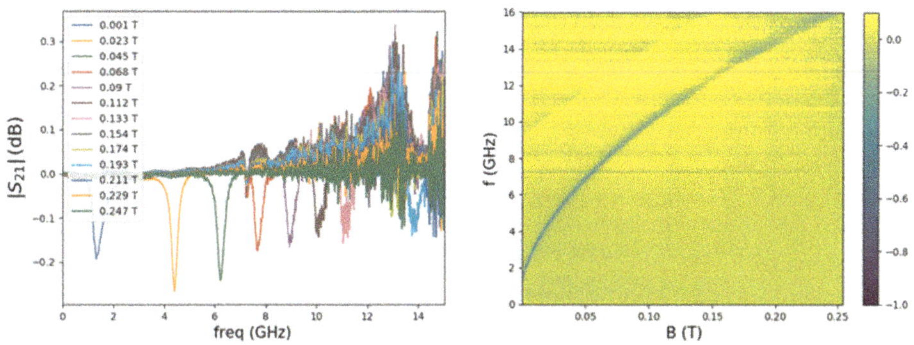

Figure 5.22. Example FMR spectra for permalloy (80% Ni; 20% Fe). Left: individual frequency sweeps at different applied magnetic fields. Right: composite resonance spectra, where colour indicates the level of RF absorption (resonance).

where M_s is the saturation magnetisation of the sample, B is the magnitude of the DC magnetic field, and γ is the gyromagnetic ratio. Thus we can extract the γ or M_s from f_0, by fitting a line to $f_0^2\ / B(B)$:

$$f_0^2 / B = \frac{\gamma^2}{4\pi^2}\mu_0 M_s + \frac{\gamma^2}{4\pi^2}B \tag{5.63}$$

Table 5.5. Comparison of microstrip and coplanar waveguides.

Property	Microstrip	Coplanar waveguide
Radiation loss at microwave frequencies, 300 MHz–30 GHz	Low	Low
Radiation loss at mm wave frequencies, >30 GHz	High	Low
Mode suppression	Difficult	Easy
Supress non quasi-TEM modes	*Vary conductor width or substrate thickness.*	*Vary conductor width, space in between, ground-signal-ground connections.*
Electrical performance variation due to PCB fabrication	Low	Moderate

The linewidth (FWHM) and slope can then be used to extract a variety of parameters such as the magnetic damping, saturation magnetisation, inhomogeneous broadening, and gyromagnetic ratio, $\gamma/2\pi$. For consideration of shape and magnetocrystalline anisotropy effects on FMR measurements see Kittel [37] and Smit [38].

For example, we can obtain an expression for the linewidth, ΔH, of the resonance spectra as a function of damping:

$$\Delta H(f_0) = \Delta H_0 + \frac{4\pi\alpha f_0}{\mu_0\gamma} \tag{5.64}$$

where α is the Gilbert damping parameter, μ_0 is the permeability of free space, γ is the gyromagnetic ratio and ΔH_0 is the inhomogeneous contribution to linewidth (an extrinsic contribution to damping). Therefore, the gradient of $\Delta H(f_0)$ is an indicator of intrinsic damping. An example of FMR data obtained from permalloy thin film is given in figure 5.22, where you can see broadening of the resonance spectra as f_0 increases.

Extensions of FMR include its application for measurement of the efficiency of spin injection in a magnetic bilayer [39]. In this scenario, additional broadening of the resonance spectra can be attributed to the spin mixing conductance, g_r:

$$g_r = \frac{2\gamma M_s t_F \Delta H}{g\mu_B f_0} \tag{5.65}$$

where $\gamma = ge/2mc$ ($g = 2$), μ_B is the Bohr magneton, M_s is the saturation magnetisation, and t_F is the ferromagnetic film thickness [40].

5.7 Examples of spectroscopy techniques in the literature

In the following sections, I will highlight several examples of research articles where spectroscopy techniques were used to characterise or identify a given material. For each article, starter questions are provided, which are aimed at focussing the reader's attention on the key aspects of the article (with respect to this chapter).

5.7.1 In 'Magnetic imaging—Snell's law for spin waves'

Stigloher *et al* demonstrated that spin waves in a magnetic thin film can be refracted and reflected, in a similar way to electromagnetic waves when incident on a surface with a different refractive index [41].

Reading through the article, try to answer the following:

1. What is a spin wave?
2. Define the term 'magnonics'.
3. What experimental techniques do the authors use?
4. How does micro-BLS differ from MOKE imaging?
5. Why did the measurements need to be time resolved?

The aim of this work was to demonstrate that spin waves can be manipulated ('steered') in a controlled fashion and that this could be modelled with an analogue of Snell's law. They showed that by introducing a change in the thin film thickness, spin waves could be efficiently refracted, accompanied by a change of wavelength.

The authors concluded that it is possible to apply Snell's law to spin waves steered in such materials if the magnetic anisotropy is incorporated. This suggests a route to predict the behaviour of more complex magnonic devices, where magnon steering will play a crucial role.

5.7.2 In 'Structural evolution and characteristics of the phase transformations between α-Fe_2O_3, Fe_3O_4 and γ-Fe_2O_3 nanoparticles under reducing and oxidizing atmospheres'

X Zhang *et al* investigated the phase transformation of α-Fe_2O_3 nanoparticles to γ-Fe_2O_3 by annealing in different atmospheres [42].

Reading through the article, try to answer the following:

1. What are the motivations for producing Fe_3O_4 and γ-Fe_2O_3 nanoparticles?
2. What is meant by monodisperse nanoparticles?
3. How was the size and morphology of the powders measured?
4. How was Raman and XPS used to distinguish between the different phases of iron oxide?
5. What additional information did Mössbauer spectroscopy provide?

The aim of this work was to find a route for production of monodisperse Fe_3O_4 and γ-Fe_2O_3 nanoparticles from α-Fe_2O_3. The authors showed that by annealing in a suitable atmosphere high quality Fe_3O_4 or γ-Fe_2O_3 could be obtained.

5.7.3 In 'A new and simple route to prepare γ-Fe_2O_3 with iron oxide scale'

Liu *et al* demonstrated a new method of preparing γ-Fe_2O_3 from iron oxide scale. This was verified with XPS, XRD, and UV–vis reflectance [43].

Reading through the article, try to answer the following:

1. What are the possible phases of iron oxide that are discussed by the authors?
2. Given the level of impurities identified in table 1 of this manuscript, would you expect these to be observable in XRD?
3. What differences in the morphology of the powders do the SEM images indicate?
4. How was XPS used to confirm the iron oxide phase?
5. What additional information did FTIR provide?

The aim of this work was to demonstrate a new, non-toxic method of refining the γ-Fe_2O_3 *phase of iron oxide (commonly used for magnetic recording) from the waste product of steel manufacture—iron scale (or rust).*

The authors showed that a relatively simple treatment could be used to produce γ-Fe_2O_3*: mixing with sodium hydrate (NaOH) and heating at* 400 °*C for* 8 *h, followed by a purifying treatment (repeated wash in deionised water and centrifuge) and heating at* 500 °*C for* 3 *h.*

5.7.4 In 'Understanding the origin of band gap formation in graphene on metals: graphene on Cu/Ir(111)'

H Vita *et al* investigated the bandstructure of graphene on Cu/Ir using ARPES [44].

Reading through the article, try to answer the following:

1. What is the motivation for studying band gap formation in graphene on metals?
2. What is meant by intercalation in this context?
3. How was photoelectron spectra (photoemission) used to confirm intercalation of the Cu in this study?
4. What is unique about the Dirac point/cone in graphene?
5. How does intercalation of the Cu layer affect the band states observed with ARPES?
6. How were the samples prepared for these measurements?

*The aim of this work was to assess the impact of Cu intercalation on the electronic properties of a graphene layer. The authors concluded that intercalation of a Cu layer between graphene and Ir(*111*) resulted in hybridisation between Cu d and graphene* π *bands, which manifested as an opening of the band gap at the Dirac point.*

5.7.5 In 'Experimental realization of a three-dimensional topological insulator, Bi_2Te_3'

Y L Chen *et al* investigated the photoemission spectra of Bi_2Te_3 doped with varying levels of Sn [45].

Reading through the article, try to answer the following:

1. What is a topological insulator?
2. Why was Bi_2Te_3 chosen for this study?
3. What is the advantage of using ARPES for characterisation of topological insulators?
4. What is the Γ point?
5. What was the impact of doping on the band gap in this system with regards to conductivity and carrier type?

The aim of this work was to investigate the formation of a 3D topological insulator through doping of the parent alloy Bi_2Te_3. The authors concluded that 0.67% doping of Sn in Bi_2Te_3 resulted in a 3D topological insulator with a large bulk band gap.

5.7.6 In 'Magnetostrictive thin films for microwave spintronics'

D E Parks *et al* investigated the formation of epitaxial thin films of Galfenol ($Fe_{81}Ga_{19}$) a magnetostrictive material. They demonstrated using FMR that it is possible to tune the microwave resonant frequency with strain as well as magnetic field [46].

Reading through the article, try to answer the following:

1. Why was Galfenol chosen for this study?
2. What are the potential difficulties of using thin film Galfenol in this context?
3. How does the linewidth measured in this study compare to previously measured values for thin film Galfenol?
4. How do the authors determine the impact of strain on the damping parameter, α?
5. How does the extrinsic damping change with increasing strain?
6. What explanations do the authors give for the strain-induced changes (or not) of α and ΔH_0?

The aim of this work was to investigate the fabrication of magnetostrictive thin films for future information and communications devices. This requires development of films with narrow resonant linewidths (as measured in FMR) whilst retaining appreciable magnetostrictive properties so that the resonant frequency could be tuned by electric and magnetic fields. The authors demonstrated that this was possible with Galfenol thin films grown epitaxially by MBE, and which showed magnetostrictive properties similar to bulk crystal, a narrow linewidth (6.7 mT at 10 GHz) and a strain tunable response of the resonance of 19.3 T per unit strain.

References

[1] Kittel C 2005 *Introduction to Solid State Physics* 8th edn (New York: Wiley) p 91

[2] Ferraro J R 2003 *Introductory Raman Spectroscopy* (Amsterdam: Academic)

[3] Maher R C *et al* 2008 *J. Phys. Chem.* A **112** 1497–501

[4] Schomacker K T, Delaney J K and Champion P M 1986 *J. Chem. Phys.* **85** 2440
[5] Wagner J *et al* 1999 *Appl. Phys. Lett.* **74** 3863
[6] Yu T *et al* 2008 *J. Phys. Chem.* C **112** 12602–5
[7] Beams R *et al* 2015 *J. Phys. Condens. Matter* **27** 083002
[8] Beugnot J *et al* 2014 *Nat. Commun.* **5** 5242
[9] Demokritov S O *et al* 2001 *Phys. Rep.* **348** 441–89
[10] Sebastian T, Shultheis K, Obry B, Hillebrands B and Shultheiss H 2015 *Front. Phys.* **3** 35
[11] Hofer F, Schmidt F P, Grogger W and Kothleitner G 2016 *EMAS 2015 Workshop, IOP Conf. Series: Materials Science and Engineering* **109** 012007
[12] Goodhew P J, Humphreys J and Beanland R 2001 *Electron Microscopy and Analysis* 3rd edn (London: Taylor and Francis)
[13] Watts J F and Wolstenholme J 2008 *An Introduction to Surface Analysis by XPS and AES* (Hoboken, NJ: Wiley)
[14] Seah M P and Dench W A 1979 *Surf. Interf. Anal.* **1** 2–11
[15] Zhang X *et al* 2013 *Cryst. Eng. Comm.* **15** 8166–72
[16] Sugiyama K *et al* 1998 *J. Appl. Phys.* **83** 4928
[17] Damascelli A, Hussain Z and Shen Z 2003 *Rev. Mod. Phys.* **75** 473
[18] Lynch D W and Olson C G 1999 *Photoemission Studies of High-Temperature Superconductors* (Cambridge: Cambridge University Press)
[19] Pan Z-H, Vescovo E, Fedorov A V, Gardner D, Lee Y S, Chu S, Gu G D and Valla T 2011 *Phys. Rev. Lett.* **106** 257004
[20] Neupane M *et al* 2016 *Nat. Commun.* **7** 13315
[21] Köver L *et al* 1995 *Surf. Interface Anal.* **23** 461–6
[22] Castaing R 1951 *Thesis University of Paris, ONERA Publication #55*
[23] Cliff G and Lorimer G W 1975 *J. Microsc.* **103** 203
[24] Stadler A 2012 *Materials* **5** 661–83
[25] Rehr J J and Albers R C 2000 *Rev. Mod. Phys.* **72** 621–54
[26] Hummer A A and Rompel A 2013 *Metallomics* **5** 597
[27] van der Laan G *et al* 1986 *Phys. Rev.* B **34** 6529–31
[28] Schütz G *et al* 1987 *Phys. Rev. Lett.* **58** 737
[29] Stöhr J and Siegmann H C 2007 *Magnetism: From Fundamentals to Nanoscale Dynamics* (Berlin: Springer)
[30] Stöhr J 1999 *J. Magn. Magn. Mater.* **200** 470–97
[31] Long G J and Grandjean F (ed) 1993 *Mössbauer Spectroscopy Applied to Magnetism and Materials Science* Vol 1 (New York: Plenum)
[32] Pennington C H and Stenger V A 1996 *Rev. Mod. Phys.* **68** 855–910
[33] Harriman J E 1978 *Theoretical Foundations of Electron Spin Resonance* (New York: Academic)
[34] Poole C P Jr. 1996 *Electron Spin Resonance: A Comprehensive Treatise on Experimental Techniques* 2nd edn (New York: Wiley)
[35] Parks D E *et al* 2013 *Sci. Rep.* **3** 2220
[36] Liu H *et al* 2018 *Nat. Mater.* **17** 308–12
[37] Kittel C 2005 *Introduction to Solid State Physics* 8th edn (New York: Wiley)
[38] Smit J and Beljers H G 1955 *Philips Res. Rep.* **10** 113

[39] Vlaminck V *et al* 2013 *Phys. Rev.* B **88** 064414
[40] Hoffmann A 2013 *IEEE Trans. Magn.* **49** 5172
[41] Stigloher J *et al* 2016 *Phys. Rev. Lett.* **117** 037204
[42] Zhang X *et al* 2013 *Cryst. Eng. Comm.* **15** 8166–72
[43] Liu J *et al* 2018 *Materials Letters* **229** 156–9
[44] Vita H *et al* 2014 *Scientific Reports* **4** 5704
[45] Chen Y L *et al* 2009 *Science* **325** 178
[46] Parks D E *et al* 2013 *Sci. Rep.* **3** 2220

IOP Publishing

Kelly Morrison

Chapter 6

Thermal characterisation

In this chapter, a broad overview of thermal measurement techniques for the study of phase transitions or commonly used properties such as specific heat and thermal conductivity of different materials will be given. We will start with a short background on such measurements including a definition of heat and temperature, followed by key details on the general methods, such as temperature and heat flux sensors. This will then be followed by more detailed treatment of each experimental technique where the aim will be to provide the reader with a broad understanding of the limiting factors (in such measurements). Finally, some examples of these approaches will be highlighted.

6.1 Some background theory

6.1.1 Temperature scales, heat and the laws of thermodynamics

Historically, the measurement of temperature was achieved by monitoring the expansion of a liquid such as mercury or alcohol with respect to two fixed points in temperature. Whilst Fahrenheit would use melting ice as one fixed point (32 °F) and the temperature of human blood as another (96 °F), Celsius, rather sensibly, used the ice point (0 °C) and the boiling point (100 °C) of water [1]. These are both relative scales that are used interchangeably across the world; in a laboratory, however, the two most commonly used temperature scales are Celsius and Kelvin. Unlike the Celsius scale, the Kelvin scale is an absolute measurement of temperature, which is defined to be zero at the point at which no molecular movement takes place. It has the same linearity as the Celsius scale (1 K = 1 °C), however, is offset by 273.15 K such that:

$$T_{\mathrm{Kelvin}} = T_{\mathrm{Celsius}} + 273.15 \tag{6.1}$$

Thus, it is often simple to convert between the two scales, and they can be used interchangeably for cryogenic (low temperatures; Kelvin) or furnace based (high temperatures; Celsius) measurements.

doi:10.1088/2053-2563/ab2df5ch6

The lesser used Fahrenheit (relative) and Rankine (absolute) scales can be defined by:

$$T_{\text{Fahrenheit}} = 1.8T_{\text{Celsius}} + 32 \quad (6.2)$$

$$T_{\text{Rankine}} = T_{\text{Fahrenheit}} + 459.69 \quad (6.3)$$

To understand why we define temperature in the way that we do, and what the physical significance of heat is, it is useful to first discuss the fundamental laws of thermodynamics. These are general observations that form the basis of our understanding of thermal physics, the most relevant of which are the zeroth and first laws:

6.1.1.1 Zeroth law: definition of temperature

If two systems are in thermal equilibrium (i.e. the same temperature) with a third system then they are in thermal equilibrium with each other.

6.1.1.2 First law: conservation of energy such as heat

When energy passes in or out of a system (generally as work or heat) the system's internal energy will change according to the principle of 'conservation of energy'; that energy is neither destroyed nor created.

The second and third laws of thermodynamics concern entropy—a measure of disorder in a system. The second law generally states that entropy will increase in a natural thermodynamic process (thus limiting the efficiency of an ideal engine), whereas the third law states that entropy will tend towards a constant low value as you approach absolute zero.

Based on these observations, temperature describes the thermal state of a material such that we could determine whether it would exchange heat with another material if it were placed in contact with it. How much heat is exchanged will depend on the heat capacity (and perhaps even the latent heat) of the two materials. The specific heat is similarly defined as the heat capacity of a material divided by its mass (i.e. the heat required to raise the temperature of a unit mass by 1°).

6.1.2 Transfer of heat

There are three basic processes that lead to transfer of heat and it is worth defining them here so that the precautions taken during measurements can be easily justified. These processes are conduction, convection, and radiation.

Conduction is defined as the transfer of heat between two materials (typically solids) that are in direct contact with one another. If we were to describe heat as the collection excitations of atoms in a material according to its current temperature, then heat could be exchanged between two materials in contact with one another by exchange of energy between atoms at the interface of these two materials. Factors that would then determine how quickly the two systems reach thermal equilibrium would be the temperature difference between them, the heat capacity of each material (i.e. how much energy is required to raise or lower their temperature by 1°), the thermal conductivity of each material (i.e. how quickly heat can be conducted) and the quality of the contact between the two materials. In general:

$$\frac{Q}{A} = \frac{\kappa}{d}\Delta T \tag{6.4}$$

where Q is the heat transferred across a material with a temperature gradient of ΔT, thermal conductivity κ, cross-sectional area A, and thickness, d.

As it is uncommon for a material surface to be atomically smooth, thus providing the perfect interface for two solids to be placed in contact with one another, it is common practice to use a 'thermal grease' to improve the thermal contact. This works well as the grease will be viscous enough to fill small gaps between the two solids, thus ensuring conduction at every point. It will then only be limited by the coverage and thermal conductivity of the grease used.

Convection is defined as the transfer of heat via an intermediary exchange gas or liquid where, rather than the heat being conducted to the next surface, it is 'carried away' by the exchange gas or liquid. As a heat transfer mechanism, convection is typically much slower than conduction, unless actively promoted by an increase of air flow or large temperature gradients between the exchange gas and the material it is cooling or heating. It can be described by:

$$\frac{Q}{A} = h_c \Delta T \tag{6.5}$$

where A is the heat transfer area of the surface, h_c is the convective heat transfer coefficient [W m^{-2} K^{-1}], and ΔT is now the temperature difference between the area of the surface and the convection medium (whether air or liquid).

In a typical thermal measurement, convection can be utilised in the form of an exchange gas to bring a sample to a given 'base' temperature, or to promote cooling if the sample is otherwise isolated. If the exchange gas is removed (achieved by pumping out the system to a pressure where the average distance between gas molecules is less than the mean free path) then it can be assumed that $h_c \sim 0$ and convective heat transfer does not play a role in the thermal processes being studied.

Radiative heat transfer is a mechanism whereby heat is exchanged with a material's surroundings by emission of energy in the form of electromagnetic radiation (photons). The rate at which energy is lost will depend on the temperature and the emissivity (a measure of how good a material is at emitting radiation). It can be defined by:

$$\frac{Q}{A} = \varepsilon \sigma T^4 \tag{6.6}$$

where σ is the Stefan–Boltzmann constant (5.6703×10^{-8} W m^{-2} K^{-4}), ε is the emissivity coefficient ($0 < \varepsilon < 1$), and T is the temperature of the radiating area, A.

Typically radiation losses are expressed in terms of the net radiative loss (as surrounding walls will also radiate heat), such that:

$$\frac{Q}{A} = \varepsilon\sigma(T_h^4 - T_c^4) \quad (6.7)$$

and T_h and T_c are the temperatures of the hot and cold surfaces, respectively.

Radiative heat transfer is often minimised during thermal measurements as it can make it difficult to efficiently maintain a specified temperature, and can also complicate the associated thermal transfer equations. This can be achieved in an experimental setting by encasing the sample environment in a radiation shield that is thermally connected to the sample temperature, such as the example given in figure 6.1.

Lastly, control of these heat transfer mechanisms can be used to achieve one of two standard measurement conditions: isothermal or adiabatic, as defined below.

Adiabatic—it is assumed that the sample is kept in 'thermal isolation' so that any heat released/absorbed will manifest. This is realised practically by placing the sample in a vacuum so that convective heat transfer is minimised. Often with specific heat measurements the sample may also be held on a platform suspended by thin wires and then placed in a vacuum (or low pressure environment), thus limiting conduction, convection and radiative paths.

Isothermal—it is assumed that a measurement is made whilst the sample is kept at the same temperature. This might be a change in state initiated by changing pressure

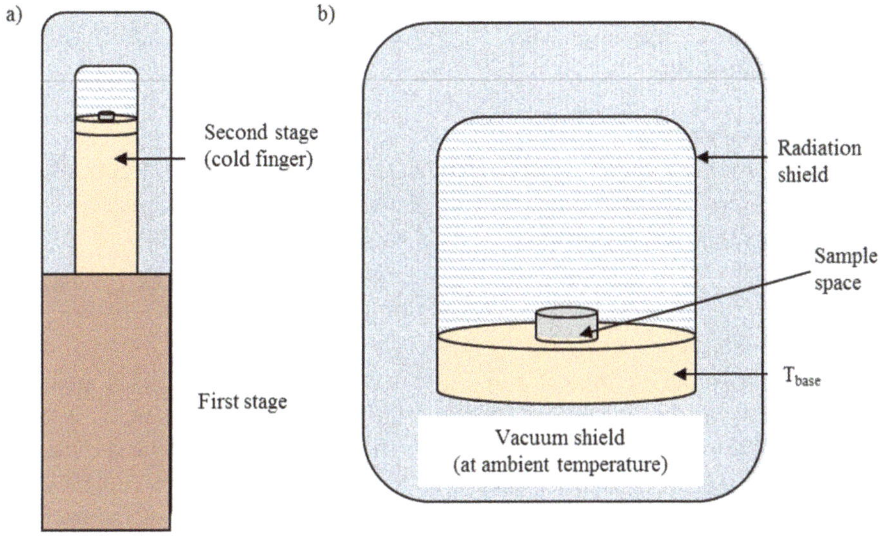

Figure 6.1. Schematic of a basic measurement set-up where a radiation shield is used to minimise radiative heat loss from the sample space. (a) Example of a two-stage cryostat measurement, and (b) the sample space. As long as the radiation shield is kept at the same temperature as the base temperature for the sample, T_{base}, radiation losses from the sample will be minimised. Careful choice of material for the radiation shield (low emissivity) will minimise radiation losses elsewhere.

or magnetic field. For the state to remain isothermal, any latent heat produced/absorbed by the material needs to be quickly equilibrated. This is typically achieved by placing the sample in contact with a thermal bath of fixed temperature, whether via conduction or use of an exchange gas, or both. Practically this could be a large (compared to sample) copper block. The advantage of using copper is that it has a reasonably high specific heat (385 J K^{-1} kg^{-1}) and that it has a particularly large thermal conductivity (401 W m^{-1} K^{-1}), so any changes in temperature can be quickly corrected for.

6.1.3 Specific heat, latent heat and phase transitions

The specific heat of a material is a parameter that allows one to determine the amount of energy (heat) required to increase the temperature of a unit mass of material by a unit amount. It can be defined as either C_p: specific heat under constant pressure, or C_v:specific heat under constant volume; the former being more widely used. It is defined by:

$$C_{p,v} = \left(\frac{\Delta Q}{m\Delta T}\right)_{p,v} \tag{6.8}$$

where m is the mass, ΔQ the heat transferred, and ΔT the temperature increase. The units for specific heat are typically either in J K^{-1} kg^{-1} (SI) or J K^{-1} mol^{-1}.

For a general material system—more commonly a crystalline solid—there is a characteristic trend of the specific heat with respect to temperature that is described by the Debye, Einstein, and Dulong–Petit models of specific heat. To summarise, at low temperatures, C_v increases with T^3, eventually reaching saturation above the Debye temperature, θ_D.

6.1.3.1 Dulong–Petit law

The Dulong–Petit law describes the tendency for the specific heat to trend towards a constant value that is largely independent of material parameters such as electron number or density when expressed in [J K^{-1} mol^{-1}]. It is defined by:

$$C_v = \frac{\partial}{\partial T}(3N_A k_B T) = 3N_A k_B \tag{6.9}$$

where k_B is Boltzmann's constant and N_A is Avogadro's number. This means that the specific heat will be dependent on the number density of atoms, as is evident when you compare the molar specific heat for single element compounds, such as in table 6.1, where $3N_A k_B = 24.94$ J mol^{-1} K^{-1}. Notice that: (1) this can be used to estimate the expected specific heat of a solid in units of J K^{-1} kg^{-1} if the number of atoms per kg is known; and (2) this law does break down for some systems, for example, where light atoms are strongly bound (such is the case for diamond or graphite).

Table 6.1. Specific heat and thermal conductivity of some common materials [2, 3].

Sample	Chemical formula	Specific heat ($J\ K^{-1}\ kg^{-1}$)	Molar mass ($g\ mol^{-1}$)	Specific heat ($J\ K^{-1}\ mol^{-1}$)	Thermal conductivity at 25 °C($W\ K^{-1}\ m^{-1}$)
Aluminium	Al	900	27	24.3	237
Apiezon N^{TM} grease	—	2300	—	—	0.194
Bismuth	Bi	130	209	27.2	8
Copper	Cu	385	63.5	24.4	401
Diamond	C	516	12	6.2	1000
Epoxy	$C_{21}H_{25}ClO_5$	1000	392.88	392.88	0.35
Glass	SiO_2	~730	60	43.8	1.05
Graphite	C	717	12	8.6	168
Ice	H_2O	2090	18	37.6	2.18
Iron	Fe	490	55.8	27.3	80
Lead	Pb	129	207.2	26.7	35
Mica	$Al_2K_2O_6Si$	880	256.24	225.5	0.71
Platinum	Pt	133	195.1	25.9	72
Polyethylene low density (PEL)	$(C_2H_4)_n$	$(1250–1429)_n$	28_n	35–40	0.33
Quartz	SiO_2	700	60	42	3
	Sand	830		49.8	
Water	H_2O	4182	18	75.3	0.606

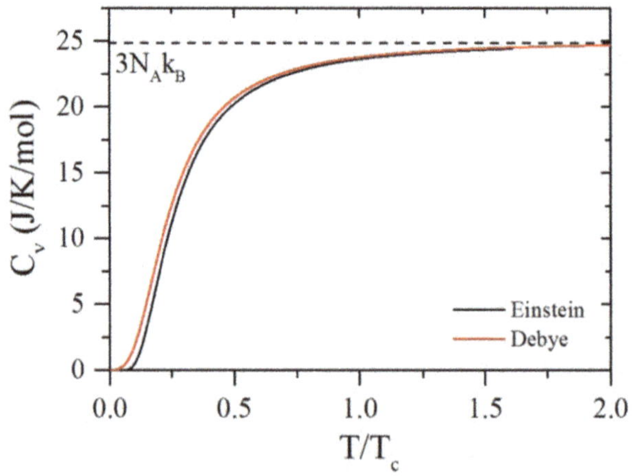

Figure 6.2. Einstein and Debye models for specific heat plotted as a function of their normalised temperature (where $T_c = 0.806\theta_E$ and θ_D, respectively).

6.1.3.2 Einstein and Debye models

Describing the low temperature behaviour of the specific heat requires a description of lattice vibrations (phonons), which we used earlier to highlight the difference between temperature and heat. The Debye model describes modes of vibration within a lattice (i.e. phonons in a box), which have a maximum value determined by the Debye frequency, ω_D:

$$\theta_D = \frac{\hbar\omega_D}{k_B} \tag{6.10}$$

where $\hbar$ (= $h/2\pi$) is the reduced Planck's constant, k_B is Boltzmann's constant, and θ_D is the Debye temperature. For Debye, C_v follows a T^3 dependence until a characteristic temperature, θ_D, is reached:

$$C_v = 9N_A k_B \left(\frac{T}{\theta_D}\right)^3 \int_0^{x_D} \frac{x^4 e^x}{(e^x - 1)^2} dx \tag{6.11}$$

where $x = \hbar\omega/k_B T$ and $x_D = \hbar\omega_D/k_B T = \theta_D/T$. For low temperatures we obtain the Debye approximation [2]:

$$C_v = \frac{12\pi^4}{5} N_A k_B \left(\frac{T}{\theta_D}\right)^3 \tag{6.12}$$

after which, the specific heat saturates at the Dulong–Petit Limit of $3Nk_B$.

The Einstein model instead determines the specific heat by assuming that the atoms in a lattice are non-interacting. In this case, the energy in the system can be described by an assembly of vibrating atoms, where:

$$C_v = \left(\frac{\partial U}{\partial T}\right)_V = 3N_A k_B \left(\frac{\theta_E}{T}\right)^2 \frac{e^{\theta_E/T}}{(e^{\theta_E/T} - 1)^2} \tag{6.13}$$

and $\theta_E = \hbar\omega/k_B$ is the characteristic temperature for the Einstein model, where ω is the frequency of vibration of each oscillator. Whilst the Einstein model fails for low temperatures (C_v will decrease as $e^{-\hbar\omega/T}$ not T^3), it does describe intermediate temperatures better than the Debye model and is often used to approximate optical phonon spectra. Finally, to compare the two models requires some consideration of how the characteristic temperatures scale to one another, but can be approximated by $\theta_E/\theta_D = 0.806$, as shown in figure 6.2.

6.1.3.3 Electronic specific heat

At low temperatures, for some metals the specific heat derived by Debye was still observed to diverge. This was largely due to an electronic contribution to the specific heat, which can be accounted for:

$$C_v = C_{\text{phonon}} + C_{\text{electronic}} \tag{6.14}$$

$$C_v = \frac{12\pi^4 N_A k_B}{5\theta_D{}^3}T^3 + \frac{\pi^2 k_B{}^2 N_A}{2E_F}T = \beta T^3 + \gamma T \tag{6.15}$$

where the first term is due to the phonon contribution derived from lattice vibrations (Debye model), and the second term (electronic contribution) is provided by electrons at the Fermi level, E_F, which are available to move. β and γ are parameters that are typically extracted from a fit of C/T as a function of T^2, which can then be used to determine θ_D and E_F. Note that derivation of this requires using Fermi–Dirac statistics, as shown in [2]. Notice also that this describes a linear temperature dependence at low temperatures that will eventually be dominated by the T^3 term.

6.1.3.4 Other contributions

So far we have largely ignored magnetic and electronic contributions to the specific heat. Whilst these can largely be considered negligible, there are situations where it becomes more apparent. For example, we have already seen that at low temperatures there may be a divergence from the Debye temperature dependence due to the electronic contribution to the specific heat. At a magnetic transition, the reorientation of magnetic spins, or development of localised magnetic moment may also manifest in a change in the specific heat. The superconducting-normal transition can also be identified by a feature in specific heat. It is for these reasons that specific heat measurements are often used to chart phase transitions, which are identified by a feature in heat capacity measurements that is not expected from the Debye or Einstein models.

6.1.3.5 Ehrenfest classification of phase transitions

The 'order' of a phase transition was originally defined by Ehrenfest by considering the derivative of the Gibb's free energy, G, of a system with respect to one of its state variables, V_{ST}. Examples of $G(V_{ST})$ include:

$$G(P, T) = U + PV - TS \tag{6.16}$$

$$G(H, T) = F - HM \quad (6.17)$$

where P is pressure, T is temperature, U is the internal energy, V is the volume, S is the entropy, H is the external magnetic field, F is the Helmholtz free energy ($F = U - TS$) and M is the magnetisation. Specifically, Ehrenfest defined the order of a phase transition to be the lowest derivative where a singularity in the free energy is observed such that a singularity in:

$\frac{\partial G}{\partial V_{ST}}$ indicates a first order phase transition

$\frac{\partial^2 G}{\partial V_{ST}^2}$ indicates a second order phase transition

For example, a first order phase transition (often identified by the presence of latent heat), is obvious by the measurement of a discontinuity in the specific heat. If this were a first order magnetic transition, then it would manifest as a step function in $M(H)$, such as the example given in figure 6.3. The difference here is due to the

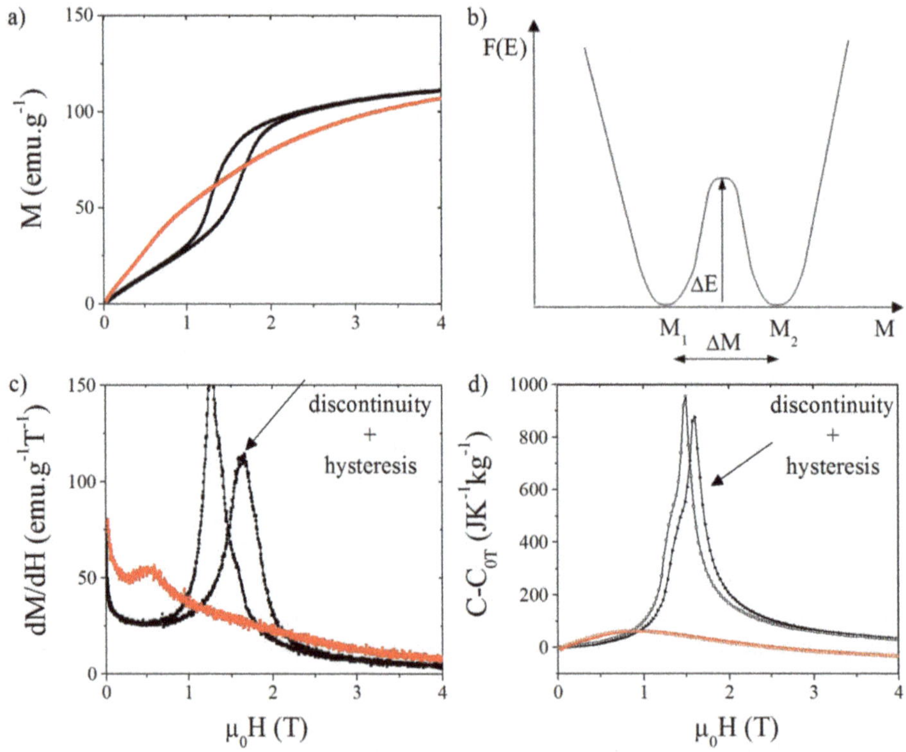

Figure 6.3. Example of the Ehrenfest classification of phase transitions for magnetic phase transitions. (a) Magnetisation as a function of applied magnetic field, H for a first order (black) and continuous (red) field driven phase transition. (b) Example schematic of the Gibb's free energy for a first order magnetic phase transition between two states, M_1 and M_2 separated by an energy barrier, ΔE. (c) the first derivative of $M(H)$ for the first order (black) and continuous (red) field driven phase transitions of part (a). (d) Corresponding specific heat measurements of the first order (black) and continuous (red) field driven phase transitions of (a) and (c).

fact that the specific heat is already a derivative of the fundamental variable, Q, as a function of temperature, whereas M is not. Higher order phase transitions could be identified by the order of differentiation that would yield a discontinuity and are now often referred to as continuous or second order phase transitions. Where there is a change of state (from solid to liquid or one crystalline state to another) there will be some manifestation of this in heat capacity measurements either by a subtle change, or by the presence of latent heat.

It should be remembered that in some material systems, the phase transition may occur as a function of several competing energies—such as thermal, magnetic and structural changes in a magnetocaloric material [4]. This can sometimes lead to complex phase transitions where there are features of both first order and second order phase transitions, or where it can move from one to another as the magnetic field is increased. Examples of this include the magnetostructural phase transition in $Gd_5Ge_2Si_2$ [5], the magnetovolume expansion in La(Fe,Si)$_{13}$ [6], the Verwey transition in Fe_3O_4 [7], or the Jahn Teller distortion in manganites [8].

6.1.4 Thermal conductivity and interfacial thermal conductance

The thermal conductivity, κ, describes the rate at which heat can be transferred along a material. For example, for a bar of cross-sectional area, A, in contact with two surfaces at temperature T_h and T_c (see figure 6.4), the rate of heat transfer, $P = dQ/dt$ [J s^{-1}], can be described by:

$$P = \frac{kA(T_h - T_c)}{L} \tag{6.18}$$

where κ is the thermal conductivity, T_h and T_c are the temperatures of the hot and cold ends, respectively, A is the contact surface area at the hot and cold ends, and L is the length of the material.

As can be seen from table 6.1, the thermal conductivity can vary widely across different materials, from as much as ~0.35 W m^{-1} K^{-1} for epoxy to 1000 W m^{-1} K^{-1} for diamond (at 25 °C).

Thermal conductance, K, describes the rate of heat transfer across the interface of two materials and has already been discussed briefly in the context of the use of thermal greases. (The inverse of thermal conductance is thermal resistance.) Whilst for two bulk materials, the thermal contact resistance is determined by the roughness of the two interfaces, for thin films a thermal boundary resistance is observed as a temperature drop across an interface. This boundary resistance is referred to in thin

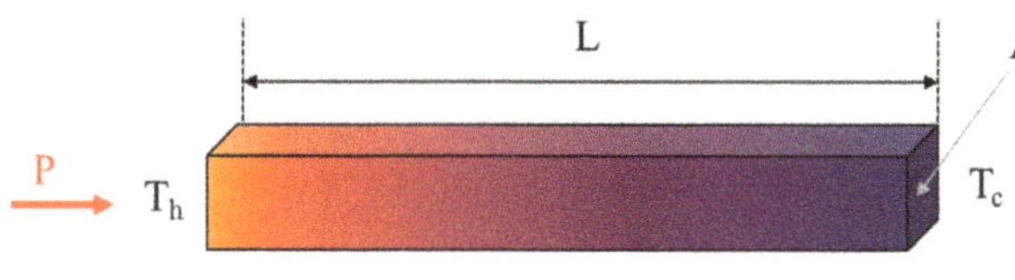

Figure 6.4. Definition of parameters used to describe the thermal conductivity of a material.

Table 6.2. Summary of commonly used temperature sensors.

Type	Temp. range	Notes
Thermocouple	1.2 K to 1543 K	Various standard curves available. Choose thermocouple based on required environment and sensitivity in the chosen temperature range.
Capacitance	1.4 K to 290 K	Good alternative sensor to the Cernox sensor for use in magnetic fields.
	Positive temperature coefficients	
Platinum	14 K to 873 K	Standard curve available and suitable for use in radiation.
Rhodium Iron	1.4 K to 420 K	Not recommended for below 77 K, suitable for use in radiation.
	Negative temperature coefficients	
Cernox®	0.1 K to 420 K	Most suitable sensor for use in magnetic fields, especially above 1 K. Suitable for use in radiation and for sub 1 K measurements.
Germanium	0.05 K to 100 K	Suitable for use in radiation and for sub 1 K measurements.
Ruthenium Oxide (Rox™)	0.01 K to 40 K	Standard curve available. Suitable for use in radiation and for sub 1 K measurements.
	Diodes	
Silicon	1.4 K to 500 K	Standard curve available.
GaAlAs	1.4 K to 500 K	

films as 'Kapitza resistance'[1] [9] and is expected due to bonding strength or difference in materials. It is defined by:

$$R_K = \Delta T / J_Q \tag{6.19}$$

where ΔT is the temperature difference across the interface, and J_Q is the heat flux.

6.1.5 Measurement of temperature and heat flux

To begin to characterise the thermal properties of a material first requires a reliable method for determination of temperature. Whilst, historically the measurement of temperature was achieved with the expansion of liquids such as mercury, this does not often lend itself well to more complex measurements. In this case, there are several different types of temperature sensors that are commonly used, as listed in table 6.2.

The choice of temperature sensor will depend heavily on the particular measurement. If, for example, it involves application of magnetic fields and low temperatures, then the Cernox sensor is the typical choice. Otherwise, the sensitivity of the

[1] Originally coined to describe the thermal resistance between a solid and liquid helium.

temperature sensor for a chosen temperature range is often the deciding factor; the typical response curves of these sensors is given in figure 6.5.

Thermocouples make use of the Seebeck effect, where a voltage is generated when a material is subject to a temperature gradient. This is utilised in practise by taking two dissimilar metals and connecting them at the temperature you intend to measure, with the voltage contacts at a reference temperature (often 0 or 20 °C). Careful choice of each metal in the pair can be used to tailor the sensitivity of the thermocouple for a given temperature range, or to be more robust for the environmental settings (high temperatures or corrosive atmosphere). Examples of commonly used thermocouples and their ranges are given in table 6.3.

Examples of sensors with a positive temperature coefficient (resistance will increase with temperature) include platinum and rhodium iron. The platinum sensor uses the well characterised temperature dependence of platinum resistivity [10] as an indicator of temperature. This can be calibrated against known reference points such as ice water (273.15 K) and boiling water (373.15 K) or liquid nitrogen (77.15 K). Sensitivity is typically of the order of 0.3–0.5 Ω $^{\circ-1}$ between 50 and 873 K.

Unlike the resistance thermometers (platinum and rhodium iron) the negative temperature coefficient sensors—Cernox, germanium, and ruthenium oxide—are particularly sensitive at lower temperatures. This is because the resistance for these sensors will increase as the temperature is lowered. The Si diode is also especially useful at low temperatures where the sensitivities of the thermocouple or Pt resistors are no longer high enough to be easily measured.

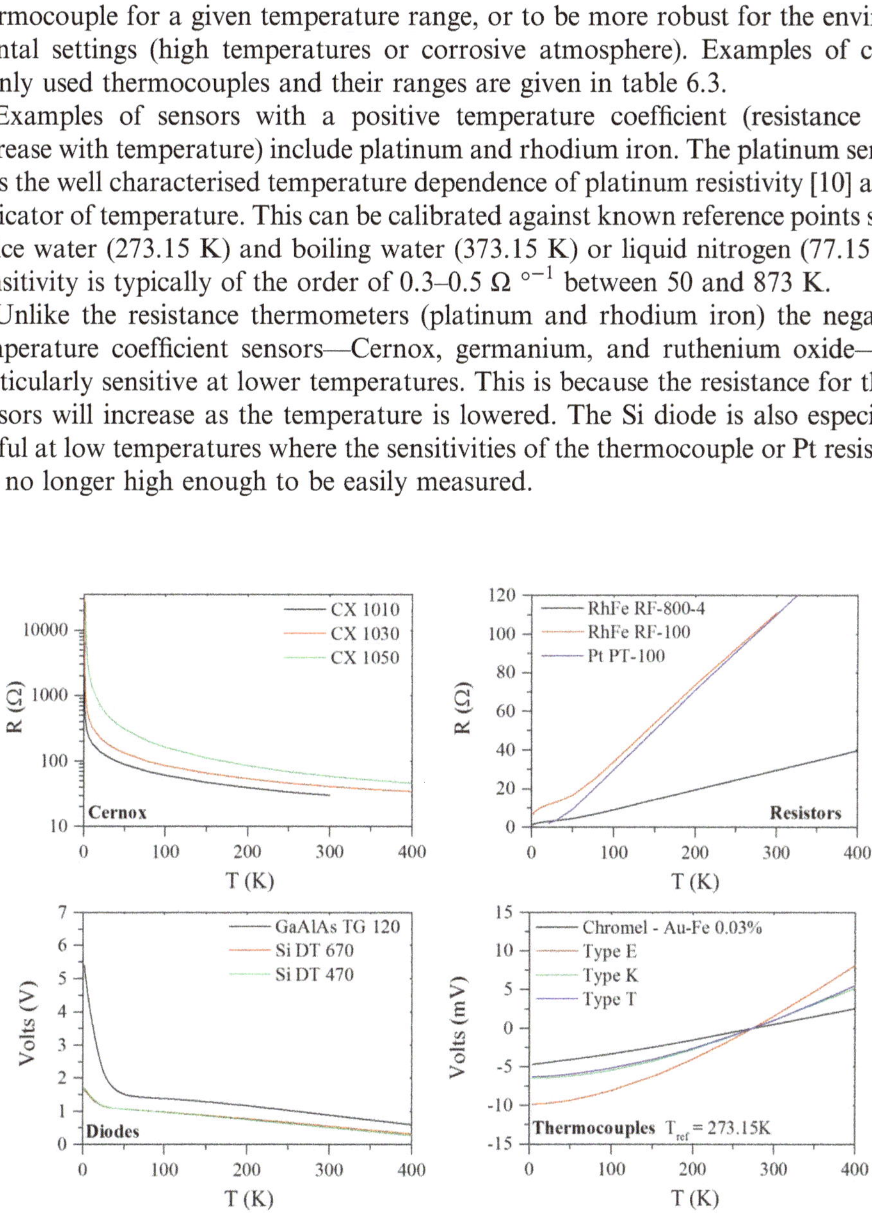

Figure 6.5. Temperature response curves for some commonly used temperature sensors.

Table 6.3. List of standard thermocouples, the metals used, temperature range and sensitivity (EMF range) as well as specific comments for each type.

Type	+ Lead	− Lead	Temp. range	EMF range (mV)	IEC 5843 colour (+/−)	Comments (bare wire)
J	Iron	Constantan (copper–nickel)	−210 to 1200 °C	−8.095 to 69.553	Black/white	*Of limited use in oxidising environment at high temperatures. Not recommended for low temperatures.*
K	Chrome(nickel–chromium)	Nickel–aluminium	−270 to 1372 °C	−6.458 to 54.886	Green/white	*Of limited use in vacuum or reducing atmosphere. Wide temperature range.*
T	Copper	Constantan	−270 to 400 °C	−6.258 to 20.872	Brown/white	*Good where moisture is present, low temperatures or cryogenic applications.*
E	Chrome	Constantan	−270 to 1000 °C	−9.835 to 76.373	Purple/white	*Of limited use in vacuum or reducing atmosphere. Highest EMF change per degree.*
N	Nicrosil(Nickel–chromium–silicon)	Nisil (Nickel–silicon–magnesium)	−270 to 1300 °C	−4.345 to 47.513	Pink/white	*Alternative to type K. Suitable for high temperatures.*
R	Platinum–13% rhodium	Platinum	−50 to 1768 °C	−0.226 to 21.101	Orange/white	*Do not insert in metal tubes; beware of contamination.*
S	Platinum–10% rhodium	Platinum	−50 to 1768 °C	−0.236 to 18.693	Orange /white	
B	Platinum–30% rhodium	Platinum—6% rhodium	0 to 1820 °C	0 to 13.820	Grey/white	
G	Tungsten	Tungsten—26% rhenium	0 to 2320 °C	0 to 38.564	No IEC standard ANSI code = brown/blue	*Beware of embrittlement. Not for oxidising atmosphere.*
C	Tungsten–5% rhenium	Tungsten—26% rhenium	0 to 2320 °C	0 to 37.066	No IEC standard ANSI code = white/red	
D	Tungsten–3% rhenium	Tungsten—25% rhenium	0 to 2320 °C	0 to 39.506	No IEC standard ANSI code = white/red	

Heat flux sensors are useful for monitoring the rate of heat flow, or total heat into a system. They are less commonly employed in thermal measurements except perhaps for the case of differential scanning calorimeters. One example of a heat flux sensor is the Peltier cell, whereby a thermoelectric effect that is proportional to heat flow (Peltier effect) is used to measure the total heat flow through a small cell (see figure 6.6).

The Peltier cell is comprised of two Al_2O_3 plates separated by an array of alternately p- and n-doped Bi_2Te_3 pillars connected together as thermocouples in series. The voltage generated at the Peltier element, V_p, is equal to the product of its Seebeck coefficient, S, and the temperature difference between the two Al_2O_3 plates, $\Delta T = T_1 - T_2$. The voltage measured by the Peltier cell, V_p, can be converted to heat flux, J_Q, if the area of the device, A, and its sensitivity, S_p (= V_p/Q), are known:

$$J_Q = \frac{V_p}{AS_p} \tag{6.20}$$

6.2 Calorimetry

Calorimetry is defined as the measurement of heat transfer in a system. This can be used to determine, for example, the specific heat of a material at a fixed pressure or volume; the latent heat of a first order phase transition (endothermic or exothermic); or the temperature at which a phase transition (whether chemical, structural or magnetic) can occur. There are various techniques that are employed, as outlined below:

1. **Adiabatic method**, where a known amount of heat is applied to the sample kept in thermal isolation. The measured temperature rise is then proportional to the heat applied and the heat capacity of the sample.

2. **Relaxation method**, as developed by Eucken [11] and Nernst *et al* [12] and the various methods that stem from this. The sample is kept in quasi-adiabatic conditions and a heat pulse of known energy is applied. The resulting temperature change or relaxation to the bath temperature is used to determine the specific heat.

3. **Heat pulse method**, where a heat pulse of known energy is applied to the sample in semi-adiabatic conditions. The slow change in temperature either side of the heat pulse is subtracted to determine the specific heat.

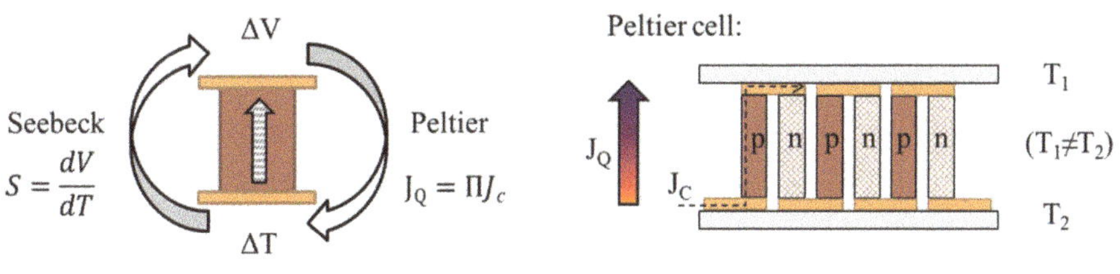

Figure 6.6. Schematic of the Peltier cell heat flux sensor. This solid state device utilises the Peltier effect to measure the heat flux across the device. To increase the sensitivity, pillars of n- and p-doped semiconductor (chosen because they have low thermal conductivity, reasonably good electric conductivity, and high Seebeck coefficient, *S*) are arranged such that heat flow passes through them in parallel, and electric current passes through them in series.

4. **Differential scanning calorimetry**, DSC. The sample is connected to a thermal bath by a well-defined heat link and the temperature of the system is scanned at a constant rate. As the temperature is scanned, small differences between the sample to be measured and a reference sample are monitored and additional power supplied to keep them at the same temperature. The difference in the power required to maintain the sample and reference cell at the same temperature is dependent on the heat capacity of the sample.

5. **AC calorimetry method**. Developed by Sullivan and Seidel [13], this techniques uses an oscillatory temperature perturbation of small amplitude that is applied to a sample linked to the thermal bath via a helium exchange gas. The amplitude and phase of the thermal response, as typically measured by a thermopile is related to the specific heat.

The choice of which technique to use often comes down to time, sample availability (i.e. how much is available) and required accuracy, as summarised in figure 6.7.

6.2.1 Adiabatic calorimetry

The traditional and arguably most accurate determination of specific heat is by adiabatic calorimetry [11, 12]. For this technique, the sample is ideally kept in adiabatic conditions and a heat pulse of known energy dQ is applied while the sample temperature is measured. The change in internal energy is equal to dQ and the usual definition for the specific heat under isobaric conditions is then:

$$C_p(T) = (dQ/dT) \tag{6.21}$$

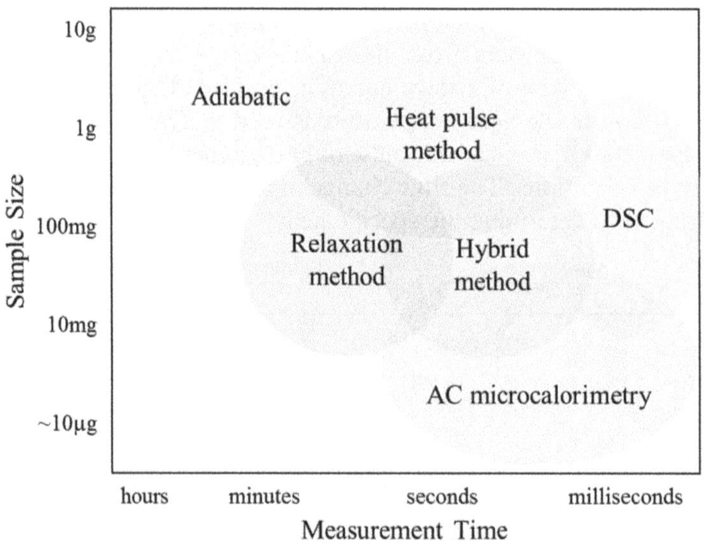

Figure 6.7. Plot illustrating approximate time constraints (for a single data point at fixed T, H) and required sample size for some of the different available calorimetry methods [12, 14–25].

Although the system is ideally adiabatic there will inevitably be a weak thermal leak to the bath temperature, T_b, and the sample temperature, T_s, will thus change. This can be characterised by an exponential decay of T_s:

$$T_s = (T_s - T_b)e^{-t/\tau} \tag{6.22}$$

where τ is the thermal relaxation time constant. The disadvantages with this technique include a required sample mass of the order of 1–10 g [26] and typically long timescales for the measurement (~1 h per temperature point) [27].

The difficulty of thermally isolating the measurement platform, such that τ is large enough for the temperature variation to be considered linear within the timescales of the measurement (e.g. $\tau \gg 500$ s), has limited the emergence of 'home-built' adiabatic calorimeters. This has led to the development of the semi-adiabatic heat pulse method.

The heat pulse method is an adaptation of the adiabatic calorimeter that has been shown to be capable of high precision measurements of specific heat for samples as small as 1 g [14]. The semi-adiabatic conditions in this case are usually provided by an adiabatic heat shield which tracks the temperature of the sample, thus eliminating any heat transfer from sample to environment. A temperature sensor and electric heater attached to the heat shield then performs the necessary feedback control [14, 28]. Whilst there may be some drift in the temperature as a function of time, as long as this has been characterised before and after the heat pulse, the specific heat can be determined from the step change in the temperature.

Other adaptations of the adiabatic calorimeter that are widely employed in the literature include the relaxation method [15, 29, 30] (which is available commercially through the Quantum Design Physical Properties Measurement System, PPMS, heat capacity puck[2]), and the hybrid method [19].

6.2.2 Relaxation and hybrid calorimeters

For the relaxation and hybrid methods there are two main differences with respect to the adiabatic heat pulse calorimeter: first, the sample size is now of the order of mg, and secondly, the thermal time constant τ is smaller, which immediately makes the measurement less time consuming.

Klaasse *et al* outlined the hybrid method in 2008 in order to treat data where 30 s $< \tau <$ 500 s, in other words bridging the gap between adiabatic ($\tau >$ 500 s) and relaxation measurements ($\tau <$ 30 s) [19]. One advantage of the method is that the starting temperature of each measurement is the final temperature of the previous measurement, removing the temperature settling stage and thus decreasing the overall time required. It was argued that as τ is finite, the heating due to heat pulse will not be linear, so to correct for this they extrapolate to halfway through the heat pulse [19], as demonstrated by the schematic given in figure 6.8(d).

[2] http://www.qdusa.com/sitedocs/productBrochures/heatcapacity-he3.pdf

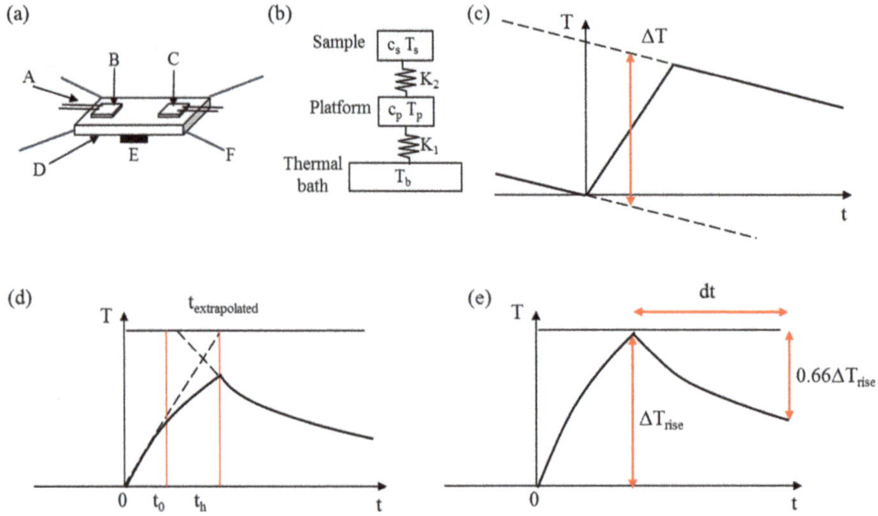

Figure 6.8. Schematic of the data extrapolation used for relaxation and hybrid methods. (a) Example of a typical experimental set-up; A–manganin wire, B–heater, C–Cernox sensor, D–gold-plated copper sheet, E–sample, and F–nylon thread. (b) Thermal circuit diagram where K_2 and K_1 describe the thermal conductivity of the thermal grease and conducting wires (manganin in this case), respectively. (c)–(e) Typical relaxation curves for the semi-adiabatic heat pulse, hybrid, and curve fitting methods, as discussed in the text.

The curve fitting method used in commercial relaxation measurements (for example, the QD PPMS) was first outlined by Hwang *et al* in 1997 [30] as a faster alternative to adiabatic calorimetry for measurement of the specific heat. Figure 6.8 shows the thermal-circuit diagram and experimental set-up of this model, which the authors applied to a sample that was suspended in a vacuum can by nylon thread to limit heat conduction to the temperature bath. Notice that now manganin wire (with almost zero temperature coefficient of resistance) supplies heat to the sample area and the temperature at the sample platform, T_p, is measured by a thin-film Cernox sensor. The curve fitting approach acknowledges that we measure the sample platform temperature, T_p, and not that of the sample, T_s, requiring a description of the various thermal contacts in the system: between sample and platform, τ_2, and between the platform and thermal bath, τ_1 [29]:

$$T_p = T_b + Ae^{-t/\tau_1} + Be^{-t/\tau_2} \tag{6.23}$$

where the variables τ_1, τ_2, and C_p are determined by fitting the relaxation curve to the characteristic functions described in [30] and [16], and the constants A, and B are proportional to the temperature differences $(T_p - T_b)_{t=0}$ and $(T_s - T_p)_{t=0}$, respectively. Further details can be found in [30].

Caution is required when using the relaxation method, especially with the commercial PPMS system, where user control is limited. There have been several discussions of the best way to use the PPMS for accurate results [17, 18, 31, 32] the outcome of which are the following general guidelines:

1. For samples with poor thermal conductivity a large mass will result in underestimation of C_p if the temperature rise, ΔT_{rise}, defined in figure 6.8(e), is too large. This was highlighted by Kennedy *et al* with measurements of zirconium tungstate [32].
2. To fully capture the latent heat it is important to ensure that ΔT_{rise} is greater than the thermal hysteresis, ΔT, and that the measurement crosses the phase transition on both heating and cooling [17].
3. At a first-order phase transition more details of the specific heat can be found by examining the relaxation *curves* as outlined by Lashley *et al* [31] and further developed by Suzuki *et al* [18]. A demonstration of this is given in figure 6.9.

6.2.3 Differential scanning calorimetry (DSC)

Complete measurement of the latent heat at a first-order phase transition is arguably best achieved by differential scanning calorimetry, DSC. The basic set-up for this technique is the symmetrical arrangement of two Peltier cells, one attached to a reference cell, the other to the sample to be measured (see figure 6.10). The temperature of the system is typically scanned at a constant rate while the differing heat flux at the sample and reference cell is measured. In addition, the Peltier can actively cool/heat the sample to maintain the same temperature. The advantages of DSC include fast data-acquisition, complete measurement of the heat-flow in and out of the system, and the use of potentially small samples (of the order of mg). A disadvantage of the technique includes difficult calibration of the sensor as it can be sensitive to the rate of change of temperature.

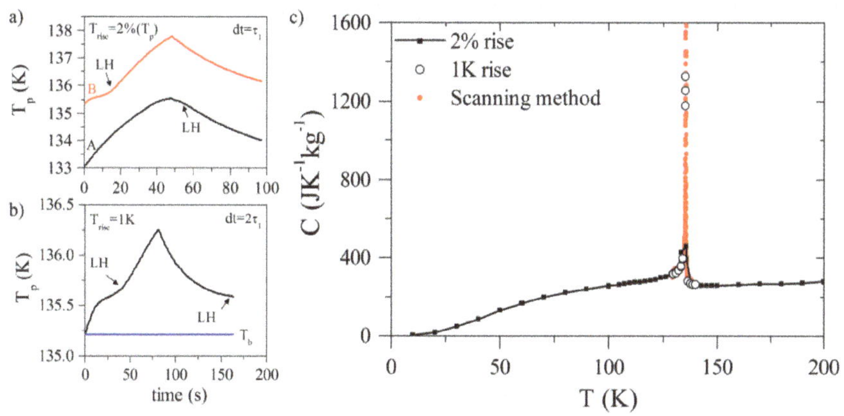

Figure 6.9. Examples of the 'scanning' or 'slope-difference' method employed to extend the relaxation technique to first order phase transitions for $DyCo_2$. (a) and (b) Raw relaxation curves close to the Curie temperature. (c) Comparison of data obtained from the scanning method, and slope-difference method with varying temperature rise. The increased number of datapoints results from analyzing the data point by point rather than fitting the entire curve.

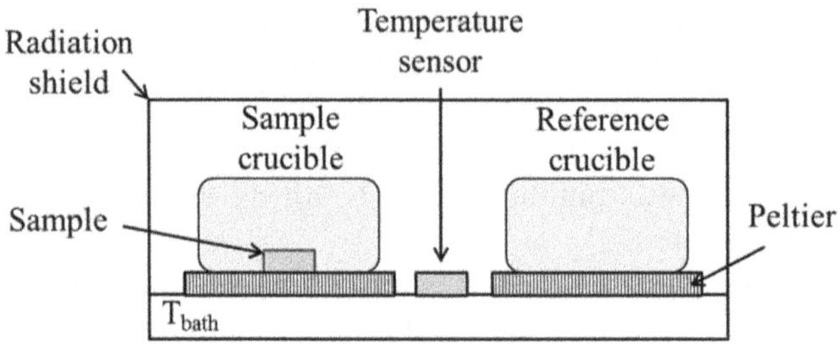

Figure 6.10. Typical arrangement of a DSC.

The specific heat can be determined if the mass of the sample, m, rate of temperature change, $\partial T_s/\partial t$, and heat supplied, Q, are known:

$$c(T_s) = \frac{-Q}{\partial T_s/\partial t} = \frac{-V_p}{S_p(\partial T_s/\partial t)} \tag{6.24}$$

Additional complications that may need to be considered are the internal relaxation timescales of the Peltier element (τ_3 effects), radiation losses, and conduction from the sample to the thermal sink as discussed in detail by Plackowski *et al* [33] and later by Basso *et al* [23]. Notice that accurate determination of the specific heat requires some knowledge of the addenda (empty calorimeter contributions), and the temperature dependence of the sensitivity, S_p. Further details of how these parameters can be determined was outlined comprehensively by Plackowski *et al* [33].

Temperature modulated DSC is an extension of this method that uses an AC signal to improve the sensitivity of the measurement. It can be considered as a hybrid between DSC and AC calorimetry.

6.2.4 Differential thermal analysis (DTA)

Differential thermal analysis (DTA) is a method of determining the transition temperatures of a material by detecting the absorption or release of heat associated with the chemical or physical changes at these temperatures. It is similar to DSC in the sense that it employs the use of a reference sample (typically Al_2O_3), however, as a quantitative measure of the specific heat or latent heat is not necessarily required, a Peltier cell to control heat flow (and thus maintain negligible temperature difference between samples) can be omitted.

A typical DTA set-up would consist of a thermocouple on the sample and reference, both of which are monitored as the sample environment is warmed up/cooled down. As the sample undergoes a phase transition a change in the $\mathrm{d}T/\mathrm{d}t$ (of the sample) will occur that allows identification of a phase transition.

6.2.5 Thermogravimetric analysis (TGA)

Thermogravimetric analysis (TGA) is typically used to monitor chemical phase transitions* (loss/gain of mass) as a function of temperature. In conjunction with DSC or DTA it can help to identify whether a chemical phase change is occurring at a specific phase transition temperature. It is also useful as a means of testing the thermal stability of a material (i.e. up to which temperature mass loss is negligible).

The main components of a TGA apparatus are: a precision balance, accurate measurement of temperature (typically a thermocouple), and a programmable furnace. The measurement is typically carried out under constant heating or constant mass loss conditions [34].

Examples of chemical phenomena for which a mass change would occur:

- Vaporisation;
- Sublimation;
- Absorption;
- Adsorption;
- Desorption;
- Chemisorption;
- Desolvation;
- Decomposition;
- Oxidative degradation;
- Solid-state reactions;
- Solid-gas reactions.

6.2.6 AC calorimetry

Sullivan and Seidel developed the AC calorimetry technique for sensitive detection of small changes in the heat capacity, coupled with a simplified measurement apparatus that is no longer sensitive to small heat leaks [13]. It works on the principle that if you supply an AC modulated heating to a sample, the resultant heating power (at twice the frequency, because $P = V^2/R$) would generate a signal that could be 'locked' into. The use of such a lock-in technique can improve the signal-to-noise ratio as any noise (or DC offsets) outside of the specified frequency range will be discarded.

One example of the AC calorimetry technique is the adaptation of a commercial Xensor SiNi membrane gauge by Minakov *et al* in 2005 for AC calorimetry of microgram samples [24]. Figure 6.11 shows a schematic of the technique. During the measurement, an AC current ($f \sim 5$ Hz) passes through a 100×50 μm resistive heater, upon which the sample is placed. He gas is used for heat transfer between the sample and temperature bath and power supplied by the AC current is chosen such that the heat transfer can be approximated by assuming a point source at the sample [24].

The specific heat of the sample can be determined using:

$$C_p = \frac{SP_h}{4\pi f V_{th}} \sin(\phi_{th}) \tag{6.25}$$

where S is the Seebeck coefficient, determined separately using a calibration routine, P_h the power supplied to the heater region at frequency $2f$, and V_{th} and ϕ_{th} are the voltage amplitude and phase of the thermopile signal measured by the lock-in amplifier.

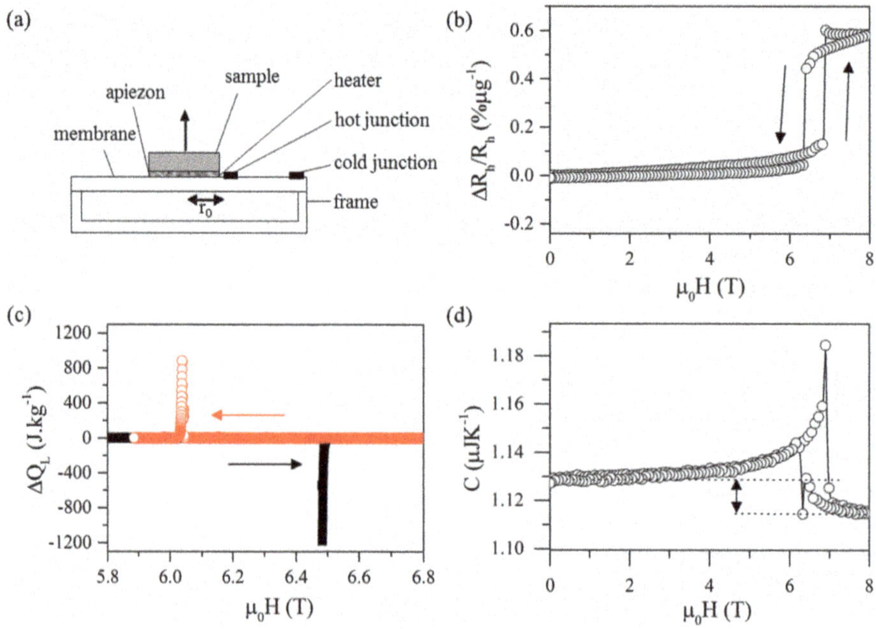

Figure 6.11. (a) AC microcalorimeter set-up. (b)–(d) AC microcalorimetry data of $CoMnGe_{0.08}Si_{0.92}$ at 190 K. (b) Strain induced change in resistance of the SiN membrane indicating onset of a volume change. (c) Latent heat and (d) specific heat, where ΔC is defined as $(C_H - C_{0T})$, as discussed elsewhere [37].

The AC nature of this technique means it is well suited to measuring specific heat, C_p, at a reversible transition. For a first-order, irreversible phase transition the measurement of latent heat is suppressed. This is because (a) the latent heat generated by sweeping H or T will not necessarily occur in phase with the ac modulation, and (b) as long as the ac temperature modulation is much less than the thermal hysteresis associated with the first order phase transition the sample will not be able to cycle between states. As a result, there may be an indication of latent heat when the sample is first driven across the phase transition but is soon averaged out in the many subsequent ac oscillations. Saruyama *et al* have discussed how the latent heat is suppressed by the ac technique in bulk measurements [35]; and this was demonstrated further in [36].

In 2008 Miyoshi *et al* showed that the ac microcalorimeter could be easily adapted to measure latent heat [25]. They argued that by pumping out the He exchange gas ($P < 0.05$ mbar) and by scanning with magnetic field at a rate of 0.5 T min^{-1}, the thermopile voltage, V_{th}, captures the sharp peaks of the latent heat, as illustrated in figure 6.11(b). A weak thermal-link still exists between the sample and the thermal bath, so the measurement is considered to be carried out under semi-adiabatic conditions. Temperature changes at the sample are thus registered as sharp increases in V_{th} as a function of magnetic field, with a characteristic decay time of approximately 1 s. Calibration of the voltage-peak-height V_{pk} into a latent heat,

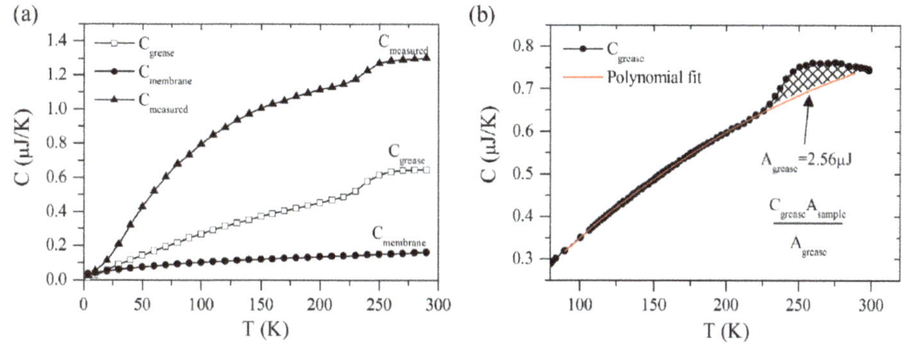

Figure 6.12. Separating addenda from specific heat of sample. (a) In this example, the addenda includes the membrane on which the sample is mounted as well as the Apiezon grease that was used to fix it to the surface. (b) As the amount of Apiezon may differ with each measurement, the glass transition at approximately 250 K can be used to determine the relative amount used.

L, is performed by comparison with a pulse of known energy E_{cal} and height V_{cal}, using:

$$L = \frac{V_{pk}}{V_{cal}} E_{cal} \tag{6.26}$$

6.2.7 Removal of addenda

With all calorimetry measurements, there will need to be consideration of the 'addenda', a term commonly used to describe erroneous additions to the measured specific heat, such as the thermal grease used to mount it. In the case of the AC microcalorimetry method described in the previous section, this addenda would encompass the thermal grease and the SiN membrane onto which the sample is mounted, as shown in figure 6.12.

Removal of the addenda can be done if empty measurements are collected. For example, the empty SiN gauge could be measured, followed by a small amount of thermal grease, before the sample itself. In this way, the three separate contributions have been measured in succession. The only difficulty might be calibrating for the exact amount of thermal grease used. For Apiezon N^{TM} this can be achieved with some accuracy, by mapping the magnitude of the glass transition at 250 K (once the membrane contribution has been subtracted).

6.3 Thermal conductivity

Measurement of the thermal conductivity is often difficult as it requires some prior knowledge of the sample's thermal properties in order to choose an appropriate measurement and geometry. Some of the more common techniques for bulk and thin film will be outlined here.

6.3.1 Steady state methods

The steady state method for thermal conductivity measurements requires two fixed points (the hot and cold bath temperatures). The assumption is that in the steady state, a constant heat flow will pass through the sample (and reference if relevant) and that the thermal conductivity can be calculated from a variation of equation (6.18). There are four common measurement geometries—absolute, comparative cut bar, radial heat flow, and parallel thermal conductance, as shown in figure 6.13.

For the absolute measurement geometry, the sample is cut such that two thermocouples can be placed a distance, L, apart. One end of the sample is fixed to the cold sink, whilst the other is attached to a heater with known power output, P. If the heat losses, Q_{loss}, are known, then the heat flowing through the sample, Q:

$$Q = P - Q_{\text{loss}} \tag{6.27}$$

and the thermal conductivity, κ is:

$$\kappa = \frac{QL}{A(T_{\text{h}} - T_{\text{c}})} \tag{6.28}$$

where A is the cross-sectional area. To minimise heat losses, thin thermocouple wires with low thermal conductivity should be used, preferably in differential configuration such that the temperature difference, ΔT, is directly measured [38]. In addition, the sample can be measured under vacuum.

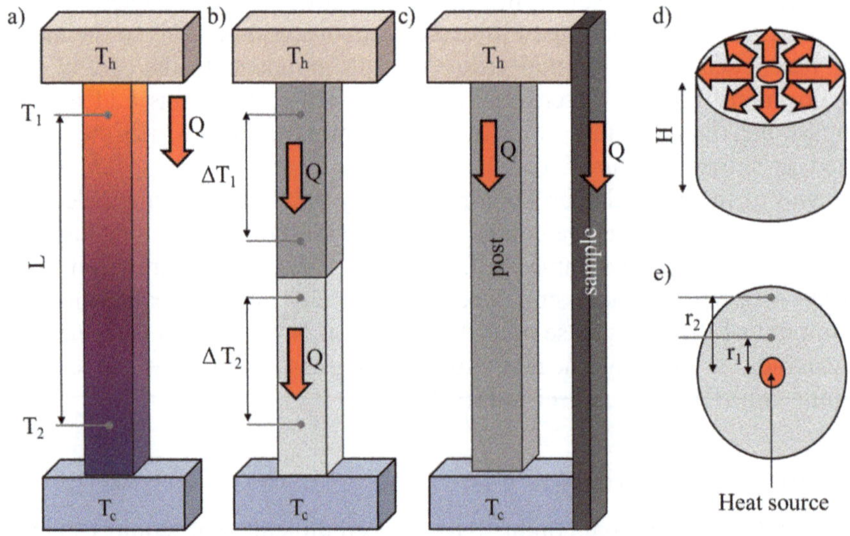

Figure 6.13. Steady state thermal conductivity measurement geometries. (a) Absolute method, where the sample is fixed to a heater (T_{h}) and cold sink (T_{c}), and the temperature measured at two points (T_1, T_2) separated by distance, L. (b) Comparative cut bar. (c) Parallel thermal conductance method, where a thin needle-like sample can be measured when connected thermally in parallel to a known reference. (d) and (e) Side and top views, respectively, for the radial method.

For the comparative cut bar method, the sample is fixed in series to a reference sample, with one end attached to the cold sink and the other to the heat source. The advantage of this method is that it enables accurate determination of the heat flow if the thermal conductivity of the reference sample is well known. This avoids requiring detailed knowledge of the heat losses (or making the assumption that heat losses are negligible) and removes the need to know the power output at the heater. In this case, the thermal conductivity of the sample will be:

$$\kappa_s = \kappa_r \frac{A_r \Delta T_r L_r}{A_s \Delta T_s L_s} \tag{6.29}$$

where $A_{s/r}$, $\Delta T_{s/r}$, and $L_{s/r}$ are the cross-sectional area, temperature difference and length of the sample and reference, respectively. Alternatively, a heat flux sensor (or transducer) could be used in place of the reference sample.

The parallel thermal conductance method is useful for small thin samples (e.g. $\sim$1 mm^2 cross-sectional area [39]), where the sample is fixed in parallel to the reference sample. A typical measurement would involve first measuring the thermal conductance of the reference bar, followed by the sample and reference in parallel, where the difference is due to the thermal conductance of the sample.

For the radial geometry, the heater is placed in the centre of a cylindrical sample, and heat flow is assumed to flow equally (radially) out from this point. In this case, the thermal conductivity is determined using:

$$\kappa = \frac{Q \ln(r_2/r_1)}{2\pi H \Delta T} \tag{6.30}$$

where r_1 and r_2 are the radial distances to the hot and cold thermocouples, as shown in figure 6.13, H is the height of the sample, Q is the heat supplied by the heater, and ΔT is the measured temperature difference.

For each of these methods, some consideration may need to be made for the additional thermal resistance (in the measurement) due to the epoxy used to attach the thermocouples to the sample(s), or, the hot and cold sinks. This is especially relevant for samples with high thermal conductivity where such thermal resistance starts to dominate the measurement. As a result, it is good practice to choose the sample size and measurement method based on the expected thermal conductivity of the sample. For example, a material with high thermal conductivity, such as copper, would require a short thin wire so that the thermal conductance ($\kappa L/A$) is increased with respect to the contact resistance. This would also result in an increase in the observed temperature difference, thus improving measurement (of this) with thermocouples. A material with low thermal conductivity, such as a ceramic, on the other hand, could be cut to have much larger cross-sectional area and produce similar thermal conductance.

6.3.2 Transient methods

Transient techniques have the advantage of faster data acquisition, and less dependence on heat losses and thermal contact resistance of the temperature sensors.

Much like the relaxation or curve fitting methods in the calorimetry section, transient measurement of the thermal conductivity relies on characterisation of the time dependant change in temperature (of a material) in response to a heat pulse, or periodic heating.

6.3.2.1 Pulsed power

The measurement setup for the pulsed power transient method is similar to that of figure 6.13(a) except that now an AC heating signal is supplied at the hot end and the heat sink is allowed to slowly vary. Similar to the hybrid method heat capacity measurement, this technique was originally designed to measure thermal conductivity and Seebeck coefficient [40].

If the AC signal has a period 2τ, then the time varying heat transfer in the system can be described by:

$$Q = C(T_h)\frac{dT_h}{dt} = R_e(T_h)I^2(t) - K\left(\frac{T_c + T_h}{2}\right)\Delta T(t) \tag{6.31}$$

where $C(T)$ is the heat capacity of the sample heater at temperature, T, R_e is the resistivity of the electric heater, $I(t)$ is the current supplied, K is the thermal conductance, T_c and T_h are the temperatures at the cold at hot sinks, respectively, and $\Delta T(t)$ is the induced temperature difference. To simplify determination of K from this, the assumption that $T_c \sim T_h$ is made (i.e. in the limit of a small temperature difference $\ll$1 K). This means that the time dependence of T_h can be ignored and $C(T_c)$, $R(T_c)$, $K(T_c)$ can be substituted for $C(T_h)$, $R(T_h)$, $K(T_h)$ such that:

$$K = \frac{RI_0^2}{\Delta T_{pp}}\tanh\left(\frac{K\tau}{2C}\right) \tag{6.32}$$

where R is the resistance of the heater, I_0 is the amplitude of the AC electric current, ΔT_{pp} the temperature difference, τ is half the heating current period, and C is the volumetric heat capacity of the heat source. Notice that the thermal conductance, K, is present in both the left and right hand sides of this equation, indicating that the solution will need to be found numerically (i.e. by fitting, or guessing the value and adjusting the guess based on the difference between the calculated and 'guessed' values of K).

6.3.2.2 Hot wire method

Similar to the radial steady state method, the hot wire method involves the determination of thermal conductivity by embedding a thin wire (as heat source) in the material to be tested [41]. Based on the assumption that heat will flow radially out from the wire, and that the wire can be considered infinitely long (i.e. do not consider the edge effects) and infinitesimally thin (i.e. can be considered as a point source), the thermal conductivity can be described by:

$$\kappa = \frac{P}{4\pi(T(t_2) - T(t_1))L}\ln\left(\frac{t_2}{t_1}\right) \tag{6.33}$$

where P is the heater power, $T(t_1)$ and $T(t_2)$ are the temperatures at times t_1 and t_2, and L is the length of the wire (such that we calculate the power per unit length from P/L).

6.3.2.3 Laser flash method

The laser flash method is a commonly employed technique due to its non-destructive nature, and commercial availability (e.g. NETZSCH or LINSEIS). This technique involves optical heating of the sample being studied with a laser or flash source, and subsequent detection of the transient heat flow on the other side of the sample using an infrared sensor [42]. The measured temperature can be described by:

$$T(t) = \frac{Q}{\rho c_p L}\left(1 + 2\sum_{n=1}^{\infty}(-1)^n \exp\left(\frac{-n^2\pi^2}{L^2}\alpha t\right)\right) \tag{6.34}$$

where q is the fluence (energy per unit area) provided by the laser pulse, ρ is the density, c_p is the specific heat of the sample, L is the sample thickness, α is the thermal diffusivity, and t is the time at which the temperature, T is measured.

Equation (6.34) can be simplified by defining:

$$V(t) = \frac{T(t)}{T_M} \tag{6.35}$$

and:

$$\omega = \frac{\pi^2 \alpha t}{L^2} \tag{6.36}$$

such that:

$$V = 1 + 2\sum_{n=1}^{\infty}(-1)^n \exp(-n^2\omega) \tag{6.37}$$

By plotting V as a function of ω it can be found that $\alpha = 1.38$ when $V = 0.5$. This means that the thermal diffusivity can be determined from the time dependant (transient) temperature profile of the sample, where:

$$\alpha = \left(\frac{1.38L^2}{\pi^2 t_{1/2}}\right) \tag{6.38}$$

and $t_{1/2}$ is the time it takes for the back surface to reach half of the maximum temperature rise. The thermal conductivity is then calculated using:

$$\kappa = \alpha\rho c_p \tag{6.39}$$

It can be seen from this that the limit of the laser flash method will be the temporal resolution of the IR sensor used in the measurement, convoluted with the length scale of the heating pulse, the thermal diffusivity of the sample, and its thickness. If the sample is too thin, for example, the heat will diffuse through the sample too fast for an appropriate $t_{1/2}$ to be determined. This is particularly relevant for thin film samples, and often results in a minimum thickness of the order of 100 μ.

6.3.2.4 Transient plane source method

In the transient plane source method, a thin metal strip or disc acts as both a plane heat source and the temperature sensor [43]. This is achieved by monitoring the current and voltage supplied to the resistor (whether a thin disc or square) such that the heating power supplied can be determined and the change in resistance of the disc (or square) can be monitored. If the resistance of the disc (or square) is well characterised then the temperature change can be calculated, and the thermal conductivity determined from:

$$\Delta T(\tau) = \frac{Q}{\pi^{1.5} rk} D(\tau) \tag{6.40}$$

where:

$$\tau = \sqrt{\frac{t\alpha}{a^2}} \tag{6.41}$$

and a is a constant that measures the size of the hot disc or square (the radius of the disc or half the length of the square), α is the thermal diffusivity, t is the time measured from the start of the transient heating, Q is the heating power, k is the thermal conductivity and $D(\tau)$ is a function that measures/describes the spatial variation of the temperature across the sample with respect to time [43]. As with other transient methods, determination of the characteristic time, τ, requires an iterative process.

6.3.3 Thin film methods

Steady state methods for measuring the thermal conductivity of a thin film have several difficulties associated with them: in particular accounting for temperature drops across the substrate (of the thin film), and minimising radiation losses (which will have a larger impact on the measurement). In the case of in-plane thermal conductivity measurements, removal of the substrate by suspending the thin film is the most comprehensive way to reduce the impact of transient heat losses (i.e. heat flow through the substrate rather than the film). For out-of-plane thermal conductivity measurements, the substrate is often chosen to have high thermal conductivity such that its impact on the measured temperature drop is minimised. Neither of these methods work well if the thermal conductivity is affected by strain at the substrate–film interface. As such, transient techniques, such as the 3ω method, are better suited.

6.3.3.1 3ω method

The 3ω method is an AC technique that was initially proposed by Cahill in 1990 [44] and can be applied to both bulk and thin film [45] samples. It works on the principle that a heater deposited directly onto the thin film to be studied can also act as a sensor if it has a large enough temperature coefficient for resistance. An example of such a heater is given in figure 6.14. Whilst for conductive thin films, some consideration would need to be made to include an electrically insulating barrier, it otherwise works well as a direct measurement of the thermal conductivity of thin films.

Similar to the AC calorimetry technique, an AC current is passed to the heater, such that:

$$I(t) = I_0 \cos(\omega t) \tag{6.42}$$

where I_0 is current amplitude, and ω (= $2\pi f$) is the angular frequency of the AC signal. As the power produced will be proportional to I^2R, there will be Joule heating present at twice the frequency of excitation:

$$\Delta T(t) = \Delta T_0 \cos(2\omega t + \varphi) \tag{6.43}$$

where ΔT_0 is the amplitude of the temperature change, and φ is the phase difference between this temperature change and the AC current. Given that the temperature difference will be measured as a change in the resistance of the strip, equation (6.43) can also be expressed as:

$$R_e(t) = R_{e,0}(1 + \alpha_R \Delta T) = R_{e,0}(1 + \alpha_R \Delta T \cos(2\omega t + \varphi)) \tag{6.44}$$

where α_R is temperature coefficient of resistance ($\Delta R_e/\Delta T$), and $R_{e,0}$ is the resistance at the start of the measurement (initial state). Given that the resistance is changing with time, and that $V = IR$, the multiple of equations (6.42) and (6.44) can then be used to determine the voltage that would be measured across the heater strip:

$$\begin{aligned} V(t) = {} & R_{e,0} I_0 \cos(\omega t) + \frac{1}{2} R_{e,0} I_0 \alpha_R \Delta T_0 \cos(\omega t + \varphi) \\ & + \frac{1}{2} R_{e,0} I_0 \alpha_R \Delta T_0 \cos(3\omega t + \varphi) \end{aligned} \tag{6.45}$$

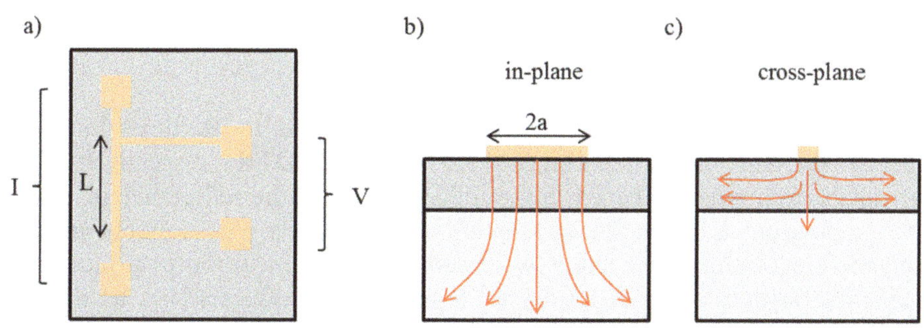

Figure 6.14. Contact geometry for the 3ω method. (a) Top side view. A strip of width, $2a$, with contact pads for signal in (I, 1ω). Voltage is measured across a length, L, of the heater strip (V, 3ω). (b) and (c) Choice of heater strip width will determine whether heat flow is predominantly in-plane or cross-plane.

where there are now three terms, two of which are proportional to the first harmonic of the measurement (ω), and the latter to the third harmonic (3ω). This latter term contains information about the thermal transport in the sample, and can be measured by a lock-in amplifier.

Determination of the film thermal conductivity then requires suitable design of the experiment. First the substrate thermal conductivity should be determined. This can be achieved using the 3ω method if the heat strip is deposited directly onto the substrate, where the temperature difference at the heater, ΔT_s can be described by [46]:

$$\Delta T_s = \frac{p}{\pi L k_s}\left[0.5\ \ln\left(\frac{\alpha_s}{a^2}\right) - 0.5\ \ln(\omega) + \eta\right] - i\left(\frac{p}{4Lk_s}\right) = \frac{p}{\pi L k_s} f_{\text{linear}}(\ln\omega) \quad (6.46)$$

where η is a constant, p/L is the peak electric power per unit length, f_{linear} is a linear function of $\ln(\omega)$, α_s is the thermal diffusivity of the substrate and we have made the assumption that the thermal penetration depth ($\alpha_s/2\omega$) is much larger than the heater half width, a. Next, the thermal conductivity of the film can be determined by deducing the temperature difference across the film from the heat flux, and known substrate thermal conductivity:

$$\Delta T_{s+f} = \Delta T_s + \frac{pd_f}{2aLk_{f\perp}} \quad (6.47)$$

where d_f is the thickness of the film, and the error can be minimised by fitting experimental data for various heater frequencies, ω, and strip width, a.

Note that the above works well on the assumption that heat flow is perpendicular to the heater plane (cross-plane conductivity). For in-plane measurements, the heater width could be reduced, such as the example given in figure 6.14. In this case, data analysis would require a more comprehensive model of the heat flow, which can be simplified with complementary measurements to isolate the cross-plane $k_{f\perp}$ first, so that it can be extracted from the in-plane measurements.

Additional analysis of 3ω measurements has since been presented, with a suggestion of analysing the results in terms of the following dimensionless variables: relative thermal conductivity $K = \kappa_f/\kappa_s$, the film's Biot number $B_{if} = hb_f/\kappa_f$, and the ratio between the effective heater half linewidth to the film thickness, $H_r = a/(b_f n_f^{1/2})$ [47].

6.3.3.2 *Transient thermoreflectance*

Transient thermoreflectance (TTR), sometimes referred to as time domain or frequency domain thermoreflectance (TDTR and FDTR, respectively), is a pump–probe technique that measures subtle changes in the reflectance of the film as its temperature is increased by a pump laser. As the measurement relies on reflectance measurements, it requires the sample to have a smooth surface, and a thin metallic layer is often deposited on top (metal transducer) [48], as shown in figure 6.15. The transducer can be chosen to have a higher reflectivity, which is sensitive to changes in temperature (dR/dT) whilst also conducting heat to the sample. If dR/dT is known then any change in the surface temperature can be

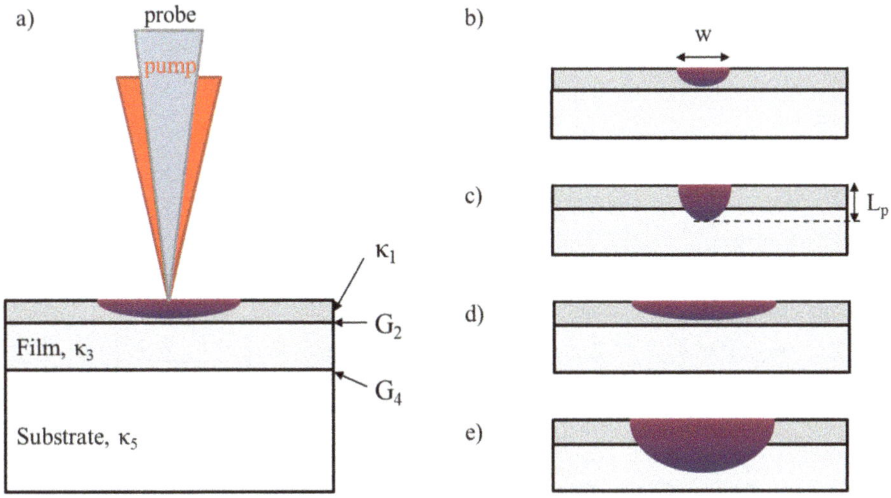

Figure 6.15. Schematic of the TTR technique. (a) A pump laser incident on the sample surface results in a temperature increase. The corresponding probe laser measures the reflectivity, which changes with temperature. (b)–(e) The thermal penetration depth, which is controlled by modulation frequency (of the pump laser), beam width, w, and the thermal conductivities $\{\kappa_1, \kappa_3, \kappa_3\}$ and conductance $\{G_2, G_4\}$.

measured; corresponding thermal models can then be developed to extract the thermal conductivity.

The advantages of this technique include:

- The ability to measure the thermal conductance of interfaces;
- Spatially resolved measurements;
- High resolution mapping.

With TTR, the pump/probe lasers can either be continuous wave (CW), or pulsed. With both a CW and pulsed laser system, an electrooptic modulator (EOM) will create a periodic signal from the CW pump laser that will periodically excite the sample to be measured such that an AC variation in the reflectivity from the probe laser can be observed. This helps to increase the sensitivity of the measurement by enabling use of a lock-in amplifier. To avoid interference, the wavelengths of these two lasers would typically be different, where part of the modulated pump laser is passed as the reference. With a pulsed laser system, a single laser source could be used alongside a delay stage (that introduces a delay between the pump and probe lasers) and second harmonic generation (SHG) crystal to introduce frequency doubling of the pump laser, in addition to modulation by an EOM. An example of such set-ups is given in figure 6.16.

The distinction between TDTR and FDTR is the variable in the measurement: for TDTR a delay is introduced between the pump and the probe and varied, whereas in FDTR the modulation frequency is varied. To understand analysis of TTR it is useful to define the thermal diffusivity, D and effusivity, ε:

$$D = \kappa / C \tag{6.48}$$

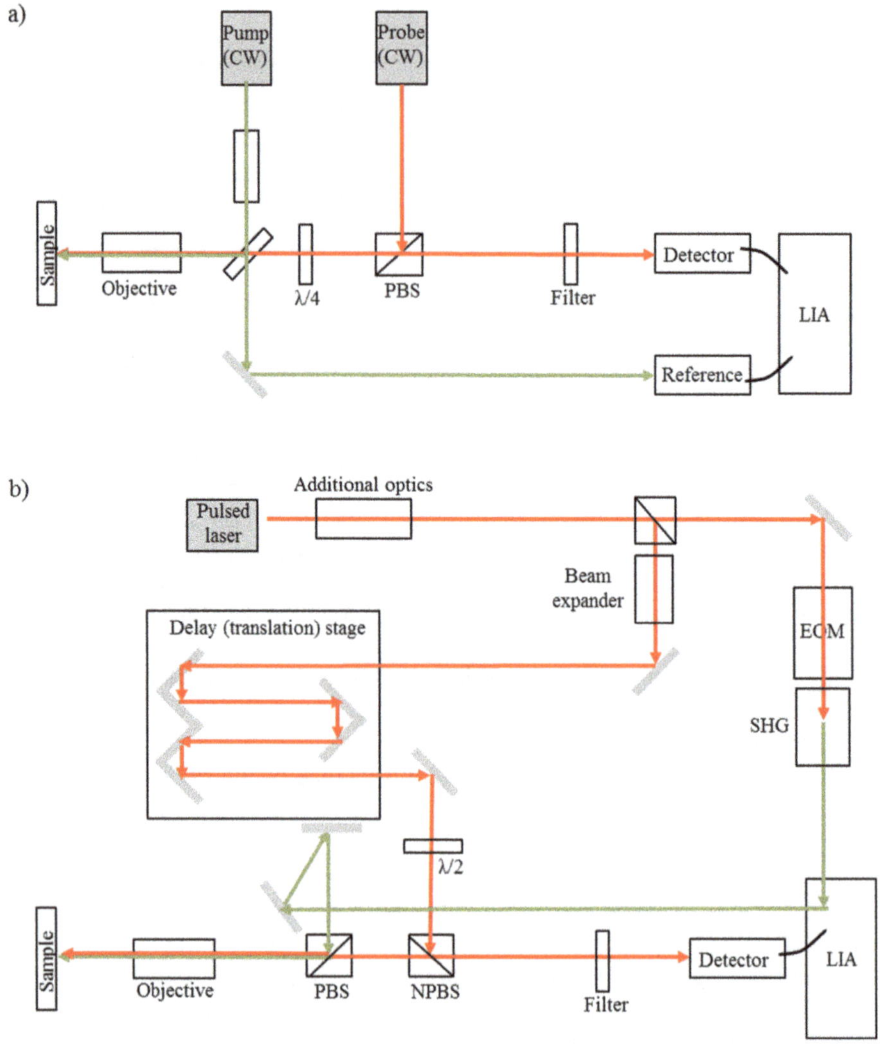

Figure 6.16. Examples of TTR measurement setups for (a) continuous wave laser, and (b) pulsed laser. EOM is the electrooptic modulator, PBS a polarising beamsplitter, NPBS a non-polarising beam splitter, SHG a second harmonic generator crystal, and LIA is a lock-in amplifier.

$$\varepsilon = \sqrt{\kappa C} \tag{6.49}$$

where κ is the thermal conductivity, and C is the heat capacity per unit volume. Whilst the diffusivity describes rate of transfer of heat of a material from the hot to the cold side, the effusivity describes the ability of a material to exchange heat with its surroundings.

Given that the temperature profile of the measurement can be defined by:

$$C\frac{dT}{dt} = \kappa\nabla^2 T \tag{6.50}$$

these measurements can also be easily represented in the frequency and spatial frequency domains by:

$$\tilde{T} = \tilde{T}_0 \, e^{i\omega t} e^{-qz} \tag{6.51}$$

$$i\omega C\tilde{T} = \kappa q^2 \tilde{T} \tag{6.52}$$

$$q = \sqrt{i\omega / D} \tag{6.53}$$

where $\tilde{T}$ is the temperature in the Fourier domain. The general solution for frequency domain thermoreflectance (FDTR) was discussed by Cahill [49] for a bulk sample:

$$\Delta T = 2\pi A \int_0^{\infty} G(u) e^{-\pi^2 (w_0^2 + w_1^2)/2} k \; \mathrm{d}k \tag{6.54}$$

where w_0 and w_1 are the diameter of the pump and probe lasers (on the sample), respectively, $G(u)$ is a function that describes the intensity profile of the laser, A is the amplitude, and k is the wavevector. Equation (6.54) effectively describes a Gaussian weighted surface temperature.

In the low frequency limit ($\omega \ll D/w_0^2$), equation (6.54) reduces to:

$$\Delta T = \frac{A}{\kappa} \int_0^{\infty} \exp(-\pi^2 k^2 w_0^2) k \; \mathrm{d}k = \frac{A}{2\sqrt{\pi} w_0 \kappa} \tag{6.55}$$

which is useful for estimating the steady state temperature rise of the sample.

In the high frequency limit ($\omega \gg D/w_0^2$), equation (6.54) reduces to:

$$\Delta T = \frac{A}{q\kappa} \int_0^{\infty} \exp(-\pi^2 k^2 w_0^2) k \; \mathrm{d}k = \frac{A}{\pi w_0^2 \sqrt{i\omega\kappa C}} \tag{6.56}$$

This limit is equivalent to describing 1D heat transfer with a uniform heat flux of $A_0/\pi w_0^2$. Equations (6.55) and (6.56) highlight a progression of ΔT from a constant to a frequency dependant value. This feature is useful for pinning down an appropriate model of the sample, where the measurement can be performed as a function of ω.

Another option is to vary the delay time (TDTR), which can be achieved with the pulsed laser based TTR measurements, and a variable translation stage (see figure 6.16). In this case, the changing thermal profile of the surface temperature in response to the pulsed heating can be probed, an example of which is given in figure 6.17. As the delay time is varied a change in phase shift and amplitude of the signal measured by the lock in amplifier will be observed.

6.3.4 Summary

Whilst there are several available methods, measurement of thermal conductivity is tricky due to the introduction of thermal interface resistance (Kapitza), which can

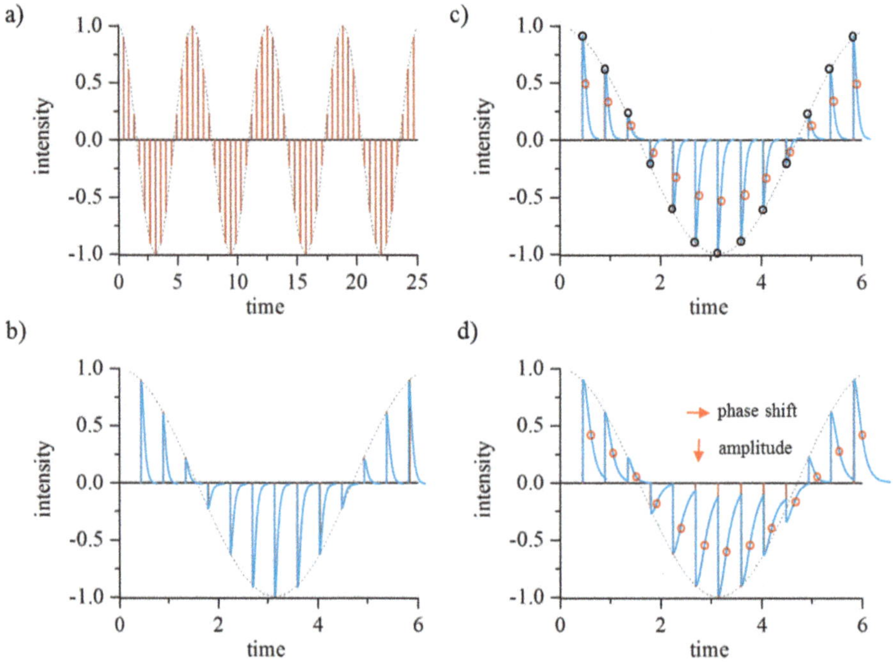

Figure 6.17. Schematic of TDTR pulsed laser measurements. (a) A modulated pump laser incident on the sample surface. (b) Resulting modulated temperature increase. (c) Schematic of the temperature rise measured by the probe laser without (black symbol) and with (red symbols) time delay. (d) Exaggerated example showing how the introduction of a time delay between pump and probe reduces the measured temperature and introduces a phase shift.

further complicate the measurement as the sample becomes smaller (thin film). Common techniques will aim to minimize radiation losses by keeping the temperature difference for the measurement small, or by using lock-in techniques such as in 3ω or TDTR. For bulk, the sample geometry may be chosen to optimise the measurement within the constraints of the chosen technique (i.e. long thin samples are ideal for high κ in steady state measurements so that a large enough temperature difference can be generated), whereas in the case of thin films, various sample geometries may be chosen in order to build up a picture of the thermal conductivity of each layer, as suitable measurements will require corresponding simulations.

6.4 Examples of thermal characterisation techniques from the literature

In the following sections, several examples of research articles where thermal techniques were used to characterise or identify a given material will be highlighted. For each article, starter questions are provided, which are aimed at focussing the reader's attention on the key aspects of the article (with respect to this chapter).

6.4.1 In 'Neutron-diffraction study of the Jahn–Teller transition in stoichiometric $LaMnO_3$'

Rodríguez-Carvajal *et al* studied the changing crystal structure of $LaMnO_3$—a perovskite—as a function of temperature [50].

Reading through the article*, try to answer the following:

1. What experimental methods were used?
2. What is Jahn–Teller distortion?
3. What is the importance of the Mn^{3+}:Mn^{4+} ratio with regards to the Jahn-Teller distortion?*
4. What information does the DTA/TGA provide?
5. What are the authors' arguments for a non-cubic primitive cell?*
6. What was the wavelength of neutrons used in this study? How does this compare to an x-ray source (Cu, Kα)?*
7. What crystal structures do the Pbnm and R3c space groups refer to? How do these differ?*

**This article was first seen in chapter 3.*

6.4.2 In 'Evaluation of the reliability of the measurement of key magnetocaloric properties: a round robin study of La(Fe,Si,Mn)H_δ conducted by the SSEEC consortium of European laboratories'

Morrison *et al* evaluated the magnetic entropy change in fragments of La(Fe,Si,Mn)H_δ fragments from the same source using different calorimetric methods [51].

Reading through the article, try to answer the following:

1. What experimental methods were used?
2. How were the samples prepared?
3. What types of calorimetry methods were presented and how do they differ from one another?
4. Why was there an offset in the position of the specific heat peak in figure 5?
5. What factors could have led to broadening of the specific heat peak in figure 5?
6. How was the temperature change determined in direct ΔT measurements?

The aim of this work was to highlight the sensitivity of the calculation of entropy change to the measurement technique. It was demonstrated that the specific heat that was measured as a function of temperature could vary significantly due to the chosen measurement or sample preparation method(s). The authors concluded that to be able to compare measurements across different laboratories and methods several key parameters such as magnetic field ramp rate, sample preparation and form (i.e. geometry), and mounting method (whether glued, held freely, or pressed into a pellet) should be provided as standard.

6.4.3 In 'Magnetic phase transitions and the magnetothermal properties of gadolinium'

Dan'kov *et al* studied the magnetic phase transition of Gd around 294 K using various magnetic and thermal methods [52].

Reading through the article, try to answer the following:

1. What experimental methods were used?
2. Define 'magnetocaloric effect'.
3. How was the Curie temperature determined from the specific heat?
4. How was the spin reorientation transition identified in specific heat measurements?
5. What considerations were made for measurement of ΔT_{ad}?
6. How was the magnetic entropy change determined from field dependant specific heat data?

The aim of this work was to determine the influence of measurement technique and sample purity on the determination of the Curie temperature, T_c, of Gadolinium. The authors showed that impurities in the Gd samples can lower T_c and broaden the phase transition even in samples designated as pure as 99.9 wt. % purity.

6.4.4 In 'Low-temperature specific heat and critical magnetic field of α-uranium single crystals'

Lashley *et al* investigated the various contributions to the low temperature specific heat measurements of single crystal uranium [53].

Reading through the article, try to answer the following:

1. What experimental methods were used?
2. Calculate the mean value and standard deviation for the spread of values reported for the electronic specific heat, γ, and Debye temperature, θ_D in table 1.
3. What are the possible sources of these differences?
4. Are the phase transitions observed in the specific heat data first order?
5. What additional contributions to the low temperature specific heat did the authors identify?

The aim of this work was to investigate the charge density wave (CDW) transitions in single crystal uranium, made possible by the absence of grain boundaries. The authors found evidence of three CDW transitions at 23, 36, and 42 K. In addition, they fit low temperature specific heat measurements in order to extract the Debye temperature and electronic specific heat.

6.4.5 In 'Thermal conductivity, electrical resistivity, and Seebeck coefficient of silicon from 100 to 1300 K'

Fulkerson *et al* investigated the electric and thermal properties of bulk silicon [54].

Reading through the article, try to answer the following:

1. What experimental methods were used?
2. What was the authors' motivation for measuring the thermal conductivity of multiple silicon samples?
3. Which thermocouple type(s) were used for the thermal conductivity measurements and why were these probably chosen?
4. What is the purpose of the multiple grooves machined into the silicon discs? (See figure 1.)
5. How was the thermal conductivity calculated?
6. How did the authors' values for thermal conductivity compare to other published results?

The aim of this work was to resolve the issue of conflicting thermal conductivity measurements of silicon at high temperatures. The authors achieved this by taking a set of polycrystalline and single crystal samples, monitoring their relative impurities and machining them specifically for high temperature measurements. They found that their results agreed well with that of Glassbrenner and Slack, and that it agreed well with what was expected from the ambipolar diffusion theory.

6.4.6 In 'Reduction of thermal conductivity by surface scattering of phonons in periodic silicon nanostructures'

Anufriev *et al* investigated the change in thermal conductivity of silicon nanostructures as their size and shape were engineered [55].

Reading through the article, try to answer the following:

1. What experimental methods were used?
2. What is the difference between coherent and incoherent scattering mechanisms?
3. How did the authors expect the thermal conductivity to change as the surface to volume ratio was decreased?
4. How were the samples made?
5. Why did the authors choose to use the μ-TDTR method?
6. How was the thermal conductivity extracted from the μ-TDTR data?

The aim of this work was to investigate the impact of systematic changes to silicon nanostructures where the thermal conductivity is expected to decrease due to the presence of additional boundaries. The authors found that not only was this the case, but additional unexpected reductions were observed due to a 'necking effect' and coherent phonon scattering.

References

[1] Boer J D 1965 *Metrologia* **1** 158–69
[2] Kittel C 2004 *Introduction to Solid State Physics* 8th edn (Hoboken, NJ: Wiley)
[3] Engineering toolbox website, https://engineeringtoolbox.com/
[4] Pecharsky V K, Gschneidner K A Jr., Pecharsky A O and Tishin A M 2001 *Phys. Rev.* B **64** 144406
[5] Pecharsky V K and Gschneidner K A Jr. 2001 *Adv. Mater.* **13** 683–6

[6] Fujita A, Fujieda S, Hasegawa Y and Fukamichi K 2003 *Phys. Rev.* B **67** 104416
[7] Westrum E F Jr. and Gronvold F 1969 *J. Chem. Thermodyn.* **1** 543–57
[8] Salamon M B and Jaime M 2001 *Rev. Mod. Phys.* **73** 583–628
[9] Kapitza P L 1941 *Phys. Rev.* **60** 354
Kapitza P L 1941 *J. Phys. (Moscow)* **4** 181
[10] Orlova M P *et al* 1966 *Metrologia* **2** 6
[11] Eucken A 1909 *Z. Phys.* **10** 586
[12] Nernst W 1911 *Ann. Phys.* **36** 395
[13] Sullivan P F and Seidel G 1968 *Phys. Rev.* **173** 679
[14] Pecharsky V K, Moorman J O and Gschneidner K A 1997 *Rev. Sci. Instrum.* **68** 4196–207
[15] Shutz R J 1973 *Rev. Sci. Instrum.* **45** 548–51
[16] Shepherd J P 1984 *Rev. Sci. Instrum.* **56** 273–7
[17] Hardy V, Breard Y and Martin C 2009 *J. Phys. Condens. Matter* **21** 075406
[18] Suzuki H, Inaba A and Meingast C 2010 *Cryogenics* **50** 693–9
[19] Klaasse J C P and Brück E H 2008 *Rev. Sci. Instrum.* **79** 123906
[20] Basso V, Küpferling M, Sasso C P and Giudici L 2008 *Rev. Sci. Instrum.* **79** 063907
[21] Jeppesen S, Linderoth S, Pryds N, Kuhn L T and Jensen J B 2008 *Rev. Sci. Instrum.* **79** 083901
[22] Marcos J, Casanova F, Batlle X, Labarta A and Planes A 2003 *Rev. Sci. Instrum.* **74** 4768
[23] Basso V, Sasso C P and Küpferling M 2010 *Rev. Sci. Instrum.* **81** 113904
[24] Minakov A A, Roy S B, Bugoslavsky Y V and Cohen L F 2005 *Rev. Sci. Instrum.* **76** 043906
[25] Miyoshi Y, Morrison K, Moore J D, Caplin A D and Cohen L F 2008 *Rev. Sci. Instrum.* **79** 074901
[26] Wunderlich B 2000 *Thermochim. Acta* **355** 43–57
[27] Tocado L, Palacios E and Burriel R 2006 *J. Therm. Anal. Calorim.* **84** 213–17
[28] Touloukian Y S and Ho C Y 1988 *CINDAS Data Series on Material Properties, Specific Heat of Solids* (New York: Hemisphere)
[29] Schwall R E, Howard R E and Stewart G R 1975 *Rev. Sci. Instrum.* **46** 1054–9
[30] Hwang J S, Lin K J and Tien C 1997 *Rev. Sci. Instrum.* **68** 94
[31] Lashley J C *et al* 2003 *Cryogenics* **43** 369–78
[32] Kennedy C A, Stancescu M, Marriot R A and White M A 2007 *Cryogenics* **47** 107–12
[33] Plackowski T, Yuxing W and Alain J 2002 *Rev. Sci. Instrum.* **73** 2755–65
[34] Coats A W and Redfern J P 1963 *Analyst* **88** 906
[35] Saruyama Y 1992 *J. Therm. Anal.* **38** 1827–33
[36] Morrison K, Lyubina J, Moore J D, Sandeman K G, Gutfleisch O, Cohen L F and Caplin A D 2011 *Philos. Mag.* **92** 292–303
[37] Morrison K, Miyoshi Y, Moore J D, Barcza A, Sandeman K G, Caplin A D and Cohen L F 2008 *Phys. Rev.* B **78** 134418
[38] Tritt T M and Weston D 2004 *Measurement Techniques and Considerations for Determining Thermal Conductivity of Bulk Materials Thermal Conductivity* ed T M Tritt (New York: Springer) pp 187–203
[39] Zawilski B M, Iv R T L and Tritt T M 2001 *Rev. Sci. Instrum.* **72** 1770–4
[40] Maldonado O 1992 *Cryogenics* **32** 908–12
[41] Franco A 2007 *Appl. Therm. Eng.* **27** 2495–504
[42] Parker W J, Jenkins R J, Butler C P and Abbott G L 1961 *J. Appl. Phys.* **32** 1679–84
[43] Gustafsson S E 1991 *Rev. Sci. Instrum.* **62** 797

[44] Cahill D G 1990 *Rev. Sci. Instrum.* **61** 802–8
[45] Cahill D G, Katiyar M and Abelson J R 1994 *Phys. Rev.* B **50** 6077
[46] Borca-Tasciuc T, Kumar A R and Chen G 2001 *Rev. Sci. Instrum.* **72** 2139–47
[47] Tong T and Majumdar A 2006 *Rev. Sci. Instrum.* **77** 104902
[48] Wang Y, Park J Y, Koh Y K and Cahill D G 2010 *J. Appl. Phys.* **108** 043507
[49] Cahill D G and Sci R 2004 *Instru.* **75** 5119–22
[50] Rodríguez-Carvajal J *et al* 1998 *Phys. Rev. B* **57** R3189
[51] Morrison K *et al* 2012 *International Journal of Refrigeration* **35** 1528–36
[52] Dan'kov S Y *et al* 1998 *Phys. Rev. B* **57** 3478–90
[53] Lashley J C *et al* 2001 *Phys. Rev. B* **63** 224510
[54] Fulkerson W *et al* 1968 *Phys. Rev.* **167** 765–82
[55] Anufriev R *et al* 2016 *Phys. Rev. B* **93** 045411

Chapter 7

Electric characterisation

In this chapter, a broad overview of electrical measurement techniques will be given. This will start with a short background on the theory of electric transport including a classification of different magneto- and thermo-electric effects. It will be followed by an overview of common sources of electrical noise, the electronics used in a typical measurement, some common transport measurement techniques for thin film and bulk samples, and a brief discussion of techniques to characterise the bandstructure. Finally, some examples of these techniques will be highlighted by selected papers, followed by questions specific to data analysis.

7.1 Introduction

The resistance, carrier concentration, band gap, and other electric characteristics of a material have, over the last century been used in a vast array of different applications, such as electronic devices (p–n junctions, transistors, diodes), sensors (photodiodes, SQuID detectors, GMR read heads), energy storage (batteries and supercapacitors), and energy harvesting (solar cells, piezoelectrics). The electric characterisation techniques such as described here are a useful step in analysing potential materials and devices, but require first a suitable grounding in electronic band theory. This will be followed by examples of different electronic devices and materials, electric phenomena (such as the thermoelectric effect) before covering the basics of a measurement.

7.1.1 Background

There are several models that can be used to describe the electronic properties of a material. Classically, the Drude model for conduction of electrons in a metal treats the electrons as a 'gas' of non-interacting particles with density $n_{3\text{d}}$ and charge q. In an electric field (i.e. if there is a potential difference across the conductor) charged particles will accelerate, with collisions that randomise the direction of motion. This results in an average drift velocity:

doi:10.1088/2053-2563/ab2df5ch7

$$v_{\mathrm{d}} = \frac{qE}{m_{\mathrm{e}}}\tau \tag{7.1}$$

where q is the charge on the electron, E is the electric field, m_{e} is the mass of the electron, and τ is the scattering rate (average number of scattering events per second). The current density, J, which describes the number of charge carriers passing through a given area per second can then be defined as:

$$J = nqv_{\mathrm{d}} = \frac{nq^2\tau}{m_{\mathrm{e}}}E \tag{7.2}$$

where n is the density of charge carriers. Similarly the conductivity, σ, which quantifies the ability of a material to conduct charge, can be defined as:

$$\sigma = \frac{nq^2\tau}{m_{\mathrm{e}}} \tag{7.3}$$

Notice that the conductivity is determined by both the scattering rate and electron density (as q and m are fundamental constants). The mobility, on the other hand, describes the ease with which charge carriers move within a material and is determined primarily from the scattering rate:

$$\mu = \frac{v_{\mathrm{d}}}{E} = \frac{q\tau}{m_{\mathrm{e}}} \tag{7.4}$$

with units of [$\mathrm{m^2\ V\ s^{-1}}$]. Finally, it can be seen that the mobility and conductivity are related by the following equation:

$$\sigma = nq\mu \tag{7.5}$$

As the scattering rate, τ, controls both the conductivity and mobility of a material it is useful to explore the impact of this on a measurement or device. For example, the mean free path, λ, is defined as the typical distance travelled before an elastic scattering event. If the electrons have an average speed of v_{d}, the collision time (also referred to as the momentum relaxation time or scattering rate) can be defined as:

$$\tau = \lambda / v_{\mathrm{d}} \tag{7.6}$$

There are then two regimes which can be described:

- Ballistic transport: where the lengthscales are short compared to λ;
- Diffusive transport: where the lengthscales are long compared to λ. In this instance the movement of charged particles looks diffusive, with a random walk stepsize of λ.

The conductivity of a material can therefore be improved by reducing the density of scattering centres within a material (such as crystal defects), or, as is the case with thin films, the thickness of the films being studied could be reduced to less than λ. As an example, silicon has a mobility of 1000 $\mathrm{cm^2\ Vs^{-1}}$, $\tau \sim 2 \times 10^{-13}$ s, $v_{\mathrm{d}} \sim 10^5\ \mathrm{m\ s^{-1}}$, and $\lambda \sim 20$ nm at 300 K.

The Drude model is, however, limited due to the various assumptions that are employed:

1. The independent electron approximation: that electrons do not interact;
2. The relaxation time approximation: that the scattering rate can be described by τ;
3. That thermal equilibrium is reached through particle collisions;
4. That collisions are instantaneous;
5. That these collisions indicate scattering of an electron with a nucleus (core).

This results in a breakdown of the model when trying to describe how the conductivity varies with temperature, in particular the heat capacity or thermal conductivity of a solid. In the former case, the electronic heat capacity determined from the Drude model is an overestimation and does not account for the variation with temperature described in chapter 6. This is the same for the description of the thermal conductivity and largely arises from ignoring the influence of phonons as well as considering the electrons to be independent particles.

An initial solution is to incorporate quantum statistics to describe the occupation of energy levels within a given phase space (with real-space volume, V_r and k-space volume, V_k): the Sommerfeld model. As discussed in chapter 1, the Pauli exclusion principle prevents the occupation of the same phase space by two identical electrons. On a basic level this means that two electrons with the same spin state cannot occupy the same energy level, but there are other characteristics of the electron (such as angular momentum) which are quantified by the quantum numbers $\{l, m, s\}$ that enable further population at each energy level. The number of electrons occupying a given phase space, N, can also be described in terms of the k-space and real-space volumes, V_k and V_r respectively:

$$N = \left(\frac{1}{2\pi}\right)^3 V_k V_r \tag{7.7}$$

If the electrons are free to move within a mean potential, V_0, they will occupy the lowest energy first, filling subsequent energy levels from the bottom up. Eventually this will fill a k-space sphere ($V_k = 4/3\pi k_F{}^3$), so that N can be described by:

$$N = 2\left(\frac{1}{2\pi}\right)^3 \frac{4}{3}\pi k_F^3 V_r \tag{7.8}$$

where the factor 2 accounts for degeneracy of the electron spin, and the Fermi wavevector, k_F, and Fermi energy, E_F, are defined in three dimensions by:

$$k_F = (3\pi^2 n)^{\frac{1}{3}} \tag{7.9}$$

$$E_F = \frac{\hbar^2 k_F{}^2}{2m_e} = \frac{\hbar^2}{2m_e}(3\pi^2 n)^{2/3} \tag{7.10}$$

where n is the number density of electrons (N/V_r), and $\hbar$ is the reduced Planck's constant ($h/2\pi$). The Fermi energy is defined as the energy difference between the lowest energy level and the highest occupied energy level at 0 K. As the temperature is increased the occupation of the higher energy levels becomes less distinct, as described by the Fermi–Dirac distribution:

$$f_D(E, T) = \frac{1}{e^{(E-\mu)/k_B T} + 1} \tag{7.11}$$

where k_B is Boltzmann's constant, and μ is the chemical potential, the energy at which the probability of occupation is 0.5. This reduces to E_F at $T = 0$, as shown in figure 7.1. An extension of this is the density of states, $g(E)$, which describes the number of different states for a given energy interval.

The Sommerfeld model may describe the temperature dependence of C_{el} and the thermal and electric conductivities of metals but still does not account for the interaction of particles, or the periodic potential of the crystal lattice. This means that it does not manage to describe non metallic materials, or the Fermi surface as measured by ARPES and discussed in chapter 5.

These problems are solved with the Bloch model, which requires inclusion of quantum mechanics and a revisit of Fourier series. That the Bloch model employs quantum mechanics to start to quantify the effect of a periodic potential on electrons within a crystalline lattice makes conceptual sense if we remember the description of a crystalline lattice from chapter 3. There, the structure of a lattice was described by a small repeating unit—the Basis—coupled with specific symmetry relations (Wyckoff sites and space group). By extension, the electronic potential that a charge carrier might see can similarly be described by a periodic potential:

$$V(\mathbf{r} + \mathbf{T}) = V(\mathbf{r}) \tag{7.12}$$

where $\mathbf{T}$ is the translation vector, which is defined by the primitive translation vectors $\mathbf{a}_1$, $\mathbf{a}_2$, and $\mathbf{a}_3$:

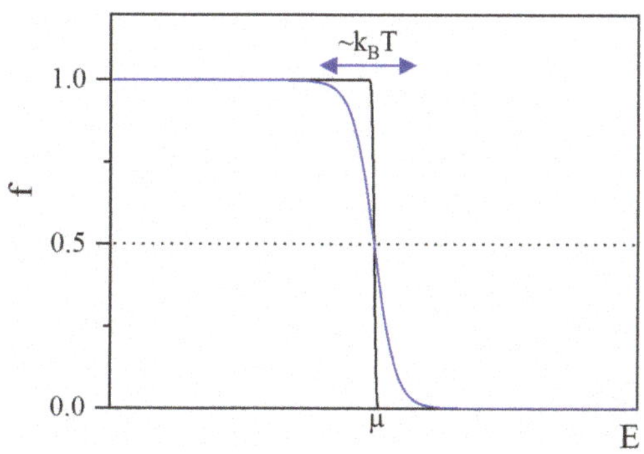

Figure 7.1. Fermi–Dirac distribution at $T = 0$ (black) and $T = 300$ K (blue). At higher temperatures there is enough available energy in the lattice for some electrons to be promoted to higher energy levels and thus blur the electron distribution.

$$\mathbf{T} = n_1\mathbf{a}_1 + n_2\mathbf{a}_2 + n_3\mathbf{a}_3 \quad (7.13)$$

and n_1, n_2, n_3 are integers. Just as the periodic crystal lattice could be described as Fourier series (see structure factor calculation in chapter 3), so too can the periodic potential:

$$V(\mathbf{r}) = \sum_{\mathbf{G}} V_{\mathbf{G}} e^{i\mathbf{G}\cdot\mathbf{r}} \quad (7.14)$$

The result of this is a periodic potential in k-space, with a volume:

$$V_k = \mathbf{A}_1 \cdot \mathbf{A}_2 \times \mathbf{A}_3 \quad (7.15)$$

Thus, all the information required to describe the periodic potential will be contained within the first unit cell of the reciprocal lattice, which is described as the first Brillouin zone (the Wigner-Seitz primitive cell of the reciprocal lattice). This periodic potential, $V(\mathbf{r})$, can then be used in the Schrödinger equation, as given in equation (7.16), which is used to mathematically describe quantum states. In this example the wavefunction, ψ is referred to as a probability distribution for the position of the electron, and energy eigenvalues, E are referred to as energy levels for a particle of mass, m, where $\nabla^2 = \frac{\partial^2}{\partial x} + \frac{\partial^2}{\partial y} + \frac{\partial^2}{\partial z}$.

$$H\psi = \left(-\frac{\hbar^2\nabla^2}{2m} + V(\boldsymbol{r})\right)\psi = E\psi \quad (7.16)$$

Analysis of equation (7.14) substituted into (7.16) finally leads to Bloch's theorem, which states that the wavefunction:

$$\psi_q(\mathbf{r}) = e^{i\mathbf{q}\cdot\mathbf{r}} \sum_{\mathbf{G}} C_{\mathbf{q}-\mathbf{G}} e^{-i\mathbf{G}\cdot\mathbf{r}} = e^{i\mathbf{q}\cdot\mathbf{r}} u_{j,\mathbf{q}} \quad (7.17)$$

where $C_{\mathbf{q}-\mathbf{G}}$ is a Fourier coefficient, and $u_{j,\mathbf{q}}$ is a function with periodicity of the lattice (q) (i.e. a plane wave multiplied by a function with periodicity of the lattice). In other words, Bloch's theorem states that by multiplying a plane wave ($e^{i\mathbf{q}\cdot\mathbf{r}}$) by a periodic wave ($u_{j,\mathbf{q}}$) we obtain a Bloch wave: a wavefunction for a particle in a periodic potential ($\psi(\mathbf{r})$).

The detailed solutions of the Schrödinger equation in the weak binding ($V(\mathbf{r}) \sim 0$, where electrons are free to move) and tight-binding ($V(\mathbf{r}) \gg \frac{\hbar^2\nabla^2}{2m}$, where electrons hardly move, occasionally hopping from atom to atom) limits are treated in *Band Theory and Electronic Properties of Solids* [1] and will not be repeated here.

To summarise however, these solutions lead to the derivation of wavevectors with crystal momentum, $\hbar\mathbf{k}$ (a quantum number that describes a Bloch state). In both cases the solutions are similar:

1) Dispersion bands are formed that electrons with a fixed energy will traverse due to the variation of the periodic potential within the Brillouin zone.

2) Away from the Brillouin zone boundary these will be similar to the free electron dispersion relationship, with the addition of **k**-space periodicity.
3) Where energy dispersions cross at Brillouin zone boundaries a bandgap will open up such that $dE/d\mathbf{k} = 0$ at the boundary and zone centre.

This has been summarised in figure 7.2. The key concept is that by including a periodic potential, the idea of an electronic bandstructure arises, which can be used to predict the electronic transport properties of a given material as a function of extensive variables such as temperature and magnetic field.

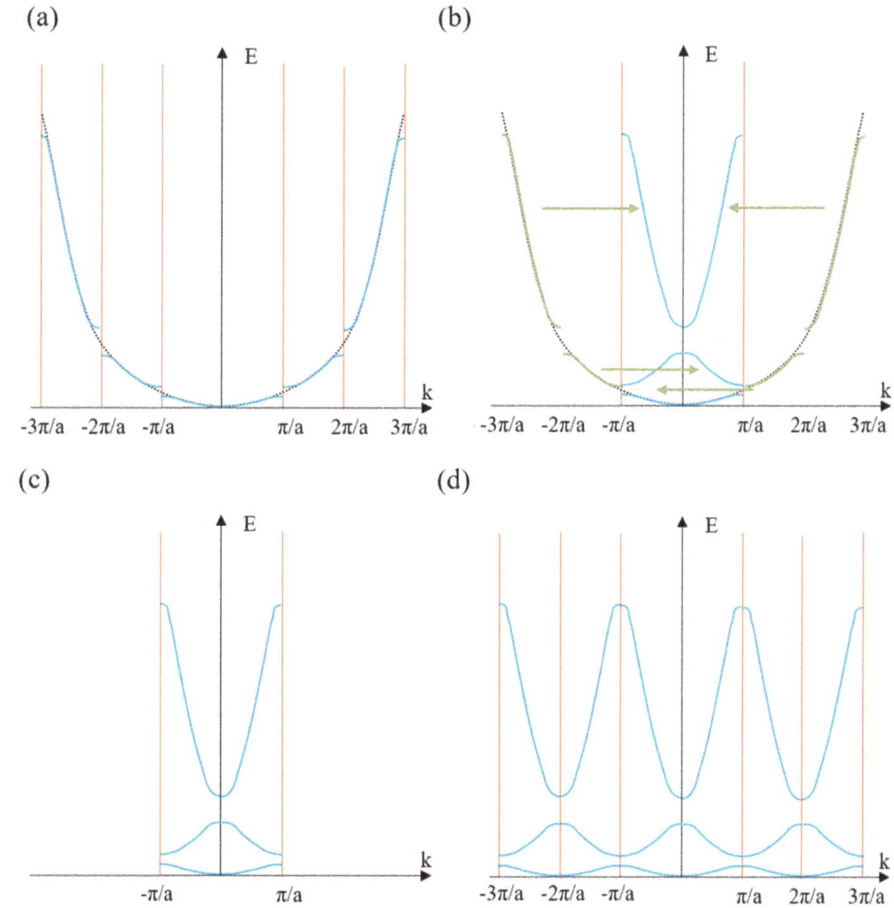

Figure 7.2. Representation of bandstructure in the extended, reduced, and periodic zone schemes. (a) The Extended zone, where the nearly free electron dispersion is shown by the dotted black lines, the Brillouin zone edges by the red lines, and the distorted nearly free electron dispersions (due to Brillouin zone edge) are shown in blue. (b) Constructing the reduced zone. The dispersion curves from the second and third Brillouin zones (green lines) are translated to the first Brillouin zone as shown (the symmetry of the periodic potential results in the bands 'folding' back on each other). (c) The resulting reduced zone representation. (d) The periodic zone, where the bands of the first Brillouin zone are repeated for each lattice translation.

7.1.2 Consequences of bandstructure

As shown in figure 7.2, by constructing the bands simply by invoking the periodicity of the lattice potential, maxima and minima will occur at zone edges. These are then occupied by electrons according to the Fermi–Dirac distribution and it is this behaviour that is described as the bandstructure. The conduction and valence bands describe the bands of energy levels closest to the Fermi level, where the valence is the lower, more populated energy level, and the conduction band is the higher, (usually) unoccupied energy level. The bandgap describes the energy difference between the top of the valence band and the bottom of the conduction band. The position of the Fermi level with regards to the valence and conduction bands, and the magnitude of the bandgap is the key determinate of metallic, insulating or semiconducting behaviour, as outlined in figure 7.3. Where an electron is excited above the valence band, this leads to a net positive charge (in the valence band) which can also act as charge carriers and is referred to as a 'hole'. This hole will act as a positive charge carrier, moving in opposite direction to the negatively charged electrons.

Finally, the effective mass, m^*, is a construct used to help describe the response of an electron to an external force. It is defined by the gradient of the dispersion curve, dE/dk:

$$m^* = \frac{\hbar^2}{\mathrm{d}^2E/\mathrm{d}\boldsymbol{k}^2} \tag{7.18}$$

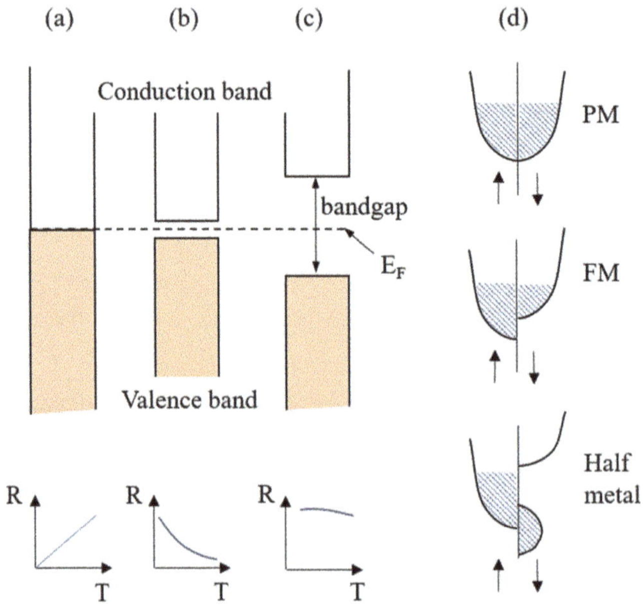

Figure 7.3. Schematic representing the density of states (in a flat band diagram), and corresponding bandgap (s), E_g, for different types of material. (a)–(c) The changing bandgap in (a) metals, (b) semiconductors and (c) insulators. (d) Spin–split density of states for magnetic materials.

And can be used to determine the density of states, $g(E)$, at a particular energy, E:

$$g(E) = \frac{1}{2\pi^2}\left(\frac{2m^*}{\hbar^2}\right)^{3/2}(E - E_0)^{1/2}$$
$$= \frac{1}{2\pi^2}\left(\frac{2}{\mathrm{d}^2E/\mathrm{d}\boldsymbol{k}^2}\right)^{3/2}(E - E_0)^{1/2} \tag{7.19}$$

where E_0 is the energy of the lowest energy state. Thus the effective mass, and density of states are related to the band curvature. For example, electrons in a periodic potential where an electric or magnetic field E/B is applied will accelerate relative to the lattice as if they had mass m^*. This results in bands that can be described as 'heavy' where the effective mass is relatively high compared to light bands, as demonstrated in figure 7.4.

7.1.3 Types of materials

Now that we have established some of the key features of electronic bandstructure we can use this to define the properties of different classes of materials:

1. **Metals** are primarily classified as conducting. This is because the Fermi level exists in the conduction band (technically this is represented as an overlap of the valence and conduction bands). As a result, electrons freely move within the material. As the temperature increases the resistance will increase because atoms have more energy and will vibrate, making it more likely that a scattering event (a source of resistance) will occur. For metals $10^{-14} < \rho < 10^{-7}\ \Omega$ m.
2. **Semiconductors** have a Fermi level between the conduction and valence bands, and an energy gap, E_g, of the order of a few eV [2]. As the temperature increases, more energy is available to electrons to move into the conduction band and the resistance will decrease. For semiconductors $10^{-6} < \rho < 10^{5}\ \Omega$ m [3].

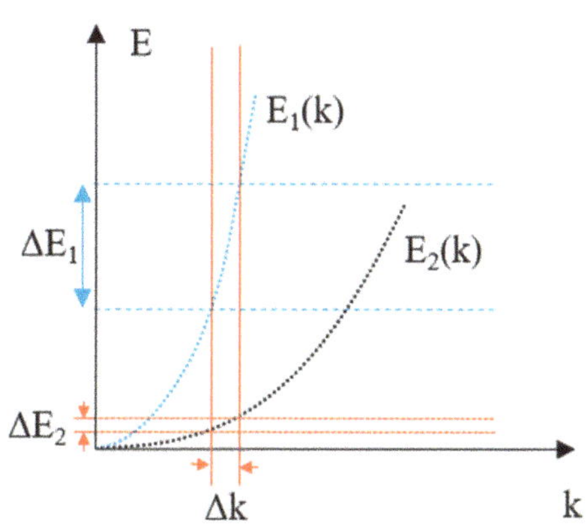

Figure 7.4. Example of heavy (E_1) and light (E_2) bands and the corresponding energy bands ΔE_1, ΔE_2 for the same wavevector interval Δk (indicating the density of states).

There are two types of semiconductor: intrinsic (naturally occurring, such as GaAs) and extrinsic (small level of doping to introduce carriers; ~1 part in 10^6 atoms). Extrinsic semiconductors are doped to change the mobility, carrier type or carrier concentration.

Extrinsic n-type semiconductors will have been doped to introduce extra electrons. This results in an excess of negative charge; Sn, As and P are good n-type dopants. Extrinsic p-type semiconductors on the other hand will have been doped to introduce extra holes. This results in an excess of positive charge; In and Ga are good p-type dopants.

Semiconductors are also described as direct (e.g. GaAs) or indirect (e.g. Si) gap. This is defined by the position of the maxima and minima in the conduction and valence bands. For direct gap semiconductors, the maxima is located at the same position of the minima with respect to k and an electron can cross the energy gap with no change in momentum (e.g. photon transition). For indirect gap semiconductors a change in momentum (k) would be required, which necessitates a two-part phonon and photon absorption for an electron to cross the gap. This is a low probability transition and as such cannot be used for light emitting structures.

3. **Semimetals** have metallic and semiconducting characteristics. In this case the valence and conduction bands overlap at the Fermi level so that the conduction band is partially filled and the valence band is underfilled. For semimetals $10^{-7} < \rho < 10^{-6}\ \Omega$ m.
4. **Half metals** are magnetic materials where the position of the Fermi level is different for different electron spins (i.e. the density of states for each spin state is not identical), as shown in figure 7.3. This can lead to the situation where the Fermi level of one spin state is between the conduction and valence bands (of that spin state) and the Fermi level for the other spin state is in the conduction band. The result of this would be spin polarised conduction electrons.
5. **Insulators** have a Fermi level between the conduction and valence bands, and an energy gap, E_g, much greater than the few eV that defines the semiconductor. (This energy gap is so large that the available thermal energy cannot overcome and promote electrons to the conduction band.) For insulators $\rho > 10^5\ \Omega$ m.
6. **Superconductors** by definition have zero resistance below their critical temperature, T_c. They are typically classified as either Type I or Type II superconductors depending on their response to magnetic field. In Type II superconductors, above a lower critical field, magnetic vortices will be introduced (see chapter 8), whereas in Type I superconductors there is only one critical magnetic field, above which the material is a normal metal. Type I superconductors can be described by BCS theory [4], where below T_c Cooper pairs are formed which can behave as bosons, forming a Bose–Einstein condensate. These Cooper pairs are a combination of spin up and spin down electrons separated by an energy gap.

7. **Topological insulators** are materials that exhibit different conductivity on the surface compared to the bulk. For example, the bulk material can be insulating whilst a surface layer is conductive. This can be summarised by the bandstructure shown in figure 7.5, where a Dirac cone is observed for surface states (linear dispersion from the valence to conduction bands) and an energy gap exists for the bulk. Examples include $Bi_{1-x}Sb_x$, Bi_2Se_3 and Bi_2Te_3 [5].
8. **Dirac and Weyl semimetals** are another example of a topologically non-trivial state where valence and conduction bands cross at a single point. In the case of Weyl semi-metals, splitting of the Dirac point occurs due to symmetry breaking (for example breaking time reversal). This results in linear dispersions with respect to k_x, k_y, and k_z [6] that can be observed in bandstructure measurements.
9. **Dielectrics** are non-conducting materials that are often used to separate the charged plates that make up a capacitor. This is due to their ability to be polarised by an electric field whilst remaining non conductive.

For comparison, the resistivity and conductivity of some common materials is given in table 7.1.

7.1.4 Bandstructure engineering

Another consequence of bandstructure is that it can be engineered by placing two different materials in contact with one another. This is the mechanism that underpins the majority of electronic devices, some examples of which are given here.

7.1.4.1 Metal–semiconductor junctions

The simplest electronic device is the Schottky diode, which can be created by placing a metal in contact with a semiconductor. The different Fermi levels, E_F, of each material results in a barrier for charge carriers at the interface: the Schottky barrier.

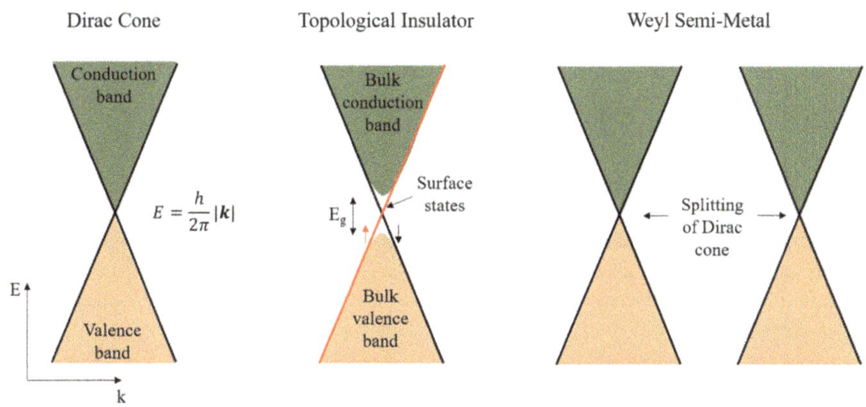

Figure 7.5. Example of characteristic features in bandstructure observed for topological materials such as topological insulators and Weyl semi metals.

Table 7.1. Resistivity of some common materials at room temperature (293 K) and their corresponding classification.

Type	Material	ρ (Ω m)	σ (S m^{-1})
Conductor	Copper	1.72×10^{-8}	5.81×10^{7}
Conductor (alloy)	Steel	20×10^{-8}	5×10^{6}
Conductor (alloy)	Manganin (Cu, Mn, Ni alloy)	44×10^{-8}	2.27×10^{6}
Conductor	Graphene	$\sim 10^{-8}$	$\sim 10^{8}$
Semimetal	Antimony	41.7×10^{-7}	2.4×10^{6}
Half metal	PtMnSb	$\sim 40 \times 10^{-8}$	2.5×10^{6}
Half metal	CrO_2	$\sim 150 \times 10^{-8}$	6.67×10^{5}
Semiconductor	Silicon	2300	4.35×10^{-4}
Semiconductor	GaAs	$\sim 10^{6}$	$\sim 10^{-6}$
Insulator	Glass	10^{10}–10^{14}	10^{-14}–10^{-10}
Insulator	Mica	10^{11}–10^{15}	10^{-15}–10^{-11}
Insulator	Wood	10^{8}–10^{11}	10^{-11}–10^{-8}

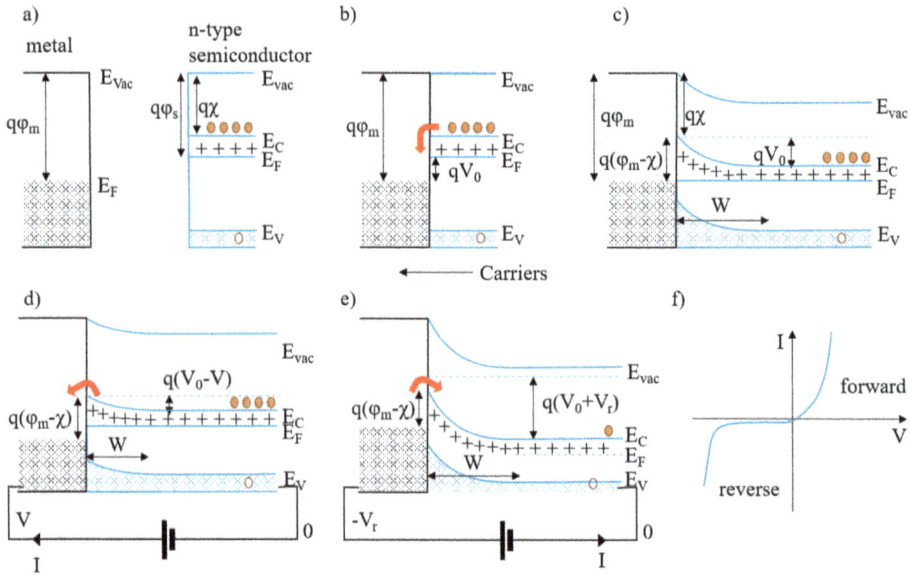

Figure 7.6. Change in bandstructure as a metal and n-type semiconductor are brought in contact with one another (Schottky barrier). (a) Initial flat band diagrams for the metal and n-type semiconductor. (b) As the two materials are brought together, the excess carriers in the semiconductor flow into the metal. (c) This movement of carriers results in an accumulation of charge (depletion region, *W*), which effectively lowers the Fermi energy. (d) and (e) The metal–semiconductor under forward and reverse bias, respectively. (f) The resulting *IV* curve.

For the case of an n-type semiconductor brought into contact with a metal, as shown in figure 7.6, this barrier results in a flow of charges from the semiconductor into the metal. This leads to a depletion of charge carriers in the semiconductor—the depletion region—which brings the Fermi levels into equilibrium. When a forward bias is applied, as shown in figure 7.6(d) the Fermi energy is once again raised and carriers can flow into the metal again. If a reverse bias is applied (figure 7.6(e)) the Fermi energy is lowered so that carriers can flow from the metal to the semiconductor, but they would need to overcome the Schottky barrier. As a result, the current that can flow as a function of applied voltage becomes non-linear as seen in figure 7.6(f). This *IV* response means that Schottky diodes can be used as rectifiers (i.e. only passes positive voltage from an alternating source). They also have applications as microwave detectors (due to high frequency response), or as infrared sensors or photodiodes if the Schottky barrier is engineered to be in a specific wavelength.

An alternative approach to creating a diode is to bring n-type and p-type semiconductors into contact with one another: the pn junction. This would, for example result in a similar *IV* curve, as shown in figure 7.6, (i.e. a diode) even though the mechanism is slightly different. In this case, the result of doping the n- and p-type semiconductors results in both excess carriers and dopants with charge (+/−). The number of n-type charge carriers will be denoted as n_n and p_n in the n- and p-type semiconductors, respectively. Similarly, the number of p-type carriers (holes) will be denoted as n_p and p_p. As the n- and p-type semiconductors are brought together there will be a flow of n_n carriers from the n-type to the p-type, and p_p carriers from the p-type to the n-type. This will eventually result in the generation of an electric field by the dopants that induces a drift current, thus reducing carrier flow from one side to another (figure 7.7) . Similar to the Schottky diode, the Fermi level will also be brought into equilibrium at the point at which no more charge flows.

The pn junction forms the basis of light emitting diodes, rectifiers and photodiodes. Extensions of this design include:

- Quantum wells, where a thin semiconductor (~10 nm) is sandwiched between two semiconductors with differing energy gaps. These can be used, for example, to build sensitive Hall probe sensors.

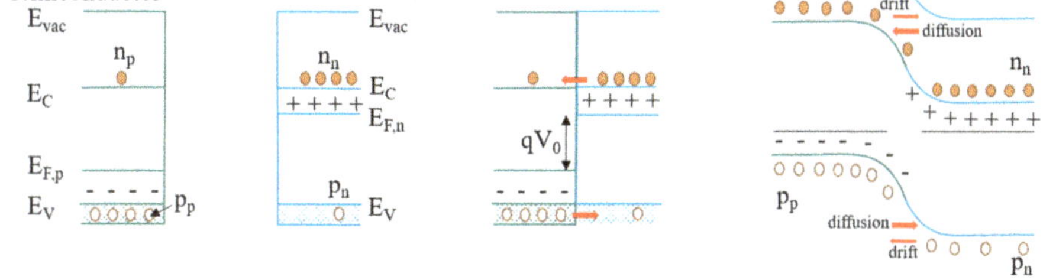

Figure 7.7. The pn junction. (a) Initial flat band diagrams of the p- and n-type semiconductors. (b) As the two materials are brought in contact with one another electrons will flow from the n-type semiconductor to the p-type and holes from the p-type to the n-type. (c) Unscreened dopants in the extrinsic semiconductors will create an electric field that induces a drift current (thus reducing carrier flow).

- Bipolar junction transistors, which are comprised of two pn junctions back to back and can act as a current amplifier.
- Metal-oxide-semiconductor-field-effect transistors (MOSFETs), which are common in digital and analogue circuits for switching or amplifying signals.
- Solar cells, where the bandstructure has been engineered to steer electron–hole pairs formed by a photoexcited electron in opposite directions. These are often constructed as a heterostructure, but the main component is a p-i-n junction, where i is an intrinsic semiconductor.

7.1.4.2 Metal–superconductor junctions

Metal–superconductor junctions can also be used to construct useful electronic devices. For example, the Josephson junction—defined as a metal sandwiched between two superconductors—exhibits tunnelling of the Cooper pairs present in the superconductor through the normal metal. Each superconductor in this junction will have a ground state described by the wavefunctions:

$$\psi_c = |\psi_c|e^{i\theta_R} \tag{7.20}$$

$$\psi_c = |\psi_c|e^{i\theta_L} \tag{7.21}$$

where $|\psi_c|$ is the magnitude and θ_L and θ_R are the phase of the left and right junctions, respectively. Tunnelling of the Cooper pairs through the normal metal junction will lead to a shift in phase $\Delta\theta = (\theta_R - \theta_L)$, which results in current flowing through the insulating layer without the need to apply a voltage.

$$I = I_c \sin \Delta\theta \tag{7.22}$$

The maximum of I_c for large $\Delta\theta$ occurs where you reach the critical current of the junction. In a DC superconducting quantum interference device (SQuID), this principle is used to measure the magnetic field with high precision. Two Josephson junctions are arranged in a ring enclosing an area through which the magnetic field is being measured (figure 7.8). The interference between the current passing

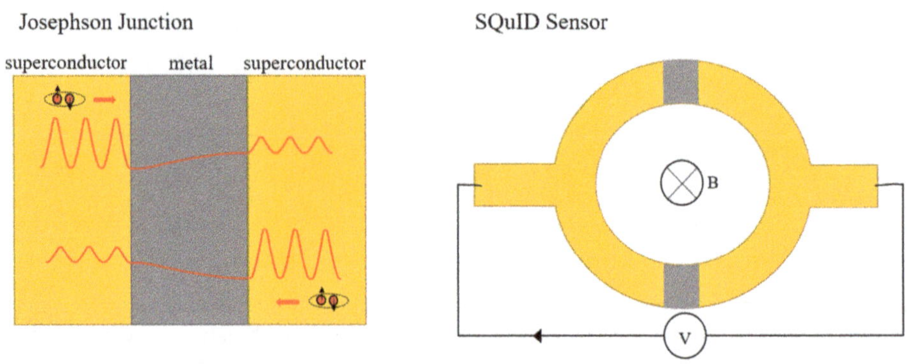

Figure 7.8. Schematic of the Josephson junction and its application in a DC SQuID sensor. The red waveform in the Josephson indicates the wavefunction of the Cooper pair as it tunnels through the metal layer.

through each path due to the presence of magnetic field will result in voltage oscillations that can be measured. This will have sensitivity of the order of one flux quanta, ϕ_0 [4]:

$$\phi_0 = \frac{hc}{2e} \tag{7.23}$$

Other applications of this type of junction include single photon detectors (by applying a dc bias), quantum computing components, and as a measurement standard. For example, the AC Josephson effect can be used to define the Volt (and consequently the Ampere).

The proximity effect, with regards to superconductors is the emergence of superconducting properties in a non-superconducting material placed in contact with a superconductor. This will extend over a characteristic lengthscale (the penetration depth) before decaying to the normal state. For example, Andreev reflection is defined as particle scattering at a normal metal:superconductor interface where normal current is converted into a supercurrent. As Cooper pairs are comprised of spin up and spin down electrons, this can be used as a sensitive measure of the spin polarisation of a metal in contact with the superconductor, which is the principle of point-contact spectroscopy measurements (see section 7.5).

7.2 Thermo- and magneto-electric effects

It has already been discussed that charge carriers will be sensitive to temperature (e.g. increased resistance for metals, increased population for semiconductors), magnetic field (e.g. Lorentz force: Hall effect) and current (e.g. Joule heating). This results in a rich variety of different magneto-electric and thermo-electric effects, as will be discussed in this section.

7.2.1 Hall effects

The Hall effect describes charge accumulation in a conductor (metal or semiconductor) where a charge current, J_c, is flowing, in response to the application of a magnetic field. The accumulation of charge results in an electric field, $\boldsymbol{E}$, which is observable as a Hall voltage, V_H. This effect arises due to the Lorentz force, $\boldsymbol{F}$, on a moving charge in a magnetic field, $\boldsymbol{B}$:

$$\boldsymbol{F} = q(\boldsymbol{E} + (v \times \boldsymbol{B})) \tag{7.24}$$

where, as will be seen in chapter 8, the internal field, B, is a function of the applied magnetic field, H, and the magnetisation of the sample, M ($B = \mu_0(M + H)$ in SI units). Extending equations (7.2) and (7.3) to three dimensions, and reminding ourselves that the crystal lattice may not necessarily be isotropic (identical along each primitive lattice vector), the current density, $\mathbf{J}$, and electric field, $\mathbf{E}$, can be described as follows:

$$\begin{pmatrix} J_x \\ J_y \\ J_z \end{pmatrix} = \begin{pmatrix} \sigma_{xx} & \sigma_{xy} & \sigma_{xz} \\ \sigma_{yx} & \sigma_{yy} & \sigma_{yz} \\ \sigma_{zx} & \sigma_{zy} & \sigma_{zz} \end{pmatrix} \cdot \begin{pmatrix} E_x \\ E_y \\ E_z \end{pmatrix} \tag{7.25}$$

$$\begin{pmatrix} E_x \\ E_y \\ E_z \end{pmatrix} = \begin{pmatrix} \rho_{xx} & \rho_{xy} & \rho_{xz} \\ \rho_{yx} & \rho_{yy} & \rho_{yz} \\ \rho_{zx} & \rho_{zy} & \rho_{zz} \end{pmatrix} \cdot \begin{pmatrix} J_x \\ J_y \\ J_z \end{pmatrix} \tag{7.26}$$

If, for example **E** = (E_x, 0, 0) and $J_x = (\sigma_{xx}E_x + \sigma_{xy}E_y + \sigma_{xz}E_z)$ (and so on for J_y and J_z), this would simplify to:

$$\sigma_{xx} = J_x/E_x \tag{7.27}$$

$$\sigma_{yx} = J_y/E_x \tag{7.28}$$

$$\sigma_{zx} = J_z/E_x \tag{7.29}$$

where we now have conductivity tensors $\{\sigma_{xx}, \sigma_{xy}, \sigma_{xz}\}$ that define the current flowing parallel or perpendicular to E_x.

Similarly, if **J** = (J_x, 0, 0):

$$\rho_{xx} = E_x/J_x \tag{7.30}$$

$$\rho_{yx} = E_y/J_x \tag{7.31}$$

$$\rho_{zx} = E_z/J_x \tag{7.32}$$

This will be useful when defining the different manifestations of the Hall effect (summarised in figure 7.9):

- **Hall effect:** Generation of an electric field (voltage) E_y in a conductor with charge current J_x and magnetic field B_z.
- **Anomolous Hall effect (AHE):** Generation of an electric field (voltage) E_y where there is spin dependent scattering, typically seen in ferromagnetic materials.
- **Quantum Hall effect (QHE):** In a two dimensional material (such as a thin film), the loss of a degree of freedom on the application of an external magnetic field can result in more pronounced quantisation of the energy levels in that material (Landau quantisation [1]).
- **Quantum anomalous Hall effect (QAHE):** In a magnetic material, the magnetisation may be large enough to observe a quantum Hall effect [7].
- **Spin Hall effect (SHE):** Generation of a spin current, J_s, from a charge current, J_c, injected into a magnetic material.
- **Inverse spin Hall effect (ISHE):** Generation of a charge current, J_c, from a spin current, J_s, injected into a metal.

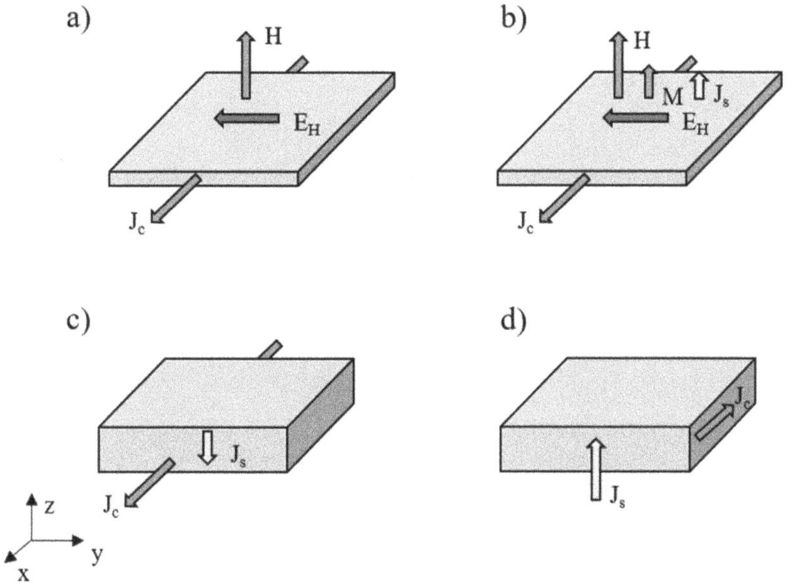

Figure 7.9. Summary of Hall effect phenomena. (a) Hall effect, where a magnetic field, H, causes charge carriers to drift due to the Lorentz force. This results in accumulation of charge, and therefore electric field, E_{H}. (b) Anomalous Hall effect, where breaking of time reversal symmetry (spin dependent scattering) results in an anomalously large Hall effect that quickly saturates. (c) Spin Hall effect, where a charge current is converted to a spin current due to spin dependent scattering. (d) Inverse spin Hall effect, where a spin current is converted to a charge current due to spin dependent scattering.

For the normal Hall effect the overall electric field in the conductor will be $\mathbf{E} = (E_x, E_y, 0)$ and it will be parametrised by the Hall coefficient, R_{H}:

$$R_{\mathrm{H}} = \frac{E_y}{J_x B_z} = \frac{\rho_{yx}}{B_z} \tag{7.33}$$

where, for the normal Hall effect:

$$\rho_{xy} = R_{\mathrm{H}} B \tag{7.34}$$

and ρ_{xy} is dependent on the available charge carriers. In the case of a semiconductor, where electrons and holes (as charge carriers) are present, we can write:

$$R_{\mathrm{H}} = \frac{\left(p\mu_{\mathrm{hh}}^2 - n\mu_{\mathrm{c}}^2\right)}{e(n\mu_{\mathrm{c}} + p\mu_{\mathrm{hh}})^2} \tag{7.35}$$

where n and p are the electron and hole charge density, respectively, and μ_{c} and μ_{hh} are the electron and heavy hole mobility, respectively [1]. In other words, the magnitude of the observed magnetoresistance due to the Hall effect is determined by the charge carrier density.

The AHE was initially labelled 'anomolous' due to the relatively large electric fields that were observed [8]. In ferromagnets, observation of the AHE mirrors that

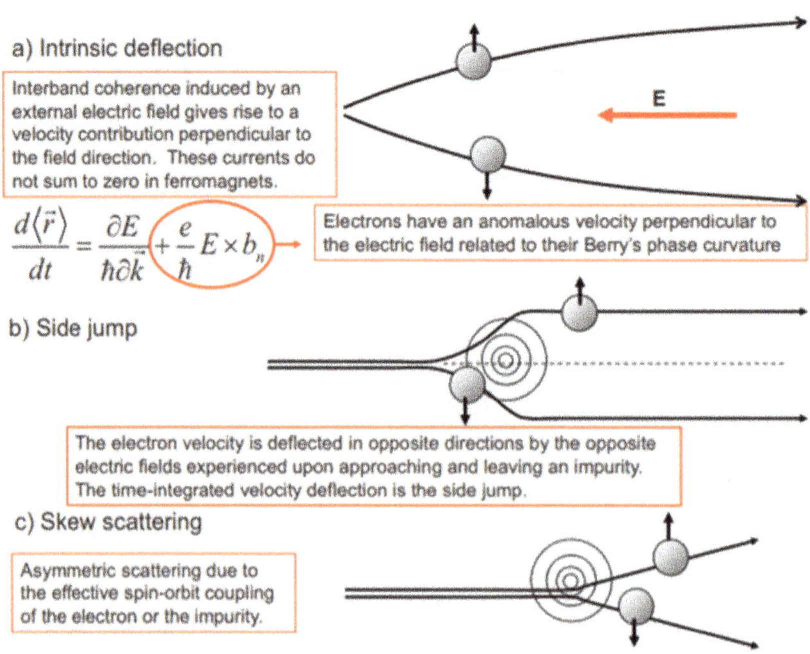

Figure 7.10. Illustration of the three mechanisms that give rise to the spin dependent scattering. Reprinted (figure) with permission from [8], Copyright (2010) by the American Physical Society.

of the magnetic hysteresis loop: the Hall voltage increases steeply as the magnetic field is initially increased, saturating at a value that mirrors the volume magnetisation of the material. In this case the resistivity, ρ_{xy} has an additional term related to the sample magnetisation, M_z:

$$\rho_{xy} = R_{\mathrm{H}}B + R_s M_z \tag{7.36}$$

where R_s is the additional Hall coefficient due to the magnetisation of the material (M_z). Unlike the normal Hall effect, where R_{H} was dependent on available charge carriers, R_s depends on several material parameters as well as the longitudinal resistivity ρ_{xx} [8]. There are three main mechanisms that have been proposed to explain the origin of the AHE, largely arising from spin dependent scattering, as summarised in figure 7.10. For more detail refer to the comprehensive review by Nagaosa *et al* [8].

The quantum Hall and quantum anomolous Hall effects are recognised by the presence of plateaus in the $\rho_{xy}(B)$ and peaks in the $\rho_{xx}(B)$ data (i.e. quantisation of the Hall effect), as shown in figure 7.11. This occurs in 2D materials such as thin films due to Landau quantisation of energy levels (where an applied magnetic field results in splitting of energy level). In an ideal system with no defects, this would manifest as distinct energy levels separated by a characteristic energy $\hbar\omega_c$, where ω_c is the cyclotron resonance. In real systems, however, the presence of defects results in slight differences in the position of each energy level, thereby broadening the global density of states. As the magnetic field is increased, the separation of these energy

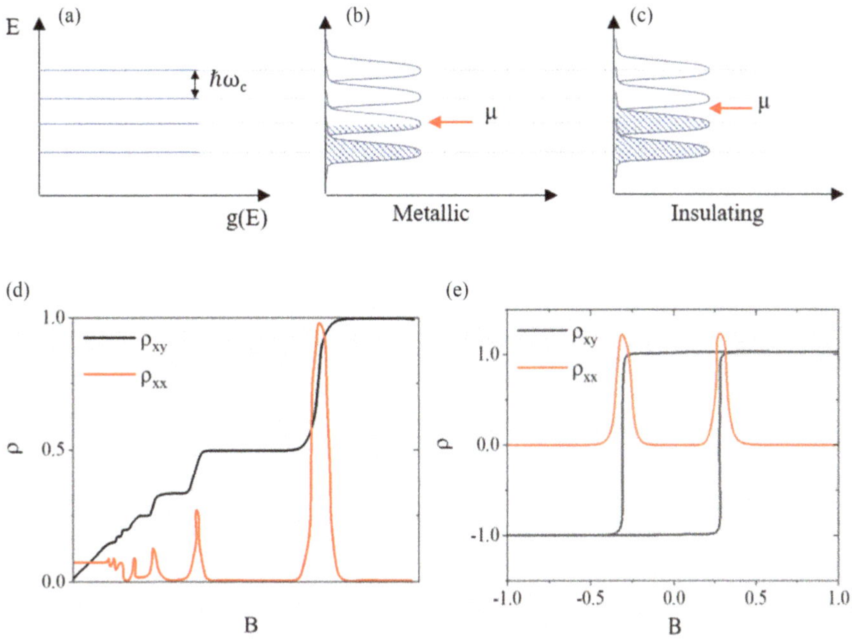

Figure 7.11. (a)–(c) Sketch of the Landau level density of states and example of metallic and insulating states where the sub band is partially or completely filled. In real samples the energy levels are broadened due to impurities; μ indicates the position of the Fermi level, which determines whether the sample is metallic or insulating. (d) and (e) Sketch of the resistivity seen in quantum Hall and quantum anomolous Hall effects, respectively. The transverse resistance, ρ_{xy}, takes on quantised values (h/e^2) as the longitudinal resistance, ρ_{xx}, vanishes.

levels increases, and the position of the Fermi level (which is fixed) will vary with respect to filled and partially filled bands. At the point where ρ_{xx} goes to zero, a whole number of Landau levels has been filled.

For example, for the Hall geometry given in figure 7.9, where current is applied in the x-direction, and magnetic field in the z-direction, the energy level, E, within each sub band, j, of electrons could be described by:

$$E(B, l, j) = \left(l + \frac{1}{2}\right)\hbar\omega_c + E_j \tag{7.37}$$

where E_j is the lowest energy level within the sub-band j, l is the orbital angular momentum quantum number, and ω_c is the cyclotron resonance defined by:

$$\omega_c = \frac{eB}{m_e} \tag{7.38}$$

where B is the magnetic field, e is the charge on an electron, and m_e is the electron mass.

In such a case, the current flowing in the x- and y-directions can be described by:

$$J_x = \sigma_{xx} E_x + \sigma_{xy} E_y \tag{7.39}$$

$$J_y = -\sigma_{xy} E_x + \sigma_{xx} E_y \tag{7.40}$$

Assuming that the material being measured is homogenous and isotropic:

$$\sigma_{xx} = \sigma_{yy} \tag{7.41}$$

$$\sigma_{xy} = -\sigma_{yx} \tag{7.42}$$

And if no current flows in the y-direction ($J_y = 0$), equation (7.40) reduces to:

$$\frac{E_y}{E_x} = \frac{\sigma_{xy}}{\sigma_{xx}} \tag{7.43}$$

Substituting into equations (7.39) and (7.40) results in the following definitions for resistivity:

$$\rho_{xx} = \frac{E_x}{J_x} = \frac{\sigma_{xx}}{\sigma_{xx}^2 + \sigma_{xy}^2} \tag{7.44}$$

$$\rho_{xy} = \frac{E_y}{J_x} = \frac{\sigma_{xy}}{\sigma_{xx}^2 + \sigma_{xy}^2} \tag{7.45}$$

As the magnetic field is varied, the separation of Landau levels will change, and thus we can move from a situation where the highest Landau level is completely filled (insulating) in which case σ_{xx} goes to zero and hence ρ_{xx} goes to zero. Experimentally, as ρ_{xx} goes to zero a plateau in the Hall voltage (ρ_{xy}) appears, as shown in figure 7.11.

Whilst ρ_{xy} is quantised in units of h/e^2 (see [1] for the derivation), the quantum Hall resistance can also be defined as:

$$R_{\mathrm{QHE}} = \frac{h}{\upsilon e^2} \tag{7.46}$$

where υ is known as the 'filling factor'[1] and is either an integer ('normal' quantum Hall effect) or a fraction of the form $p/(2p + 1)$ (fractional quantum Hall effect) [1].

In the anomalous quantum Hall effect, the magnetisation of the sample itself can result in Landau quantisation, hence the resistivity will follow the behaviour of the magnetic hysteresis loop. This manifests as ρ_{xx} peaking when there is a change in the magnetisation direction and ρ_{xy} switching from one plateau to another, as shown in figure 7.11.

The spin Hall effects arise due to spin dependent scattering in a material. They are in this way, manifestations of the processes observed in AHE and shown in figure 7.10 (spin–orbit interaction, skew scattering and side-jump scattering), but

[1] As it describes the number of full Landau levels.

Table 7.2. Some spin Hall angles for different elements (data taken from [9]).

Element	Spin Hall angle, θ_{SH} (%)
Al	0.02
Au	0.25–11
Bi	>0.8
Cu	0.22
Mo	−0.05 to −0.8
Nb	−0.87
Pd	0.64–1
Pt	1.3–11
Ta	−0.37 to −12
W	−33

without the external magnetic field. They are particularly useful as a way to generate or detect spin currents in potential spintronic materials or devices [9]. The spin angle, θ_{SH}, as defined by:

$$\theta_{SH} = \frac{\sigma_{xy}^{s}}{\sigma_{xx}^{c}} \frac{e}{\hbar} = \frac{J_s}{J_c} \tag{7.47}$$

is used to quantify the efficiency of conversion from spin current, J_s, to charge current, J_c, in the spin Hall and inverse spin Hall effects, where $\sigma^{s}{}_{xy}$ and $\sigma^{c}{}_{xx}$ are the spin and charge conductivity tensors, respectively. This property can be measured by spin torque FMR [10, 11] or non-local spin injection [12], however, there is some ambiguity in the actual value of θ_{SH} due to the impact of impurities on the measurement [9]. Some examples of good spin Hall metals is given in table 7.2.

7.2.2 Thermoelectric effects

The thermoelectric effects described here, whilst not exhaustive, give an example of the interplay of currents, thermal gradients, and magnetisation (figure 7.12):

- **Nernst effect:** Generation of an electric field in a conductive sample with temperature gradient perpendicular to applied magnetic field.

$$|N| = \frac{E_x/B_y}{\mathrm{d}T/\mathrm{d}z} \tag{7.48}$$

- **Anomolous Nernst effect:** Generation of an electric field in a magnetised conductive sample with temperature gradient perpendicular to applied magnetic field.
- **Seebeck effect:** Generation of a charge current, J_c, (commonly observed as a voltage) in a material subjected to a temperature gradient, ΔT. This effect is

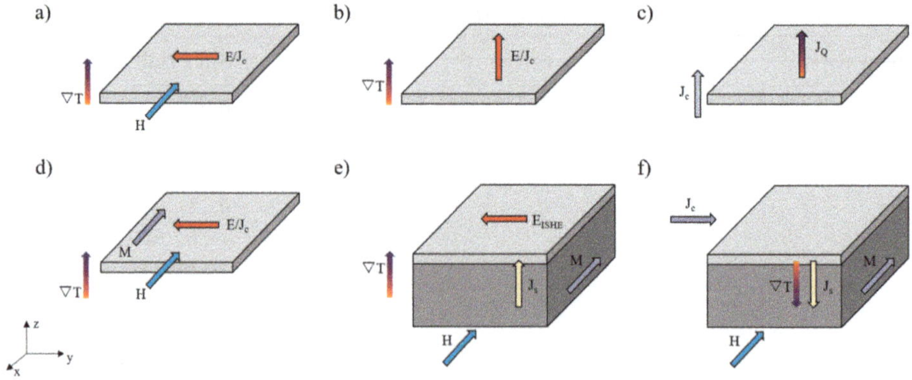

Figure 7.12. Summary of thermoelectric phenomena defined in the text. (a) Nernst effect. (b) Seebeck effect. (c) Peltier effect. (d) Anomalous Nernst effect. (e) Spin Seebeck effect (longitudinal geometry). (f) Spin Peltier effect.

more commonly employed in thermocouples where two materials are connected and held at different temperatures. The difference in the Seebeck coefficients for these materials, S, as defined in equation (7.49), results in a measurable voltage.

$$S = \left(\frac{\Delta V}{\Delta T}\right)_{Jc=0} \tag{7.49}$$

- **Peltier effect:** Generation of a thermal gradient (or heat flow, J_Q) across a material due to an applied voltage. It is defined by the Peltier coefficient, Π, where:

$$\Pi = \left(\frac{J_Q}{J_c}\right)_{\Delta T=0} \tag{7.50}$$

- **Spin Seebeck effect (SSE):** Demonstrated experimentally in 2008 by Uchida *et al* [13], this phenomena is defined as the generation of a spin current in a magnetic material subjected to a temperature gradient (i.e. spin dependent thermal transport of charge carriers in a magnetic material). In general, the spin Seebeck effect is observed by placing a thin layer of normal metal (such as Pt) on top of the magnetic material in order to detect the spin current generated via the inverse spin Hall effect.
- **Spin Peltier effect:** Generation of a thermal gradient in a magnetic material where a spin polarised current is injected (i.e. the inverse of the spin Seebeck effect [14]).

7.3 Basics of an electric measurement

7.3.1 Measurement of current

The ideal ammeter has negligible resistance so that a current can pass through relatively unperturbed. Historically this was achieved with coil based analogue galvanometers, but this has since evolved to digital ammeters, where the voltage is measured across a shunt resistor. The Galvanometer was used to detect current by the deflection of a needle due to the Lorentz force on a conducting coil at right angles to a magnetic field. Two classic examples are the tangent and D'Arsonval Galvanometers. In the case of the former, a single coil was aligned perpendicular to the Earth's magnetic field. The deflection would be described by:

$$\tan \theta = \frac{nI}{2Hr} \tag{7.51}$$

where θ is the angle of deflection, I is the current, n is the number of turns (of the coil), r is the radius of the coil, and H is the magnetic field strength perpendicular to the plane of the coil. Notice that the deflection is not linearly proportional to the current. An improvement to this was the D'Arsonval galvanometer, where the coils are wound around an iron core held between two radial magnets such that the magnetic field seen by the coil is the same as it is rotated. In this case, the deflection would be defined by:

$$I = \frac{k\theta}{BAN} \tag{7.52}$$

where I is the current flowing through the coils, k is the spring constant, θ is the angle of deflection, B is the magnetic flux density, A is the area of the coil(s) and N is the number of turns. Notice that the deflection is now linearly proportional to the current, which makes it more sensitive to the detection of small currents.

The definition of the Ampere, was originally the force exerted on two infinitely long conducting wires separated a distance of 1 m apart in vacuum (2×10^{-7} N m^{-1}). As part of the SI redefinition voted in on November 16th 2018 by the General Conference on Weights and Measures, the Ampere will be defined by the charge of an electron as $1.602\,176\,634 \times 10^{-19}$ Coulombs ($C = A \cdot s$). In practice this will likely be achieved by design of a semiconducting device that counts single electrons flowing through or the AC Josephson junction described in section 7.1.4.

7.3.2 Measurement of voltage

The ideal voltmeter has infinite resistance so that minimal current is diverted away from the main circuit. A common method employed in digital voltmeters is the integrating analogue-to-digital convertor, which averages an input voltage over time, and compares to a reference voltage. (Before this, a sensitive galvanometer coupled with a large known resistor in series with the circuit would have been used.) As the analogue-to-digital convertor is designed to measure direct current (DC)

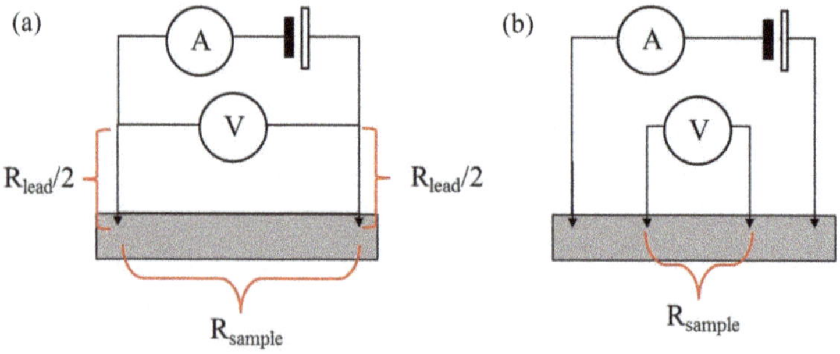

Figure 7.13. (a) 2-probe and (b) 4-probe configuration for a volt-ammeter measurement. The advantage of the 4-probe configuration is the removal of contact lead resistance from the total measurement.

voltage, a rectifier such as the diodes described in section 7.1.4 is required to convert alternating current (AC) to DC before measurement.

7.3.3 Measurement of resistance

The resistance of a material, measured as a function of temperature will depend on various factors, such as whether: it is semiconducting, insulating or metallic; Ohmic or non-Ohmic; or in thin film, bulk, polycrystalline or single crystal form. Contributions to resistance include: phonons; screened Coloumb impurity scattering; spin-disorder scattering, for example, in simple alloys such as Au_3Mn where disorder of local spins in a metal results in temperature dependent scattering above T_c or T_n; the Kondo effect, where scattering of electrons occurs due to magnetic impurities leading to a minima in the resistivity for metallic samples as the temperature approaches 0 K (rather than ρ approaching zero); many body scattering, where a minima in the resistivity is observed; spin-wave scattering; and spin fluctuation scattering.

The simplest measurement of resistance, for example, would be the 2-probe method, where contact is made across the sample to be measured at two points. The current is passed through these contact leads, and the voltage is measured (figure 7.13(a)). The disadvantage of this is that the resistance measured will include the resistance of the contact leads and if the sample is conductive (i.e. $R_{sample} \sim R_{leads}$) this becomes an appreciable error. The 4-probe method, on the other hand, where voltage and current are measured or passed separately, as shown in figure 7.13(b), measures the resistance of the sample only.

There are three common methods for the measurement of resistance (the first step in most electrical characterisation techniques):

1. The resistance bridge methods (e.g. Wheatstone and Kelvin double bridge).
2. The substitute method: where the voltage across a known resistor is measured.
3. The volt-ammeter method: where the current flowing in a circuit is measured simultaneously to the voltage generated across the resistor to be measured.

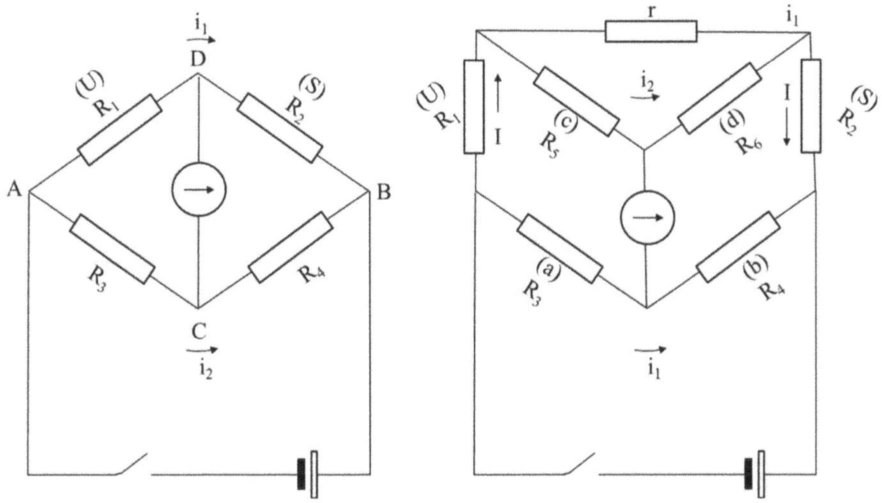

Figure 7.14. Two commonly used resistance bridges. Left: the Wheatstone bridge. Four resistors connected to a current source (battery in this case), and pairs in parallel. When the resistances R_1/R_2 and R_3/R_4 are balanced, the galvanometer will read zero deflection. Right: the Kelvin double bridge. In this configuration, the resistances *a*/*b* and *c*/*d* are balanced so that the current flowing through the unknown resistor (*U*) can be more precisely determined (accounting for small contact resistance, *r*).

The resistance bridge methods (as seen in figure 7.14), require the use of a sensitive galvanometer to monitor the 'balance' of current flow across the circuit as the resistance of a section is adjusted (e.g. by utilising a sliding wire). For the Wheatstone bridge, for example, there are six conductors, which are balanced when no current flows through the galvanometer. As a result of the circuit geometry, the following equations hold:

$$i_1R_1 = i_2R_3 \tag{7.53}$$

$$i_1R_2 = i_2R_4 \tag{7.54}$$

$$R_1/R_2 = R_3/R_4 \tag{7.55}$$

where i_i is the current flowing through the *i*th branch, and R_i is the *i*th resitor, as shown in figure 7.14. Therefore, to determine the resistance of a sample, you only need to know the resistance of one of $\{R_1,R_2,R_3,R_4\}$ and the ratio of two others. If a sliding contact were implemented along ACB, for example, the ratio of R_3 and R_4 could be easily changed (whilst the total resistance is fixed). This could be used to balance the circuit against the sample (e.g. R_1) and a known resistor (R_2). This implementation is limited to a range of approximately 1–100 000 Ω, due to the resistance of the contacts (for low *R*) and the sensitivity of the galvanometer (for high *R*).

For measurement of low resistance samples (~1 mΩ), where the contact resistance starts to become appreciable with respect to the sample, the Kelvin double bridge can be used. In this case, an additional pair of resistors is added (as seen in

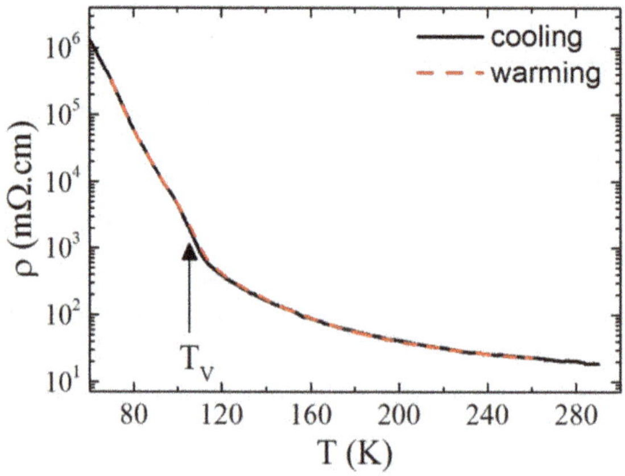

Figure 7.15. Log plot of the resistivity across the semiconductor–insulator Verwey transition in Fe_3O_4 at the Verwey temperature, T_V.

figure 7.14). The galvanometer is first balanced across these four resistors (R_3–R_6) before adjusting R_2.

Once balanced:

$$R_1/R_2 = R_3/R_4 = R_5/R_6 \tag{7.56}$$

The substitute method, on the other hand, simply requires knowledge of a known resistor, which is connected in series with the resistor to be measured. As the current flowing through both will be the same, the voltage measured across the known resistor can be used to determine the resistance of the sample. The advantage of this technique is that it only requires the current to be constant for the millisecond it takes to perform the measurement.

The volt-ammeter method is the most commonly employed technique in research laboratories due to the prevalence of digital voltmeters and ammeters. For a stable current source, I, the voltage read across the sample to be measured, V (as shown schematically in figure 7.13) can be used to determine the resistance, R, by Ohm's law:

$$V = IR \tag{7.57}$$

Of note is the choice of probe current for these measurements. Too high and the sample may heat up (Ohmic heating, $P = I^2R$), too low and the voltage generated may be too small to accurately measure with the voltmeter used. In addition, samples with particularly large changes in the resistance as a function of temperature or field (such as the metal–insulator Verwey transition in Fe_3O_4, as shown in figure 7.15) will require a change in the probe current during the measurement.

7.3.4 Sources of noise

The limitation of any measurement is the noise or background that is observed. Random fluctuations in a measurement may occur due to an unstable current source or power supply, but more often than not, it is due to intrinsic sources of noise such as Shot, Johnson, and $1/f$ noise, as outlined below.

- Johnson–Nyquist, or 'white' noise is the result of thermal fluctuations:

$$V_{\text{noise}}(\text{rms}) = \sqrt{4k_B TR\Delta f} \tag{7.58}$$

where k_B is Boltzmann's constant, T is the temperature, R is the resistance, and Δf is the bandwidth of the measurement. As Johnson noise is proportional to the temperature it can be used in thermometry [15] or for an accurate measurement of the Boltzmann constant [16]. Note that to reduce the impact of thermal noise semiconductor detectors are often cooled with liquid nitrogen.

- $1/f$ noise, also referred to as 'flicker' or 'pink' noise is recognised as a frequency (f) dependent noise that decreases in intensity as f is increased:

$$V_{\text{noise}}(\text{rms}) \propto \frac{1}{f} \tag{7.59}$$

This is shown schematically in figure 7.16. To limit the impact of $1/f$ noise a high pass filter or lock-in amplifier could be used. This would only work for alternating current (AC) measurements.

- Shot noise arises from the finite nature of charge (i.e. the electron) and is observed in circuits where there is a device with a barrier such as the pn junction discussed in section 7.1.4. (Resistors, for example, will not exhibit shot noise because current transport in them is diffusive and they do not have an energy barrier.) As these electrons cross the barrier, statistical variation will occur that is quantised by the charge on an electron, q:

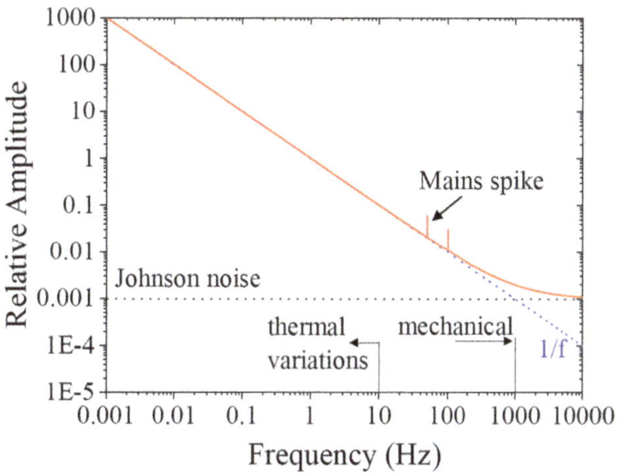

Figure 7.16. Sketch of the contributions to electrical noise as a function of frequency. For frequencies greater than 10^5 Hz radio, TV and radio can also contribute. The noise due to thermal and mechanical fluctuation will depend on the environment.

$$I_{noise}(\text{rms}) = \sqrt{2qI\Delta f} \tag{7.60}$$

where I is the rms AC or DC current supplied, and Δf is the bandwidth of the measurement. A common analogy used to describe shot noise is to picture rainfall on a tin roof. Whilst the noise generated by this is continuous (much like an electric current), each raindrop is discrete.

In addition to these fundamental sources of noise there are others that are highly sensitive to how the measurement circuit is assembled. For example, inductive coupling between wires in proximity to one another due to the magnet field produced by a current flowing through a wire. Other sources can include capacitive coupling, resistive coupling, ground loops, noise from the mains supply (50 Hz in the UK), mechanical vibrations or radio pick-up.

Routes to limit the impact of these sources of electrical noise for low level measurements include: the use of a lock-in amplifier (thus operating at higher frequencies, and enabling limitation of the bandwidth); implementation of twisted pairs and/or shielding (such as is found in coaxial cables) to limit cross-talk; use of a pre-amplifier to amplify the signal (where the main component is the bipolar junction transistor, BJT, described in section 7.1.4); and altering the measurement environment to limit mechanical vibrations and temperature fluctuations.

Note that for AC measurements additional consideration must be made for contributions to inductance or capacitance in the circuit that could result in large delay times.

7.3.5 The lock-in amplifier

For measurements where there is an appreciable DC offset or background noise to the measurement, it is common to employ a lock-in technique. This requires the signal to be perturbed by an AC signal (such as the source current), so that a parameter (such as voltage) can be monitored with respect to a signal generated at the same frequency (or harmonics thereof). The lock-in amplifier will then extract a signal in a defined frequency range.

The output of the LIA is the product of two sine waves:

$$\begin{aligned} V_{LIA} &= V_{sig} V_{ref} \sin(\omega_{sig} t + \theta_{sig}) \sin(\omega_{ref} t + \theta_{ref}) \\ &= \frac{1}{2} V_{sig} V_{ref} \cos((\omega_{sig} - \omega_{ref}) t + \theta_{sig} - \theta_{ref}) \\ &- \frac{1}{2} V_{sig} V_{ref} \cos((\omega_{sig} + \omega_{ref}) t + \theta_{sig} + \theta_{ref}) \end{aligned} \tag{7.61}$$

where V_{LIA}, V_{sig} and V_{ref} are the voltages of the lock-in amplifier output, measurement signal, and reference signal, ω_{sig} and ω_{ref} are the frequencies of the signal and reference, respectively, and θ_{sig} and θ_{ref} are the phase of the signal and reference, respectively. When passed through a low pass filter, only a DC component (i.e. at $\omega_{ref} = \omega_{sig}$) will be recovered:

$$V_{LIA} = \frac{1}{2} V_{sig} V_{ref} \cos(\theta_{sig} - \theta_{ref}) \tag{7.62}$$

i.e. a DC signal proportional to the reference amplitude.

In a dual phase lock in, a second reference at a phase difference of 90 degrees is used to extract the signal magnitude and phase. You then have access to two signals that enable determination of magnitude, R, and phase, θ of the AC signal being measured:

$$X = V_{sig} \cos\theta \tag{7.63}$$

$$Y = V_{sig} \sin\theta \tag{7.64}$$

$$R = \sqrt{X^2 + Y^2} = V_{sig} \tag{7.65}$$

$$\theta = \tan^{-1}(Y/X) \tag{7.66}$$

As the LIA is an AC technique, the signal that is output will be the average over several cycles (of measurement), as determined by the time constant. In addition, it is common for various filters to be available to further reduce the noise of the signal. The aim of a filter (whether low-pass, band-pass, or high-pass) is to select a frequency range of interest and 'throw away' everything else (such as the $1/f$ contribution to noise). In an LIA this may be represented by a filter 'order', which is equivalent to several filters (e.g. a RC circuit) connected in series. As the filter order is increased the cut-off with respect to frequency becomes sharper, but the response time increases. As a result, a compromise between the noise floor and measurement response time is often found such that artificial hysteresis is not introduced into the measurement.

Aside: A Decibel, dB, is the unit that results from a logarithmic comparison of two amplitudes, A_1, A_2:

$$d = 20 \log(A_2/A_1) \tag{7.67}$$

e.g. if $A_2 = 2A_1$, then $d = 20 \log 2 = 6.02$ dB

The best instruments have a dynamic reserve of 120 dB and are therefore capable of measuring a signal with background noise up to 1 million times the amplitude (of the signal to be measured) [17].

Applications of the lock-in amplifier include:

- AC measurements (such as the AC microcalorimetry described in chapter 6).
- Impedance spectroscopy.
- Ferromagnetic resonance (in a simplified setup to replace the network analyser).
- Vibrating sample magnetometry.

7.4 Measurement of resistivity

The (electric) resistance of a given material is defined as a measure of the difficulty with which an electric current can be passed through. The higher the resistance the more difficult it is for current to flow. It is most commonly referred to in relation to Ohm's law for a conductor:

$$V = IR \tag{7.68}$$

where V is the voltage applied, I is the current flowing, and R is the resistance of the conductor. This relationship will only hold for Ohmic conductors (such as metals). Non-Ohmic conductors such as semiconducting or insulating materials will exhibit a non-linear IV curve, as shown in figure 7.17, where the act of passing a current through the material, may, for example increase the temperature (of the material) and thus the resistance.

Resistance depends on geometry. It should be familiar enough that copper wires with a large diameter will have a lower resistance than copper wires with a small diameter. The analogy here would be to imagine trying to pass the same amount of water through two different pipes. Resistivity, on the other hand, is a material property that is independent of the sample geometry (normalised per unit volume). For a sample with a given resistance (R), cross-sectional area (A), and length (l), the resistivity will be:

$$\rho = \frac{RA}{l} \tag{7.69}$$

Conversely, the conductance (G) and conductivity (σ) of a material are the inverse of the resistance and resistivity, respectively:

$$G = \frac{1}{R} \tag{7.70}$$

$$\sigma = \frac{1}{\rho} \tag{7.71}$$

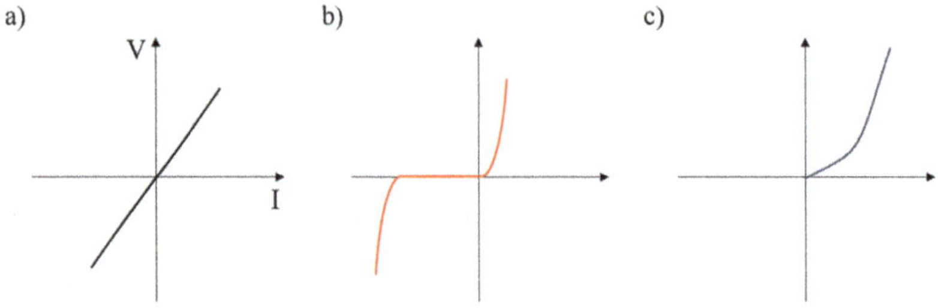

Figure 7.17. *IV* plots for (a) Ohmic and (b) and (c) non-Ohmic conductors. (b) Semiconductor diode (pn junction). (c) Bulb, or resistive element heating as current is passed.

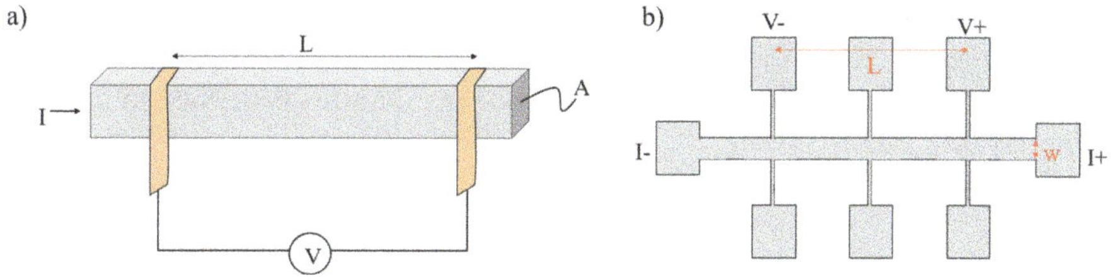

Figure 7.18. (a) Schematic of a simple bulk resistivity measurement. (b) Schematic of the Hall bar geometry for resistivity measurements.

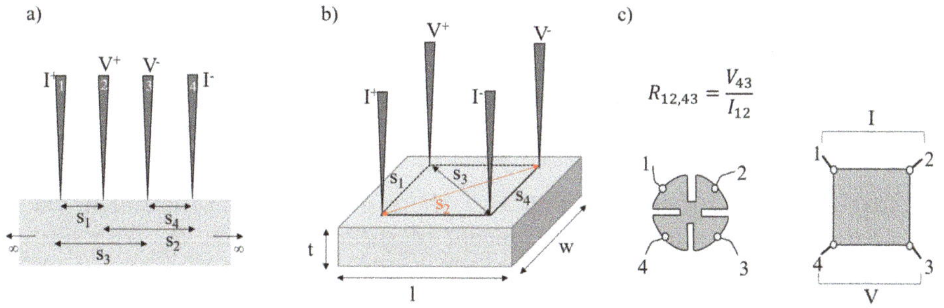

Figure 7.19. 4 point probe configurations. (a) In-line 4-point probe for the semi-infinite sheet. (b) Square 4-point probe configuration. (c) Recommended Van der Pauw connections.

To measure the resistivity, therefore, simply requires careful design of a resistance measurement so that the area, A, and length L are well known (see equation 7.69). For bulk samples this is achieved by cutting and polishing the sample to a specific shape (more often the bar seen in figure 7.18). For thin film samples, it would require knowledge of the film thickness, t, and width of the contacts, w (often achieved by patterning a Hall bar, as seen in figure 7.18(b)).

7.4.1 The 4-point probe technique

As discussed in the previous section, if the geometry of a sample is well known, and uniform enough to enable the bulk resistivity measurement shown in figure 7.18 then the resistance and resistivity of a sample can be easily measured. There are cases, however, where the sample is either not uniform, or is a thin film, both of which require a different approach. The 4-point probe method was designed to tackle this problem over 100 years ago, the idea being that 4-point contacts are arranged in a line or square on the sample, as shown in figure 7.19, and the current and voltage sourced/measured in a similar way to bulk resistivity measurements. For a semi-infinite 3D bulk sample, if the contact separations, s, are equal ($s = s_1 = s_2 = s_3 = s_4$) and in-line, the resistivity will be:

$$\rho = 2\pi s\frac{V}{I} \tag{7.72}$$

where V is the voltage measured, I is the current sourced, and the 2π arises from considering the spread of the current within the sample.

If the probes are arranged in a square, where $s_1 = s_4$ and s_2 is defined in figure 7.19 and is equal to $\sqrt{2}s_1$:

$$\rho = \frac{2\pi s_1}{2 - \sqrt{2}} \frac{V}{I} \tag{7.73}$$

Similarly for an infinite 2D sheet (i.e. thickness of the film is less than s) in-line:

$$\rho = \frac{\pi}{\ln 2} \frac{V}{I} \tag{7.74}$$

And in the square configuration:

$$\rho = \frac{2\pi}{\ln 2} \frac{V}{I} \tag{7.75}$$

Note that for the 2D sheet measurements the resistivity is not dependent on the probe distance. This results in the concept of sheet resistance, R_s, with units of Ω/square (to distinguish from a resistance measurement). In this case, if the thickness of the film, t, is known, then the resistivity, ρ, can be determined:

$$\rho = R_s t \qquad [\Omega \text{ m}] \tag{7.76}$$

In practice, for the 4-point probe measurements there will be a correction factor due to geometry of the sample:

$$\rho = f\frac{V}{I} \tag{7.77}$$

where f takes into account finite thickness, alignment of probes in proximity of sample edge, and finite lateral width, as discussed in more detail in a review article by Miccoli [18].

7.4.1.1 Van der Pauw measurements

The Van der Pauw measurement is a useful extension of the 4-point probe method for determining the resistivity of a thin film sample of non-uniform geometry. It was outlined by Van der Pauw in 1958 [19] in order to test the Hall effect and resistivity of randomly shaped discs or samples.

To measure R_s, a current is passed along one edge of the film and the voltage is measured along the opposite (parallel) edge. Contacts are then swapped amongst eight possible configurations, as shown in figure 7.19, and the voltage logged again. The combination of these eight measurements will account for any differences due to thermal voltages and geometry and is used to determine R_s from equations (7.78) to (7.80):

$$R_A = (R_{12,43} + R_{21,34} + R_{43,12} + R_{34,21})/4 \tag{7.78}$$

$$R_B = (R_{23,14} + R_{32,41} + R_{14,23} + R_{41,32})/4 \tag{7.79}$$

$$e^{-\pi R_A/R_S} + e^{-\pi R_B/R_S} = 1 \tag{7.80}$$

If $R = R_A \sim R_B$ then equation (7.80) reduces to:

$$R_s = \frac{\pi R}{\ln 2} \tag{7.81}$$

which is the limit of the 4-point probe measurement of an infinite 2D sheet described above. The above equations describe the determination of the sheet resistance, R_s, using the iterative method (equation (7.80)). Alternatively, a correction factor described by the function, f, can be used, where f is defined by:

$$\cosh\left(\frac{\ln 2}{f}\cdot\frac{(R_A/R_B) - 1}{(R_A/R_B) + 1}\right) = \frac{1}{2}\exp\frac{\ln 2}{f} \tag{7.82}$$

and:

$$R_s = \frac{\pi}{\ln 2}\frac{R_A + R_B}{2} f\left(\frac{R_A}{R_B}\right) \tag{7.83}$$

$$\rho = \frac{\pi d}{\ln 2}\frac{R_A + R_B}{2} f\left(\frac{R_A}{R_B}\right) \tag{7.84}$$

When this method is employed the user typically refers to a chart such as that given in figure 7.20, which shows f as a function of R_A/R_B.

The van der Pauw method works best when the following assumptions are made:

- Contacts are at the edge of the sample.
- The contacts are small compared to the sample size.
- The sample has a homogenous thickness.
- The surface does not have isolated holes (connected surface).

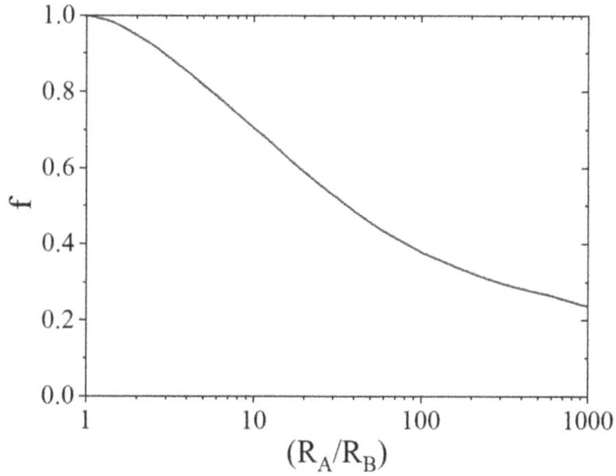

Figure 7.20. Van der Pauw correction factor, *f*.

In general, the impact of contact size [20] and sample geometry [21], when these conditions have not been met is an increase in uncertainty.

7.4.2 Magnetoresistance measurements

For some materials there is a significant change in the resistance under applied magnetic field and with careful design, this magnetoresistance can be enhanced with magnetic multilayers. This is used in conventional GMR read heads (GMR = giant magneto resistance), which read the '0' or '1' of a magnetic hard disc and won Fert and Grunberg the 2007 Nobel Prize in Physics.

To measure the magnetoresistance, *MR*, a comparison to the zero field resistance, R_0, is made as a function of applied magnetic field, as defined by:

$$MR = \frac{R_0 - R_H}{R_0} \tag{7.85}$$

7.4.3 Hall measurements

When doing Hall measurements we are interested in the voltage accumulation perpendicular to the flow of current when a magnetic field is applied perpendicular to the plane of the film. To do this in detail a Hall cross would be fabricated, as seen in figure 7.18, where the magnetic field, *B*, is out of plane. Alternatively, Hall measurements can also be carried out using the Van der Pauw method, however, this time the voltage and current pairs are crossed (e.g. V_{13}, I_{24}) rather than parallel (e.g. V_{12}, I_{34}).

To determine the carrier concentration from Hall measurements we need to know the field applied, *B* [G], current passed, *I* [A], and average transverse voltage measured.

From the Van der Pauw configuration the average voltage measured is $|\Sigma V_i|/8$:

$$\left|\sum V_i\right|/8 = (V_{24P} - V_{24N} + V_{42P} - V_{42N} + V_{13P} - V_{13N} + V_{31P} - V_{31N})/8 \tag{7.86}$$

where V_{24P} and V_{24N} indicate voltage measured across contacts 2–4 with magnetic field in a positive (P) or negative (N) direction, respectively.

The carrier concentration, n_s (or p_s depending on carrier type) [cm^{-2}], is then:

$$n_s = \frac{8 \times 10^{-8} IB}{q|\Sigma V_i|} \tag{7.87}$$

where a positive value of ΣV_i would indicate positive carriers (holes), p_s, and a negative value of ΣV_i would indicate negative carriers (electrons), n_s. In this example the units of field are Gauss (see chapter 8 for more on magnetic units).

If the sheet resistance, R_s [Ω/□] is known then the mobility, μ [$cm^2\ V^{-1}\ s^{-1}$], can be determined:

$$\mu = \frac{1}{q n_s R_s} \tag{7.88}$$

7.5 Bandstructure measurements

In chapter 5, angle resolved photoemission spectroscopy (ARPES) was outlined as a method of measuring the bandstructure of a single crystal sample. The basic principle was the measurement of the energy of electrons emitted from the sample when a known wavelength (of light) was incident on it from a specific angle. By measuring a single crystal, enough information is obtained to build a 3D model of the band structure around the Fermi surface, for symmetric (e.g. simple cubic) and asymmetric (e.g. hexagonal) structures. The advantage of this technique is that it can give detailed information on the valence electron band dispersion and the Fermi surface, however, it will not necessarily yield information on unoccupied states [22]. Inverse photoemission spectroscopy, on the other hand, where a collimated beam of low energy electrons is incident on the sample, can provide some information on the unoccupied states through analysis of subsequent radiative decay of these electrons after capture [23]. Spin polarised photoemission, where the spin of the electrons emitted is measured by observing the asymmetry of diffraction off of a single crystal, can also be employed to measure the spin polarisation close to the Fermi level.

Other methods that can be employed to measure features of the bandstructure include:

1. Tunnelling spectroscopy, which can provide information on the density of states close to the Fermi energy [24–26].
2. Point contact Andreev reflection measurements (PCAR), which can provide information on the spin polarisation close to the Fermi level [27].
3. Optical (absorption) measurements, which can provide information on the size of the bandgap in insulators.
4. Quantum oscillation measurements that observe phenomena such as the de Haas–van Alpen and Shubnikov–de Haas effects, which can provide information on the Fermi surface.

7.5.1 Point contact spectroscopy

PCAR is a technique that involves measurement of the conductance of a superconducting tip in contact with the sample to be measured as a bias voltage is applied. With the highest transition temperature among metals, Nb, ($T_c = 9.3$ K) is an ideal candidate for the point contact tip. This technique is analogous to STM and scanning tunnelling spectroscopy (STS) in that as a tip approaches the sample surface the current passing from tip to sample is measured. (A micrometer attached to the tip can be used to move it into contact.) Where it differs is that the tip is driven into contact with the surface, combined with sensitivity to spin polarisation in the metal being measured. The measurement also requires temperatures <9.3 K for the tip to be superconducting and thus enable observation of Andreev refection. This can be achieved by immersing the tip (and sample) in liquid helium.

Andreev reflection is described as ‘supercurrent conversion at a superconductor–metal interface’ or a process whereby a dissipative electric current is converted to a dissipationless supercurrent due to proximity with a superconductor [28]. It is a two-electron process whereby an incoming electron is transferred to the superconductor

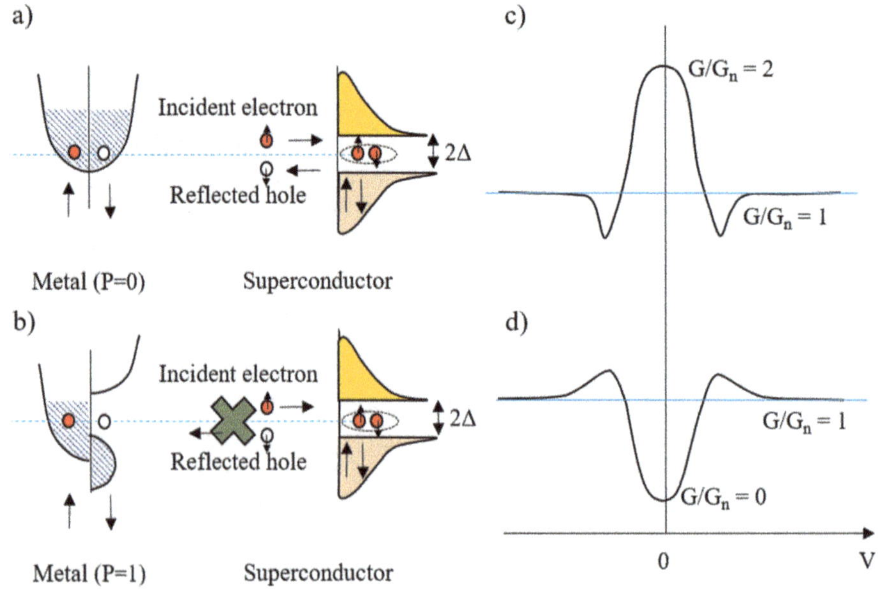

Figure 7.21. Schematic of the basic principles of PCAR. (a) Density of states for a normal metal where $P = 0$ (e.g. Cu:Nb) and Andreev reflection is unhindered by spin minority population at the Fermi energy. Red circles are the electrons and open circles are the holes. (b) Density of states for a half metal where $P = 100\%$ (e.g. CrO_2:Nb). Here, there is no supercurrent conversion due to the unavailability of suitable state for the reflected hole. (c) and (d) Sketch of the expected conductance spectra for (a) and (b), respectively. The solid line indicates the position of the conductance spectra in the normal state.

if a second electron is transferred through the interface forming a Cooper pair (reflection of a hole), as shown in figure 7.21. For a perfect contact this process will result in a measured conductance that is twice that of the normal metal (G_n) [4].

It is this feature of the conductance that is used to probe the spin polarisation of the ferromagnet being studied. If Andreev reflection occurs (assuming a perfect interface between the superconductor and ferromagnet) then a conductance, G, with a value twice that of the normal state, G_n, is expected to be seen. If there is splitting of the density of states for spin up and spin down electrons, the polarisation, P, can be defined by:

$$P = \frac{N_\uparrow(E_F) - N_\downarrow(E_F)}{N_\uparrow(E_F) + N_\downarrow(E_F)} \tag{7.89}$$

where $N_{\uparrow\downarrow}(E_F)$ is the spin dependent density of states. As shown in figure 7.21, if P is non-zero (i.e. there is spin polarisation at the Fermi energy) then the conductance will drop. For a fully spin polarised material the conductance will drop to zero as the formation of Cooper pairs through the interface is no longer allowed. An example of this is given in [27] where the density of states in the normal metal and superconductor are shown for Cu (normal metal) and CrO_2 (ferromagnetic metal) alongside corresponding IV and conductance measurements. In these measurements, G/G_n

approaches 2 for the Nb:Cu measurement, and zero (at $V \sim 0$) for the CrO_2:Nb interface.

Ultimately, as the Cooper pairs that leak into the normal metal they will be influenced by impurities, tunnel barriers and boundaries. As a result, determining the spin polarisation from conductance spectra requires extensive fitting of the data [29, 30].

7.5.2 Quantum oscillations

At low temperatures (typically <10 K) and high magnetic fields (typically >5 T) it may be possible to observe quantum oscillations, which can be useful for determining the bandstructure. Examples of these quantum oscillations include:

- The de Haas–van Alpen effect—oscillations of magnetisation.
- The Shubnikov–de Haas effect—oscillations of magnetoresistance.
- Oscillations in sample length, temperature, thermal conductivity, Peltier effect and thermoelectric voltage.

In general, the amplitude of the oscillations will increase with increasing magnetic field and decreasing temperature as the electron population described by the Fermi–Dirac distribution becomes more defined. The frequency of the oscillations can be used to infer information about the Fermi surface, as discussed in more detail in [1] and [2].

7.6 Examples of electric characterisation techniques in the literature

In the following sections, I will highlight several examples of research articles where electric characterisation techniques were used to investigate a given material. For each article, starter questions are provided, which are aimed at focussing the reader's attention on the key aspects of the article (with respect to this chapter).

7.6.1 In 'Negative Hall coefficient of ultrathin niobium in Si/Nb/Si trilayers'

Zaytseva *et al* investigated the structural and transport properties of Nb thin films sandwiched between Si [31].

Reading through the article, try to answer the following:

1. How were the films in this study produced?
2. Why study the transport properties of Nb thin films as a function of thickness?
3. Why deposit a layer of Si either side of the Nb layer?
4. Why would T_c depend on film thickness?
5. Why is lithography required for the 'Hall bar' structure?
6. What can the XRD patterns tell us about the structure/quality of the films?

The aim of this work was to investigate the effect of film thickness on the superconducting properties of ultrathin Nb. The authors show that the superconducting transition temperature, T_c, decreases significantly as the film thickness drops below 10 nm and attribute this to the impact of boundary scattering on the relaxation rate on the carriers in the trilayer.

7.6.2 In 'Simultaneous detection of the spin-Hall magnetoresistance and the spin-Seebeck effect in platinum and tantalum on yttrium iron garnet'

Vlietstra *et al* demonstrate simultaneous measurement of the spin Seebeck and spin Magnetoresistance effects using a lock-in detection technique [32].

Reading through the article, try to answer the following:

1. Define the inverse and normal spin Hall effects (V_{ISHE} / V_{SHE}).
2. What is the spin Hall angle, and how does it affect the observed V_{ISHE}?
3. What is the spin Seebeck effect?
4. Why does the SMR signal scale linearly with current whereas the SSE signal scales quadratically with current?
5. What is the impact of the Oersted field on the measurement?

The aim of this work was to demonstrate the applicability of current-induced heating for measurement of the spin Seebeck effect. The authors showed, using a lock-in technique that it was possible to simultaneously measure the spin Hall magnetoresistance and the spin Seebeck effect as low and high field contributions to the measured signal, respectively. In addition, they observed a second harmonic signal that they attributed to current-induced magnetic fields.

7.6.3 In 'Structural, electronic, and magnetic investigation of magnetic ordering in MBE-grown $Cr_xSb_{2-x}Te_3$ …'

Collins-McIntyre *et al* investigated the impact of Cr doping on Sb_2Te_3 thin films [33].

Reading through the article, try to answer the following:

1. What experimental methods have been used in this paper?
2. Define the Hall effect. How does the quantum Hall effect differ from this?
3. Why did the authors choose to study Cr doped Sb_2Te_3?
4. What crystal structure does this material form in? How did the authors determine this?
5. What are the magnetic contributions for this sample?
6. What are the key differences between XRR and PNR data?

The aim of this work was to explore the theorised development of long range magnetic order in Cr doped Sb_2Te_3 due to valence electrons (the van Vleck mechanism). The authors concluded it was possible to dope the films to $x = 0.42$ without any significant change to the structural quality (i.e. phase segregation). They showed that long-range magnetic order did exist in the doped films, with an increase in the Curie temperature, T_c, as Cr was increased.

7.6.4 In 'Transport magnetic proximity effects in platinum'

Huang *et al* investigated manifestation of magnetic proximity effects in magneto-resistance measurements of Pt deposited onto YIG [34].

Reading through the article, try to answer the following:

1. Define the inverse spin Hall effect.
2. What experimental techniques were used here?
3. What was the purpose of the patterned devices shown in figure 1(b) and (c)?
4. Define AMR. What is the relevance of this measurement here?
5. Why was YIG used as the magnetic layer in this study?

The aim of this work was to demonstrate the potential contamination of magneto-thermal measurements where Pt thin films are used as detectors of spin current (by the inverse spin Hall effect). The authors showed, by measuring the magnetresistance of Pt, Cu, and Py thin films deposited on the insulating YIG and non-magnetic Si substrates that a clear magnetic signal was induced in the Pt films.

7.6.5 In 'Topology-driven magnetic quantum phase transition in topological insulators'

Zhang *et al* observed a magnetic quantum phase transition in Cr-doped $Bi_2(Se_xTe_{1-x})_3$ thin films [35].

Reading through the article, try to answer the following:

1. What are the key features of a topological insulator?
2. What experimental methods are described here?
3. Identify the key differences between magnetotransport measurements in Te and Se doped BiCr.
4. Why is ARPES a useful tool for investigating topological materials?
5. What are the key features of figure 3? In particular, identify the Dirac point.

The aim of this work was to explore the interplay of topological surface states and magnetic ordering. The authors demonstrated a shift in the Dirac cone seen with ARPES measurements to a distinct band gap on Cr doping, accompanied by a loss of ferromagnetism. This demonstrates the tunability of a topological-magnetic material that could be useful in future spintronic applications.

7.7 Questions

7.7.1 Extracting information from bulk measurements

1) ***Resistivity***

(a)–(c) For the measurement geometries shown in figure Q7.1, calculate the resistance and resistivity using the values give in table 7.3.

(d) Identify the differences between the measurement for (a) and (b). Would this lead to significant error in this instance?

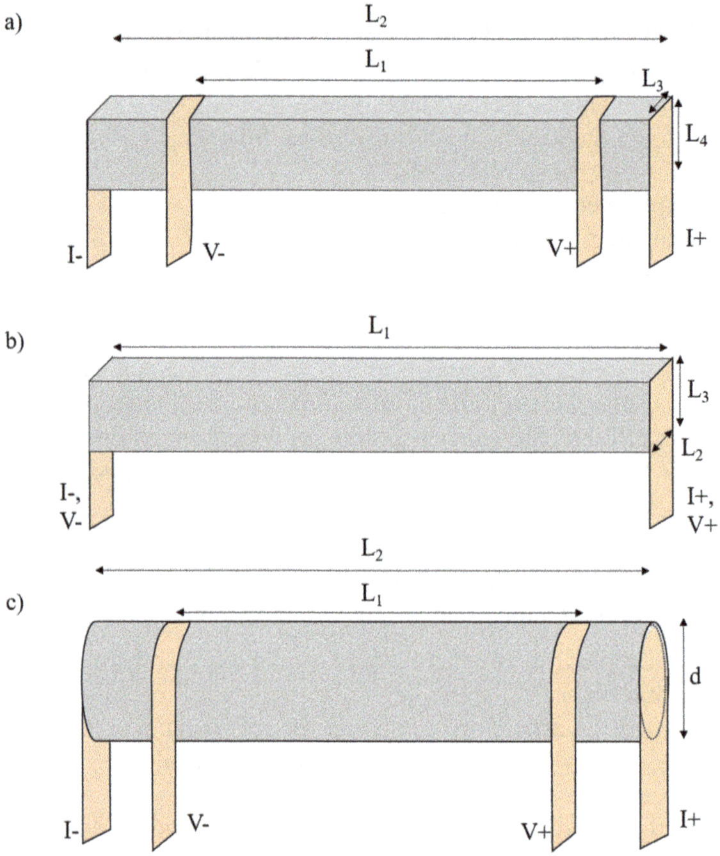

Figure Q7.1. Geometry of some resistivity measurements.

Table 7.3. Summary of resistivity measurements.

	I	V	L_1	L_2	L_3 (d)	L_4
(a)	0.1 mA	4.3 mV	5 mm	7 mm	2 mm	2 mm
(b)	0.1 μA	52.7 mV	7 mm	—	2 mm	2 mm
(c)	0.1 A	67.8 mV	26 mm	30 mm	(100 μm)	—

2) ***Thermal transport measurements***

Steady state thermoelectric measurements of a bulk sample were obtained for the following sample geometry shown in figure Q7.2. If for a single measurement, the values shown in table 7.4 were obtained:

a) Determine the Seebeck coefficient, S.
b) Estimate the thermal conductivity, κ.
c) Calculate the resistivity, ρ.

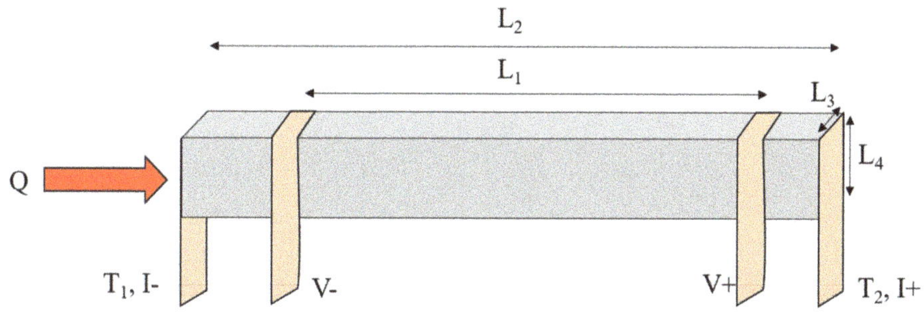

Figure Q7.2. Geometry of a Seebeck measurement.

Table 7.4. Summary of Seebeck measurements.

Q	I	T_1	T_2	V	L_1	L_2	L_3	L_4
0.02 W	0.1 mA	290 K	295 K	170 mV	5 mm	7 mm	2 mm	2 mm
0.02 W	0 mA	290 K	295 K	52 mV	5 mm	7 mm	2 mm	2 mm

d) Why might the Seebeck coefficient need to be defined with reference to another material?

Conventional thermoelectric generators use the Seebeck effect to convert a temperature difference intro useful power. This is parametrised by the figure of Merit, $zT = S^2T/\rho\kappa$.

e) Calculate zT for this sample.

7.7.2 Extracting information from thin film measurements

1) ***Sheet resistivity***

a) Determine the sheet resistance for the data in table 7.5.

Table 7.5. Van der Pauw measurements of a thin film.

Current (mA)	Voltage (mV)
$B = 0$ G	
$I_{21} = 0.1$	$V_{34} = 0.756$
$I_{12} = 0.1$	$V_{43} = 0.757$
$I_{32} = 0.1$	$V_{41} = 14.25$
$I_{23} = 0.1$	$V_{14} = 14.23$
$I_{43} = 0.1$	$V_{12} = 0.756$
$I_{34} = 0.1$	$V_{21} = 0.758$
$I_{14} = 0.1$	$V_{23} = 14.23$
$I_{41} = 0.1$	$V_{32} = 14.25$

Table 7.6. Van der Pauw and Hall measurements of a thin film.

Current (mA)		Voltage (mV)
	$B = 0$ G	
$I_{21} = 0.1$		$V_{34} = 0.756$
$I_{12} = 0.1$		$V_{43} = 0.757$
$I_{32} = 0.1$		$V_{41} = 0.756$
$I_{23} = 0.1$		$V_{14} = 0.762$
$I_{43} = 0.1$		$V_{12} = 0.756$
$I_{34} = 0.1$		$V_{21} = 0.758$
$I_{14} = 0.1$		$V_{23} = 0.756$
$I_{41} = 0.1$		$V_{32} = 0.762$
	$B = 10\,000$ G	
$I_{13} = 1$		$V_{24} = 0.23$
$I_{31} = 1$		$V_{42} = 0.25$
$I_{42} = 1$		$V_{13} = 0.22$
$I_{24} = 1$		$V_{31} = 0.24$
	$B = -10\,000$ G	
$I_{13} = 1$		$V_{24} = -0.24$
$I_{31} = 1$		$V_{42} = -0.26$
$I_{42} = 1$		$V_{13} = -0.23$
$I_{24} = 1$		$V_{31} = -0.25$

b) If the assumption was made that $R_A \sim R_B$, how would the value of R_s differ?

2) ***Hall measurements***

a) Determine the sheet resistance, carrier concentration (and type), and mobility for the data in table 7.6.

b) If the assumption was made that $R_A \sim R_B$, how would the value of R_s differ?

c) If the film was 54 nm thick, determine the resistivity and carrier concentration density.

References

[1] Singleton J 2001 *Band Theory and Electronic Properties of Solids, Oxford Master Series in Condens. Matter Phys.* (Oxford: Oxford University Press)

[2] Kittel C 1986 *Introduction to Solid State Physics* 6th edn (New York: Wiley) p 185

[3] Grahn H T 1999 *Introduction to Semiconductor Physics* (Singapore: World Scientific)

[4] Tinkham M 1996 *Introduction to Superconductivity* 2nd edn (New York: McGraw Hill)

[5] Hasan M Z and Kane C L 2010 *Rev. Mod. Phys.* **82** 3045

[6] Gooth J, Shierning G, Felser C and Nielsch K 2018 *MRS Bull.* **43** 187–92

[7] Chang C-Z *et al* 2013 *Science* **340** 167–70

[8] Nagaosa N *et al* 2010 *Rev. Mod. Phys.* **82** 1539–92

[9] Hoffman A 2013 *IEEE Trans. Magn.* **49** 5172–93

[10] Wang Y *et al* 2014 *Appl. Phys. Lett.* **105** 152412
[11] Berger A J *et al* 2018 *Phys. Rev.* B **98** 024402
[12] Valenzuela S O and Tinkham M 2006 *Nature* **442** 176–9
[13] Uchida K *et al* 2008 *Nature* **455** 778–81
[14] Sola A *et al* 2019 *Sci. Rep.* **9** 2047
[15] White D R *et al* 1996 *Metrologia* **33** 325
[16] Qu J *et al* 2015 *Metrologia* **52** S242
[17] *Zurich Instruments White Paper: Principles of Lock-in Detection and State of the Art* November 2016
[18] Miccoli I *et al* 2015 *J. Phys. Condens. Matter* **27** 223201
[19] Van der Pauw L J 1958 *Philips Res. Rep.* **12** 1–9
[20] Chwang R, Smith B J and Crowell C R 1974 *Solid-State Electronics* **17** 1217–27
[21] Ramadan A A, Gould R D and Ashour A 1994 *Thin Solid Films* **239** 272–5
[22] Damascelli A, Hussain Z and Shen Z 2003 *Rev. Mod. Phys.* **75** 473–541
[23] Fauster T and Dose V 1986 Inverse photoemission spectroscopy *Chemistry and Physics of Solid Surfaces VI Springer Series in Surface Sciences* vol 5 (Berlin: Springer) 483–507
[24] Jordan K *et al* 2006 *Phys. Rev.* B **74** 085416
[25] Zandvliet H J W and van Houselt A 2009 *Annu. Rev. Anal. Chem.* **2** 37–55
[26] Krenner W, Kühne D, Klappenbergr F and Barth J V 2013 *Sci. Rep.* **3** 1454
[27] Soulen R J *et al* 1998 *Science* **282** 85–8
[28] Pannetier B and Courtois H 2000 *J. Low Temp. Phys.* **118** 599
[29] Upadhyay S K, Palanisami A, Louie R N and Buhrman R A 1998 *Phys. Rev. Lett.* **81** 3247
[30] Strijkers G J, Ji Y, Yang F Y, Chien C L and Byers J M 2001 *Phys. Rev.* B **63** 104510
[31] Zaytseva I *et al* 2014 *Phys. Rev. B* **90** 060505(R)
[32] Vlietstra N *et al* 2014 *Phys. Rev. B* **90** 174436
[33] Collins-McIntyre L J *et al* 2016 *Europhysics Letters* **115** 27006
[34] Huang S Y *et al* 2012 *Phys. Rev. Lett.* **109** 107204
[35] Zhang J *et al* 2013 *Science* **339** 1582

IOP Publishing

Kelly Morrison

Chapter 8

Magnetic characterisation

In this chapter, the underlying theory for different types of magnetic order will be presented. This will be complemented by a broad overview of the unit systems typically used, as well as several characterisation techniques. Starting with the fundamentals of magnetic characterisation (global observations) we shall then follow with a discussion of the magnetic interactions that lead to such long-range magnetic order. Examples of magnetometry and magnetic imaging methods will then be presented. The aim will be to provide the reader with a broad understanding of magnetism and magnetic units required to start characterising the broader family of magnetic materials. Finally, some examples of these approaches will be highlighted by selected publications followed by questions specific to data analysis.

8.1 Fundamentals of magnetism

Whilst the physical origin of magnetism is often misunderstood, its basic properties—the ability to attract certain metals from a short distance, and the tendency to align with the Earth's magnetic field—have made magnets a valuable commodity over the centuries. For example, the magnetic ore, lodestone, is a form of iron oxide (Fe_3O_4) that occurs naturally as a magnet. Until the 18th century, where manufacture of permanent magnets from steel was first industrialised (based on observations from William Gilbert's *De Magnete*, which was published in Latin in 1600 and later translated) [1, 2], lodestone was commonly used as the source material for simple applications such as the compass.

The first compass was likely Chinese in origin [1, 3], where descriptions of a magnetic compass needle have been found from as early as AD 1080 compared to the first European mention in AD 1190 [1]. To build an accurate compass would require a magnetised needle that could be suspended by a fine thread of thin silk. As lodestone is so brittle (and thus difficult to shape) this needle was often some form of

doi:10.1088/2053-2563/ab2df5ch8

iron, that required 're-magnetising' by stroking with a lump of lodestone at frequent intervals.

The key observations made by Gilbert are now commonly accepted knowledge for a permanent magnet:

- That it has an 'affinity' for magnetite or more metallic iron.
- That it can exert a force (attraction) on another material without touching it.
- That there are two 'poles', one of which will attract, one of which will repel.
- That dividing the lodestone (or magnet) in two produces two smaller magnets.
- That an artificial magnet could be produced by stroking an iron rod with one end (pole) of a lodestone (or other permanent magnet).

It was not until the observations of Oersted and Faraday that the interplay between electricity and magnetism was fully unravelled, and electromagnets were developed. The result of this work can be found in nearly every aspect of everyday life, from the small magnets that we use as clasps on bags, or note holders on fridges, to the permanent magnets that are integral to the design of motors (cars, washing machines, vacuum cleaners) and generators (wind turbines). Indeed, even transformers require a common magnetic core in order to step up (or down) an alternating current (AC). As such, there are several areas of research focussed on magnetic materials optimised for different applications, all of which require a consistent approach to measurement of the magnetic moment: the strength of the magnet itself. In order to characterise the magnetic properties of these materials there are several key concepts and units that need to be first understood. These include:

1) Strength of a magnetic field
   - **magnetic flux density (induction),** ***B***—relates to the magnetic field (vector).

     Commonly used units include Tesla [T] and Gauss [G].
   - **magnetic field strength,** ***H***—relates to the magnetic field (vector).

     Commonly used units include Amperes per metre [A m^{-1}] and Oersted [Oe].

     Where B and H are scaled versions of one another

     In SI units [T, A m^{-1}]

$$B = \mu_0(M + H) \tag{8.1}$$

In CGS units [G, emu cm^{-3}, Oe]

$$B = H + 4\pi M \tag{8.2}$$

μ_0 is the permeability of free space and M is the magnetisation of the material that exists at the point where B or H is measured.

2) Magnetic strength of a material
   - **magnetisation, σ, *M***—quantifies the magnetic strength of a material (magnetic moment per unit volume or mass).

Commonly used units include emu cm^{-3}, emu g^{-1} and $Am^2\ kg^{-1}$.

- **magnetic moment, *m***—describes the magnitude of the magnetic force (in a given direction). This is often referred to as the dipole moment.
  Commonly used units include emu, Joules per Tesla [$J\ T^{-1}$] and Am^2.

$$M = m/V = B/\mu_0 \tag{8.3}$$

Historically the magnetic moment was defined as the moment generated perpendicular to a current carrying loop.

3) Response of a magnetic material to magnetic field

- **magnetic susceptibility, χ_p, χ_v**
  Commonly used units include emu cm^{-3}, emu g^{-1}, or cm g^{-1}.

$$\chi = M/H \tag{8.4}$$

$$\chi = \lim_{B\to 0} \frac{\mu_0 M}{B} \propto \begin{cases} \dfrac{1}{T + T_N} & \text{Antiferromagnet} \\ \dfrac{1}{T - T_C} & \text{Ferromagnet} \end{cases} \tag{8.5}$$

- **magnetic permeability, μ, μ_r**—where $\mu_r = \mu/\mu_0$ is the relative permeability, and μ_0 ($= 4\pi \times 10^{-7}$) is the permeability of free space.

$$\mu = B/H \tag{8.6}$$

$$\mu_r = 1 + \chi \tag{8.7}$$

8.1.1 Some units

The trouble with magnetic measurements is that even to this day, the use of standard (SI) units is not always the default. Quite often, the historically defined CGS (centimetre-gram-second) units such as 'emu' or 'Oersted' will be used, requiring conversion from one unit system to another. The most common error in analysing magnetic data is to either forget to convert to SI units (when using equations defined in this way), or to do it incorrectly. This is further complicated by the different way that induction (*B*) and field strength (*H*) scale to one another in these unit systems, as seen in equations (8.1) and (8.2).

A detailed table indicating the conversion ratio from CGS to SI units for various magnetic quantities is given in appendix B. Of note, however, are the following useful observations:

M: 1 emu g^{-1} (CGS) = 1 $Am^2\ kg^{-1}$ (SI)
B: 1 Gauss (CGS) = 10^{-4} Tesla (SI); conversely 1000 Gauss = 0.1 Tesla
H: 1 Oersted (CGS) = $10^3/4\pi$ ($A\ m^{-1}$) (SI)

The shorthand for Gauss, Tesla and Oersted are G, T, and Oe, respectively; by definition, 1 G = 1 Oe.

8.1.2 Types of magnetic order

Whilst most are familiar with bar magnets such as iron, there are several different types of magnetism that can often be determined by their behaviour under changing magnetic fields or temperature. For example, the relative permeability, μ_r, measures the magnetic response of a material to an applied magnetic field (compared to free space) and can be used to define three basic magnetic states:

1. Ferromagnetism (FM), $\mu_r \gg 1$
2. Paramagnetism (PM), $\mu_r \sim 1$
3. Diamagnetism, $\mu_r < 1$

FM describes the magnetic state in a material—a ferromagnet (FM)—with strong magnetic exchange. In order words, local magnetic moments in the material show the tendency to align with respect to one another and to strongly reorient themselves with respect to an external magnetic field. This can result in a material that acts as the archetypal permanent magnet: with the ability to attract ferrous metals and align itself to the Earth's magnetic field. In addition to FM, another common type of magnetic order can arise where layers of atoms align antiparallel with respect to one another. This is known as antiferromagnetism (AFM).

PM describes a weakly responsive material—a paramagnet (PM)—where localised magnetic moments will weakly align themselves with an applied magnetic field.

Diamagnetism describes a magnetic state whereby, the material—a diamagnet—shows the tendency to align anti-parallel to an applied magnetic field. The result of this is a combined reduction in magnetic flux density, B, within the material. The Meissner state in superconductors is a common example of this: when cooled below their superconducting temperature the superconductor will generate a magnetic field that opposes any external magnetic field (below a critical value) by generating screening currents in the surface of the material. This opposing magnetic field will act on the external magnetic field as a force pushing against it, thus leading to magnetic levitation (when the gravitational force is compensated by the magnetic repulsion). Other examples of diamagnetic materials include pyrolytic graphite, which is strongly diamagnetic along its c-axis (perpendicular to easily cleaved planes) and water (exhibited quite nicely by videos of strawberries or raspberries floating in a large magnetic field).

FM, PM and AFM are commonly characterised by inverse magnetic susceptibility measurements (by mass, χ_p, or volume, χ_v), where a linear dependence with respect to temperature is observed in the paramagnetic state (known as the Curie–Weiss law). This is summarised in figure 8.1; some values of permeability and susceptibility for common materials are also given in table 8.1.

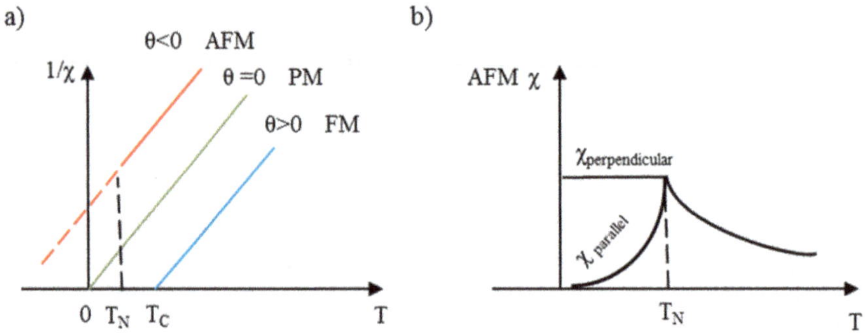

Figure 8.1. (a) Magnetic susceptibility, χ, behaviour for the three classic magnetic systems: ferromagnets, antiferromagnets and paramagnets. T_N is the Néel temperature below which antiferromagnetic order appears, T_C is the Curie temperature, below which ferromagnetic order appears. Whilst this linear relationship for $1/\chi$ might in general be valid, note that close to T_C or T_N it could diverge slightly due to magnetovolume or other coupling effects. (b) Magnetic susceptibility for an antiferromagnetic where the magnetic field is applied parallel or perpendicular to the direction of spontaneous magnetisation.

Table 8.1. Relative magnetic permeability and corresponding susceptibilities for some common materials (some data taken from http://www.engineeringtoolbox.com/), where μ = magnetic permeability, μ_r = relative permeability, χ_p = magnetic susceptibility, χ_r = relative susceptibility.

Material	μ (H m^{-1})	μ_r (unitless)	χ_p (10^{-8} m^3 kg^{-1})	χ_v (unitless)
Aluminium	$1.256\,65 \times 10^{-6}$	1.000 022	0.811	2.2×10^{-5}
Bismuth	$1.256\,43 \times 10^{-6}$	0.999 834	−1.70	-16.6×10^{-5}
Cobalt-iron	2.3×10^{-2}	18 000	$\sim 2.25 \times 10^5$	17 999
Copper	$1.256\,629 \times 10^{-6}$	0.999 994	−0.107	-0.6×10^{-5}
Hydrogen	$1.256\,6371 \times 10^{-6}$	1	−2.49	0
Iron (99.8% pure)	6.3×10^{-3}	5000	—	4999
Iron (99.95% pure, annealed in H)	0.25	200 000	—	199 999
Mu metal	6.3×10^{-2}	50 000	—	49 999
Neodymium based magnet (NdFeB)	1.32×10^{-6}	1.05	—	0.05
Permalloy (80% Ni:20% Fe)	1.0×10^{-2}	8000	—	7999
Pyrolitic graphite	-4.9×10^{-5}	0.9996	—	-4×10^{-4}
Superconductors	0	0	-10^8	−1
SiO_2 (glass)	$\sim 4\pi \times 10^{-7}$	~0.999 988 22	−0.62	-1.178×10^{-5}
Vacuum	$4\pi \times 10^{-7}$	1	0	0
Water	$1.256\,627 \times 10^{-6}$	0.999 992	−0.8	-0.8×10^{-5}

More complex magnetic states such as ferrimagnetism, helimagnetism, metamagnetism, skyrmions and spin glasses arise from the competition between different types of magnetic exchange, geometric constraints, and magnetic anisotropy, as described in more detail later.

8.1.3 Magnetic domain theory

Even given a known ferromagnet (such as iron) it is not always apparent that it is magnetic, i.e. not all iron will be able to pick up ferrous items such as the odd spare screw. This is because there are various competing interactions within a magnetic material that can lead to the formation of magnetic domains: large areas of locally magnetised material that may not necessarily be aligned with respect to each other. The result of this is that overall (globally) the material appears non-magnetic, whereas locally it is. It is for this reason that the active element of a compass needed to be recharged with (lodestone) until stable permanent magnets could be reliably produced.

The formation of domains is controlled by five competing energies, where the system will want to find the lowest energy state, as summarised by:

$$E = E_{\text{magnetostatic}} + E_{\text{exchange}} + E_{\text{anisotropy}} + E_{\text{SA}} + E_{\text{Zeeman}} \tag{8.8}$$

where each energy term can be defined as follows:

Magnetostatic, $E_{\text{magnetostatic}}$: the energy required to form external magnetic (stray) fields.

Magnetic exchange, E_{exchange}: the energy between two neighbouring spins, where alignment or anti-alignment is favoured for ferromagnets and antiferromagnets, respectively. This is defined by the exchange integral, J (section 8.2.2).

Magnetocrystalline anisotropy, $E_{\text{anisotropy}}$: where the relative positions of elements within a crystal lattice are not necessarily isotropic (same in all directions). As such, the magnetic exchange integral, J, may vary dependent on the crystal axis direction. This results in what are termed 'easy' and 'hard' axes, where atoms may prefer to align (or not).

Shape anisotropy, E_{SA}: where demagnetising effects cause the shape of the material to affect the preferred magnetisation direction (see section 8.1.4).

Zeeman, E_{Zeeman}: where the application of a magnetic field will favour a given magnetisation direction.

Whilst the magnetic exchange interaction encourages alignment of neighbouring spins, asymmetry with regards to the exchange interaction along different directions within the crystal lattice (anisotropy), or the energy required to form external magnetic fields (magnetostatic) can encourage competing magnetisation directions. As a result, where there is no external magnetic field a magnetic material may form closure domains such as the example given in figure 8.2. Notice, that even after the external magnetic field is removed, the sample does not immediately return to the original closure domain. This is because it requires energy to reorient the spins in the magnetic material leading to field history or *hysteresis*.

There are several different types of magnetic domain patterns that can form depending on the complexity of the magnetic exchange interactions, anisotropy, and geometry. For example the competition between the different energy terms in face centred cubic Fe thin films can lead to complex maze-like domains [4]. These will typically be observed with some form of magnetic imaging such as the magneto optic Kerr effect (MOKE), as discussed in section 8.4. For more information on magnetic

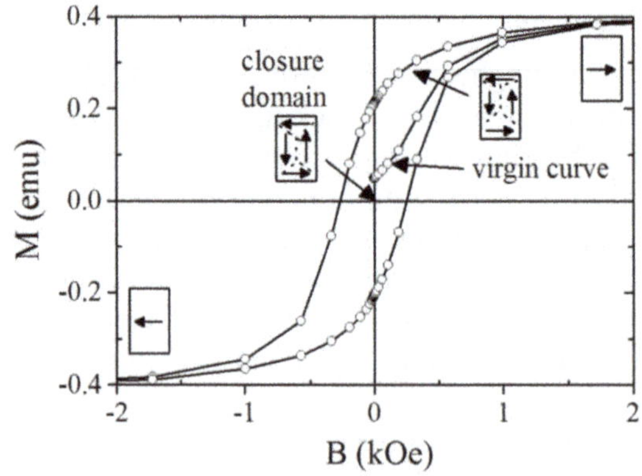

Figure 8.2. Magnetic domains in a ferromagnet under applied magnetic field. The 'virgin curve' refers to the first time a magnetic field is applied, where the material changes from a closure domain (no observable magnetisation) to magnetised in one direction (saturation).

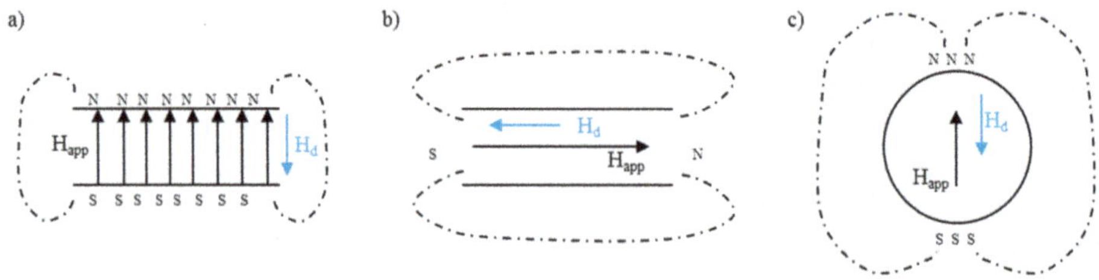

Figure 8.3. Demagnetising field in three common geometries: (a) thin plate (out of plane), (b) thin plate (in plane), and (c) sphere. Magnetic poles form at the surface of a material where there is a change in magnetisation as a magnetic field, H_{app}, is applied. This leads to an opposing magnetic field, H_{d}.

domains the reader is referred to the comprehensive body of work by Hubert and Shäfer [5].

8.1.4 Demagnetisation effects

Whilst the application of a magnetic field may cause the magnetic moments within a material to align (with respect to the field direction), the total field inside the sample will not simply be a sum of the applied field and the magnetisation of the material (in relevant units). This is because a demagnetising field is generated that varies with the shape of the material, and its orientation with respect to the applied field.

To understand why this is, you need to consider what happens at the surface of the material, as shown in figure 8.3. In each geometry, the application of a magnetic field will result in alignment of magnetic moments, and consequently the appearance of 'North' and 'South' poles at the surface. If these 'poles' are separated by a short distance (narrow sample with field applied perpendicular to the long plane) then the field generated—the demagnetising field—will be relatively large (figure 8.3(a)). If these 'poles' are separated by a longer distance (narrow sample with field applied parallel to long plane) then the demagnetising field will be smaller (figure 8.3(b)).

The magnitude of the demagnetising field, H_d, will of course be dependent on the strength of the magnetisation in the material, M, as well as the geometry with respect to the applied magnetic field. It is defined by equation (8.9), where N is the demagnetisation factor.

$$H_d = -NM \tag{8.9}$$

In SI units H_d and M will be in units of A m^{-1}, and typical values of N for the examples given in figure 8.3 are (a) 0.9, (b) 0.1 and (c) 0.33. *Note that in CGS units the values of N are multiplied by a factor of 4π as the scaling between H and B is different.*

With regards to magnetic measurements, the impact of the demagnetising field is to broaden the apparent response of the magnetic material to applied magnetic fields. An example of uncorrected data is given in figure 8.4. As this is largely dependent on the geometry of the material it is an intrinsic effect, and should ideally be corrected for. To determine the actual magnetic field within the sample one would use:

$$H_{in} = H_{app} - NM \tag{8.10}$$

8.2 Magnetic interactions

At this point, we have to some extent discussed different types of magnetic order (PM, diamagnetism, FM and AFM), as can be defined by the magnetic susceptibility, χ, or relative permeability, μ_r. There is however a complex mix of magnetic order, as shown to some extent in figure 8.5. This often results from the competition of the different magnetic exchange interactions that will be outlined here.

The response of these materials to magnetic fields (i.e. during a measurement) are as follows:

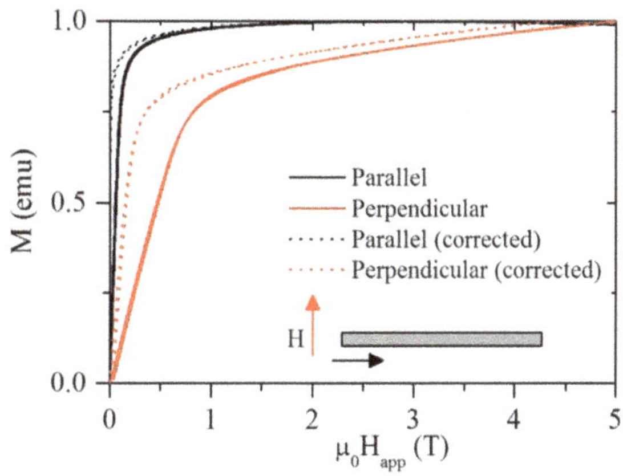

Figure 8.4. Measurement of the same sample mounted in the parallel and perpendicular (with respect to field) geometries. Dotted lines show data that has been corrected for demagnetising field. The difference between corrected measurements is due to magnetic anisotropy.

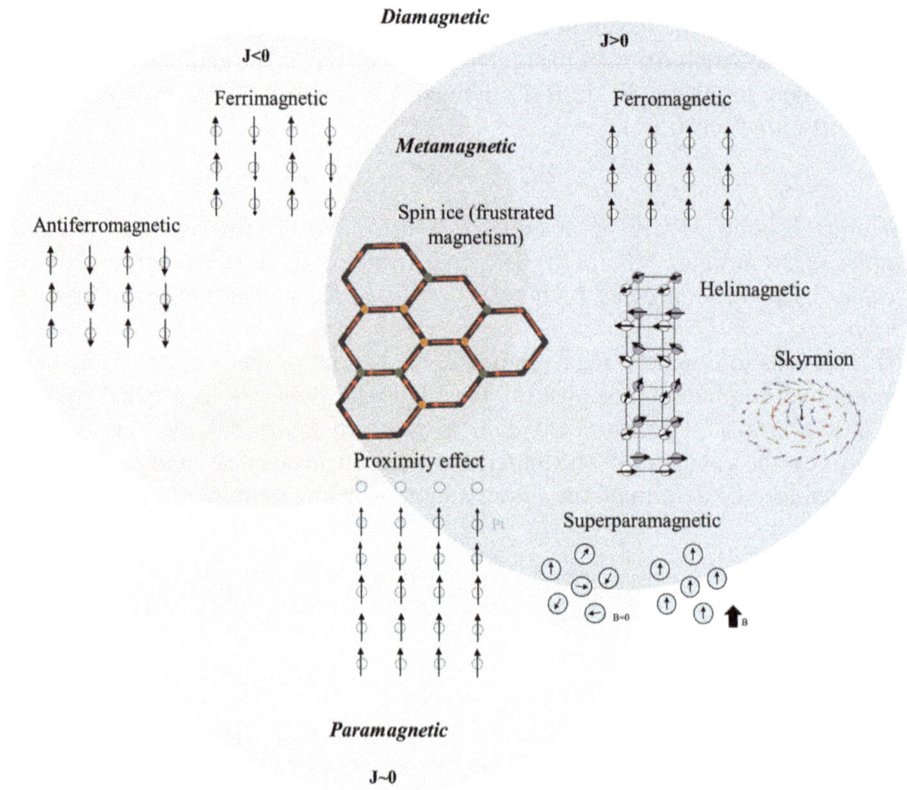

Figure 8.5. Interplay of magnetic exchange interactions leads to different manifestations of magnetic order.

- **Diamagnetic:** response of the material is to generate a magnetic field that opposes the applied magnetic field.
- **Paramagnetic:** spins will weakly align with an applied magnetic field.
- **Ferromagnetic:** magnetic moments will strongly align with an applied magnetic field.
- **Antiferromagnetic:** whilst the material may respond weakly to an applied magnetic field, the magnetic moment on individual atoms could be quite large. This arises due to an exchange interaction that favours anti-alignment of magnetic moment in a material (typically resulting in anti-aligned layers). This reduces the overall magnetic moment as on average these individual moments will largely cancel each other out.
- **Ferrimagnetic:** similar to AFM, where the individual 'anti-aligned' layers have different magnetic moments, therefore, the average magnetic moment is higher.
- **Frustrated magnetism:** material systems where there is competition between two or more magnetic interactions (e.g. antiferromagnetic exchange versus ferromagnetic exchange). Whilst the ground state might be, for example, antiferromagnetic order, this may not be observed due to the chosen field or

temperature history of measurements (see zero field cooling, field cooled cooling and field cooled warming measurements). Spin glasses such as observed in RMnGa (where R is a rare earth) are a common example of this [6]. Other examples of frustrated magnetism include the artificial spin ice systems, where careful design of sample geometry can also result in magnetic frustration [7].

- **Metamagnetic:** material systems where there is a well-defined transition from one magnetic state to another, usually via a first order phase transition (a discontinuity is observed in magnetometry measurements). This is often the result of coupling to another parameter such as crystal structure. For example, the metamagnet $Gd_5Ge_2Si_2$ exhibits a magnetostructural phase transition where the magnetism changes from antiferromagnetic to ferromagnetic in conjunction with a monoclinic to orthorhombic structural phase transition [17].
- **Helimagnetic:** material which exhibits magnetic order that is canted by some angle along a specific crystal axis. This can be commensurate (an integer multiple of unit cells) or incommensurate.
- **Proximity effect (incipient FM):** magnetism induced in a material placed in proximity to another magnet. A common example of this is Pt, which is close to the Stoner criterion—a condition required for FM to arise in a simple material system. The induced magnetic moment typically decays exponentially from the interface between the magnetic and non-magnetic samples.
- **Superparamagnetic:** observed in nanoparticles small enough for single magnetic domains to form. Whilst the particles themselves might be randomly aligned (no long-range order) they will respond strongly to magnetic field much like a ferromagnet but without the hysteresis associated with magnetic domain reorientation.
- **Skyrmions:** a topologically protected magnetic state that resembles a magnetic vortex structure (see figure 8.5).

8.2.1 Magnetic dipolar interaction

The simplest magnetic element is the dipole moment (the existence, or not, of magnetic monopoles will not be the focus here). This describes the classical idea that you have a North and South pole, and that these poles will attract or repel each other dependent on their respective orientations.

The energy associated with two magnetic dipoles, $\boldsymbol{\mu}_1$, $\boldsymbol{\mu}_2$, separated by distance, $\mathbf{r}$ is given by:

$$E = \frac{\mu_0}{4\pi r^3}\left[\boldsymbol{\mu}_1 \cdot \boldsymbol{\mu}_2 - \frac{3}{r^2}(\boldsymbol{\mu}_1 \cdot \mathbf{r})(\boldsymbol{\mu}_2 \cdot \mathbf{r})\right] \tag{8.11}$$

This can of course also be written as:

$$E = \frac{\mu_0}{4\pi r^3}\left[\mu_1\mu_2 \cos\theta - \frac{3}{r^2}(\mu_1 r \cos\alpha)(\mu_2 r \cos\beta)\right] \tag{8.12}$$

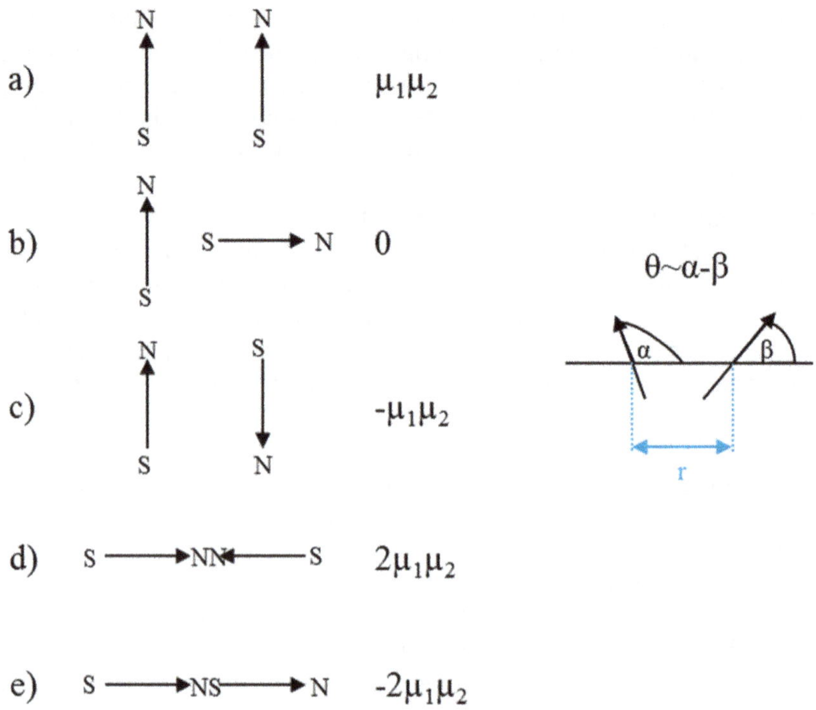

Figure 8.6. Dipole arrangements and the corresponding magnetic dipolar interaction energy, E (normalised by $\mu_0/4\pi r^3$).

where θ, α and β describe the angle between vectors $\boldsymbol{\mu_1}$ and $\boldsymbol{\mu_2}$, $\boldsymbol{\mu_1}$ and $\mathbf{r}$, and $\boldsymbol{\mu_2}$ and $\mathbf{r}$, respectively. A schematic of this for various examples is given in figure 8.6. For example, where the dipoles are aligned (figure 8.6(a)), $\theta = 0°$, $\alpha = \beta = 90°$, and equation (8.11) reduces to:

$$E = \frac{\mu_0}{4\pi r^3}[\mu_1\mu_2] \tag{8.13}$$

In order to assess the relative magnitude of this interaction we can try inserting some sensible numbers. It is useful at this point to remember that the magnetic unit of reference—the Bohr magneton, μ_B—is defined as:

$$\mu_B = \frac{2h}{4\pi m_e} = 9.274 \times 10^{-24}\,[\mathrm{A\ m^{-2}}] \tag{8.14}$$

and that the magnetic moment of a magnetic atom such as Fe can be as high as 4 μ_B. So, it is reasonable to assume for a pair of dipole moments that μ_1, $\mu_2 = 1\ \mu_B$ and $r = 1$ Å. Substitution of these values into equation (8.11) yields an energy of 8.6×10^{-24} J, which is equivalent to ~1 K. This tells us that whilst the dipole moment will act to align two magnetic dipoles, the exchange energy is too low to be stable for temperatures much greater than 1 K. In other words, the magnetic dipolar interaction is unlikely to result in long-range magnetic order.

8.2.2 The exchange integral

If the magnetic dipole moment is not enough to support long-range magnetic order at temperatures >1 K, what is? For this question we need to start to consider the impact of quantum mechanics: that electrons have spin, and that this spin couples with the angular momentum of the nucleus (for background reading refer to [8, 9]). It stands to reason, therefore, that these electrons could also drive magnetic ordering, typically during what we would call an exchange (of electrons). The first step in determining the energy involved in electron exchange (and whether this is influenced by its magnetic state) is to express each electron state as a wavefunction, ψ, that describes the probability of the electron being at a particular position and with specific spin. The simplest case to start with is the two-electron model.

Let us first denote electron 1 as being in state $\psi_a(\mathbf{r}_1)$, and electron 2 in state $\psi_b(\mathbf{r}_2)$. We need to consider what will happen if we exchange these electrons (1 and 2), and how we might construct a master wavefunction for them based on this: otherwise referred to as *particle exchange symmetry*. A summary of the concepts considered during this process is given in figure 8.7. The result of this consideration is the singlet and triplet wavefunctions, ψ_S and ψ_T:

$$\psi_S = \frac{1}{\sqrt{2}}[\psi_a(\mathbf{r}_1)\psi_b(\mathbf{r}_2) + \psi_a(\mathbf{r}_2)\psi_b(\mathbf{r}_1)]\,\chi_S \tag{8.15}$$

$$\psi_T = \frac{1}{\sqrt{2}}[\psi_a(\mathbf{r}_1)\psi_b(\mathbf{r}_2) - \psi_a(\mathbf{r}_2)\psi_b(\mathbf{r}_1)]\,\chi_T \tag{8.16}$$

where χ_S, χ_T denote the combined spin of the singlet and triplet wavefunctions and $\chi_S = 0$, $\chi_T = 1$[1].

Once we have defined a wavefunction for the system then we can calculate the energy associated with exchange of electrons, as long as the Hamiltonian, $\hat{H}$, is known. For the singlet state (symmetric exchange):

$$E_S = \int \psi_S^* \hat{H} \psi_S \mathrm{d}\mathbf{r}_1 \mathrm{d}\mathbf{r}_2 \tag{8.17}$$

For the triplet state (antisymmetric exchange):

$$E_T = \int \psi_T^* \hat{H} \psi_T \mathrm{d}\mathbf{r}_1 \mathrm{d}\mathbf{r}_2 \tag{8.18}$$

We can also use this to determine the difference in energy between the two states. This is often expressed as the exchange integral, J:

$$J = \frac{E_S - E_T}{2} = \int \psi_a^*(\mathbf{r}_1)\psi_b^*(\mathbf{r}_2)\hat{H}\psi_a(\mathbf{r}_2)\psi_b(\mathbf{r}_1)\mathrm{d}\mathbf{r}_1\mathrm{d}\mathbf{r}_2 \tag{8.19}$$

[1] Recall the calculation of the coherent neutron scattering length, b_c, which was spin dependent, in chapter 3.

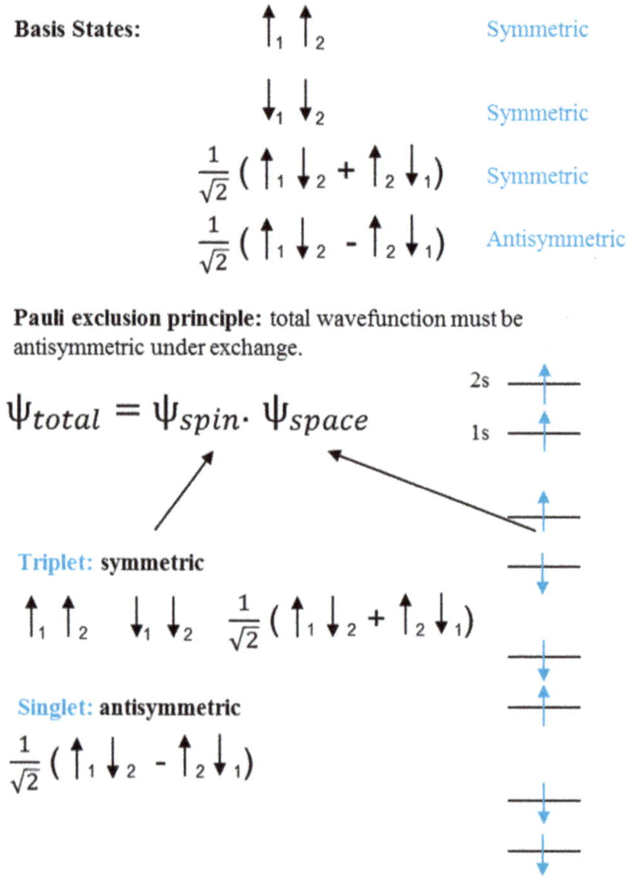

Figure 8.7. The antisymmetric wavefunction: singlet and triplet states. The Pauli exclusion principle requires the wavefunction to be antisymmetric under exchange—this can be achieved with a combination of symmetric/antisymmetric spin states and antisymmetric/symmetric space states.

There are several different Hamiltonians that are commonly used to describe magnetic exchange, one of which is the Heisenberg model, where the energy exchange between two neighbouring spins, S_i, S_j, is considered:

$$\hat{H} = -\sum\nolimits_{ij} J_{ij} \boldsymbol{S}_i \cdot \boldsymbol{S}_j = -2\sum\nolimits_{i>j} J_{ij} \boldsymbol{S}_i \cdot \boldsymbol{S}_j \tag{8.20}$$

Here, the summation indicates that the energy from neighbouring spins is considered for all spins in the system. The solution of equations (8.17)–(8.20) is not trivial, however, there are some conclusions that we can draw from it:

1. If $J < 0$, then $E_S < E_T$. This means that the singlet state is lower energy and leads to antiferromagnetic (anti-aligned) exchange.
2. If $J > 0$, then $E_S > E_T$. This means that the triplet state is lower energy and leads to ferromagnetic (aligned) exchange.

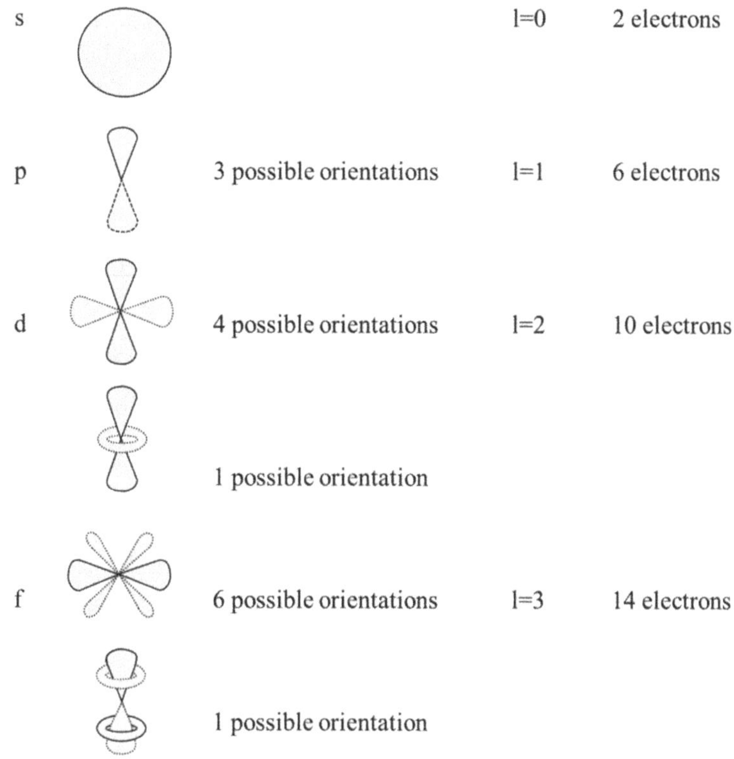

Figure 8.8. Sketch of s, p, d, f electron orbitals and their relationship with the fill levels (of electrons) per shell.

8.2.3 Direct exchange

Much like the dipole interaction, it can be imagined that magnetic order might arise from a direct exchange of electrons. In this case, the exchange of electrons between neighbouring atoms would lead to magnetic order, due to the interaction between the electron spins and the angular momentum of the atoms they move between. For this to occur, however, there needs to be overlap of neighbouring magnetic orbitals (see figure 8.8), such as with transition metal (d band) elements: Ni, Co, Fe, Mn.

More often than not, however, direct exchange is not enough to completely describe the magnetic interaction in a real system. For example, if the system is metallic, then the conduction band will result in a delocalisation of electrons. This leads to itinerant (i.e. delocalised) magnetism that will no longer be sufficiently described by the direct exchange mechanism. If, on the other hand, there is not enough orbital overlap between magnetic atoms (as is the case with the 4f shell of the Lanthanide rare earth series), then direct exchange can also not occur.

8.2.4 Indirect metallic exchange (RKKY interaction)

The Ruderman, Kittel, Kasuya and Yosida (RKKY) interaction [10], describes magnetic exchange mediated by electrons in the conduction band of a material (ionic solid). In this case, a localised magnetic moment (for example an atom with nuclear magnetic moment, or localised d- and f-shell orbitals), spin polarises a conduction electron in proximity to it, which in turn couples to the next (neighbouring) magnetic moment a distance, *r*, away. As this exchange interaction relies on conduction electrons it will be influenced by the shape of the Fermi surface.

For a spherical Fermi surface (Gd is a good example and is ferromagnetic until $T_c = 293$ K), the exchange integral, *J*, that describes the RKKY interaction is given by:

$$J_{\mathrm{RKKY}}(r) \propto \frac{\cos(2k_{\mathrm{F}}r)}{r^3} \tag{8.21}$$

The cosine function here suggests that *J* can oscillate between positive and negative values. This means that the type of interaction (ferromagnetic or antiferromagnetic) will depend on separation of the magnetic atoms.

For example, *ab initio* calculations of the exchange energy for MnP has been shown to exhibit an oscillation between AFM and FM ground states, as demonstrated by figure 3 in [11], and summarised here in figure 8.9. This has been suggested to be largely due to the RKKY interaction. This type of interaction is particularly important in rare earth compounds such as $Gd_5Ge_2Si_2$ and $Nd_2Fe_{14}B$, where the magnetic electrons in the 4f shell are shielded by 5s and 5p electrons. This is because direct exchange (between the magnetic atoms) will be weak (shielded), leaving indirect exchange via the conduction band to drive magnetic order.

8.2.5 Indirect ionic exchange (superexchange)

There are situations where a typically non-magnetic atom, such as oxygen, can mediate a magnetic interaction: otherwise called superexchange. This is a short-ranged magnetic exchange interaction that typically favours antiferromagnetic exchange. For example, for a Mn–O–Mn sub-unit, electrons could be shared between the Mn and O atoms, in which case it is energetically favourable for them to delocalise over the whole structure. If you consider the arrangement of spins in this system—as shown in figure 8.10—you should see that for delocalisation to occur, the spin of electrons on the same energy level (of the Mn atoms) would need

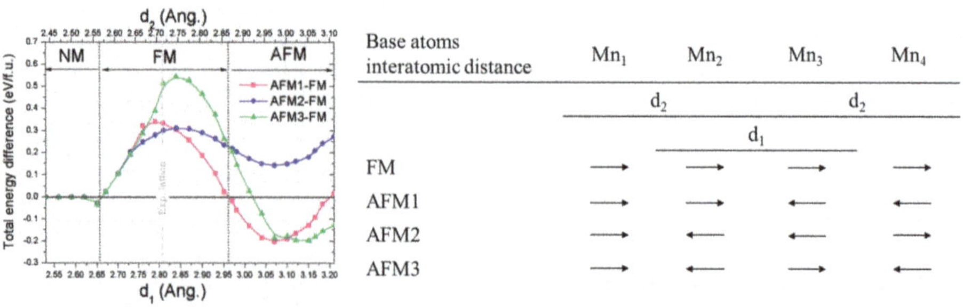

Figure 8.9. Left: Difference between energies calculated for the AFM and FM states of MnP as a function of the Mn–Mn separation(s). Right: corresponding magnetic configurations for the FM and AFM states. Reprinted (figure and table) with permission from [11]. Copyright (2010) by the American Physical Society.

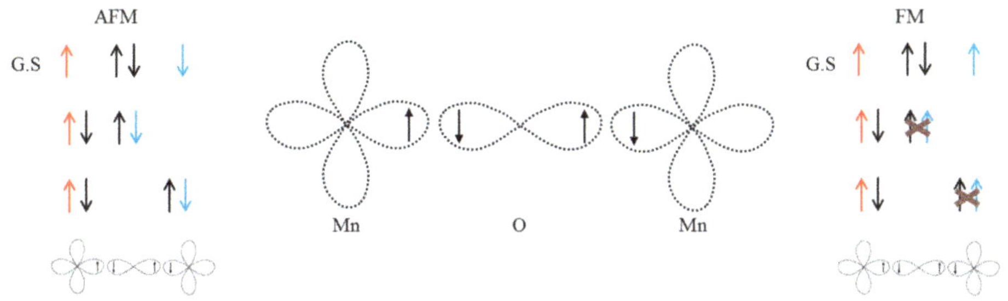

Figure 8.10. Example superexchange across Mn–O–Mn with one unpaired electron. The spin configurations on the left and right show the ground state (G.S) and example excited states for antiferromagnetic (AFM) and ferromagnetic (FM) exchange, respectively. The Pauli exclusion principle forbids excited states where two electrons occupy the same energy level with the same spin state (indicated by the red crosses). As a result, delocalisation of the electrons across the Mn–O–Mn unit is only possible for antiferromagnetic exchange between the two Mn atoms.

to be opposite (AFM ordering) so that at any one time they could occupy that energy level on one of the Mn atoms.

Indirect exchange in ionic solids such as $LaMnO_3$ typically has an exchange integral approximated by:

$$J \sim \frac{t^2}{U} \tag{8.22}$$

where U is the energy cost of an excited state and t is a hopping integral that is proportional to the bandwidth of the conduction band.

8.2.6 Double exchange

Another type of magnetic exchange that can occur in the Mn–O–Mn system is ferromagnetic double exchange. This results from crystal field splitting of the energy levels, such as is observed in manganites where the d_{z^2} and $d_{x^2-y^2}$ orbitals are grouped into the e_g energy levels, and the d_{xy} d_{xz} d_{yz} orbitals are grouped into the t_{2g} energy levels, as shown in figure 8.11(a) and discussed in [9]. Double exchange is similar to superexchange, in that it is indirect, and arises largely from delocalisation of electron orbitals. The difference, however, is that now, different valencies of the magnetic atoms (Mn) enables ferromagnetic exchange: it costs less energy if the electrons that are hopping from one Mn e_g level to the next (via the O atom) do not flip spin.

For example, if we substitute Ca ([Ar] $4s^2$) for La ([Xe] $5d^1$ $6s^2$) into $LaMnO_3$ (where Mn = [Ar] $3d^5$ $4s^2$, and O = [He] $2s^2$ $2p^4$), this results in a mixture of Mn^{3+} (for La) and Mn^{4+} ions (for Ca). In such a situation, there will be electrons that are free to hop between atoms (i.e. to a neighbouring site). If, however, there is strong exchange between the energy levels on each atom that favours alignment (in this case the e_g and t_{2g} levels), then these available spins will need to be aligned between Mn atoms for there to be no energy cost associated with such hopping. In other words,

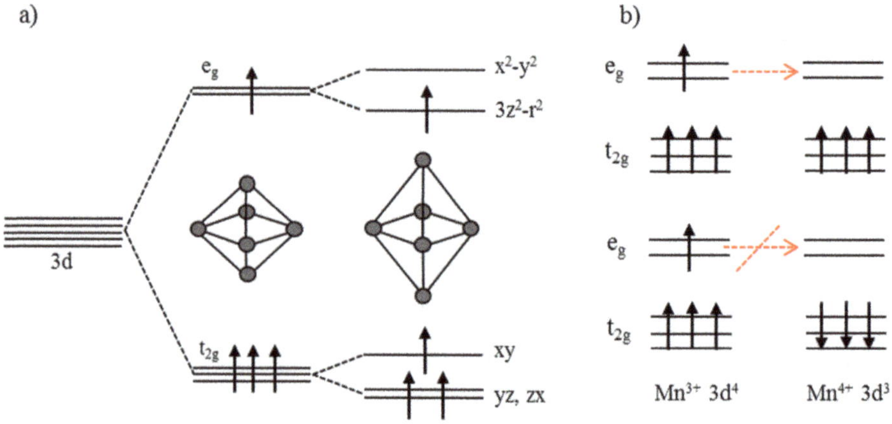

Figure 8.11. Double exchange in a manganite. (a) Sketch of crystal field splitting in a manganite crystal, where the Jahn–Teller distortion (right hand side of (a)) results in further splitting of the energy levels. (b) Sketch of the possible spin configurations for the double exchange mechanism for FM (top) and AFM (bottom) alignment. Where the Mn atoms are AFM aligned, hopping of the e_g electron between Mn^{3+} and Mn^{4+} would cost energy due to the change in ferromagnetic exchange between the t_{2g} and e_g electrons.

the e_g electrons would be able to hop to a neighbouring site only if there is a vacancy of the same spin, as shown in figure 8.11.

Other examples of systems that exhibit double exchange are the ferrites (Fe^{2+}, Fe^{3+}, where Fe = [Ar] $3d^6$ $4s^2$) where the magnetic moment on Fe^{2+} is 4 μ_B per atom, and the multiferroics RMn_2O_5 (where R is a rare Earth such as Eu or Ho).

8.2.7 Anisotropic exchange (Dzyaloshinky–Moriya interaction)

Finally, there are instances where an anisotropic arrangement of magnetic atoms can lead to a perturbation of the exchange interaction (of an otherwise symmetric arrangement). This is known as the Dzyaloshinksy–Moriya interaction (DM interaction, or DMI).

The Hamiltonian for this case is given by:

$$\hat{H} = D \cdot \boldsymbol{S}_i \times \boldsymbol{S}_j \tag{8.23}$$

where D describes the crystal field symmetry (if the crystal field has an inversion symmetry with respect to the centre between two magnetic ions D vanishes), and the force due to $\boldsymbol{S}_i \times \boldsymbol{S}_j$ will be at right angles in a plane perpendicular to D. As the force experienced by the spins will be at right angles (to each other), this interaction tends to cant the magnetic moment (see helimagnetism and speromagnetism).

Another mechanism proposed by Fert in 1990 is the formation of a DMI at the interface of a non-magnetic metal and a thin ferromagnetic film [12]. There is currently a lot of interest in the application of this exchange interaction, in particular for the emergence of weak FM (which was measured with x-rays [13]), or the emergence of Skyrmions (topological magnetic vortices) [14]. More recently, the

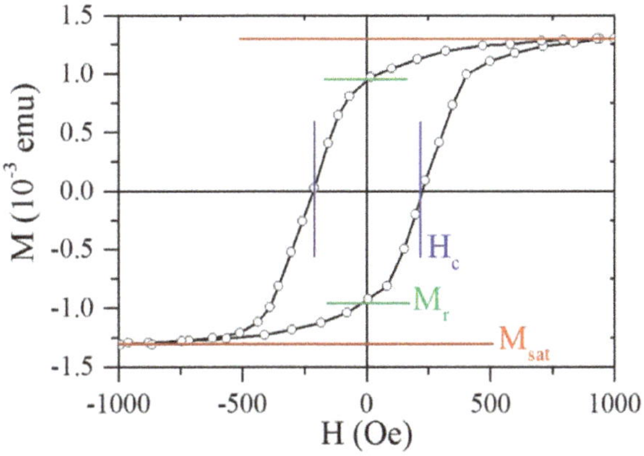

Figure 8.12. Key features of a magnetic hysteresis loop for a ferromagnet.

(interfacial) DM interaction has also arisen as a mechanism for driving out of plane magnetisation in asymmetric thin film stacks such as Pt:Co:Pt [15].

8.3 Magnetometry

There are several ways to determine the magnetisation, or magnetic moment, of a material under investigation. For example, Faraday's law of induction:

$$V = -\frac{d\Phi_B}{dt} \tag{8.24}$$

or the Lorentz Force:

$$\boldsymbol{F} = q\boldsymbol{E} + q\boldsymbol{v} \times \boldsymbol{B} \tag{8.25}$$

can both be used to determine B, and hence the magnetisation, M ($B = \mu_0(H + M)$ in SI units). This is the case with the extraction method, vibrating sample magnetometers (VSMs), and the torque magnetometer. In most labs, the most commonly employed techniques are the VSM or SQuID (superconducting quantum interference device) magnetometers, in part due to their reliability and ease of calibration, but also largely because they are commercially available. Before we start, it is useful to define the key features of a magnetic hysteresis curve for both ferromagnets and superconductors, as shown in figures 8.12 and 8.13.

As outlined in section 8.1, even for a ferromagnet, it is not necessarily the case that it will be naturally 'magnetised' due to the formation of magnetic domains, which act to lower the total energy of the system. At very low fields, these same domains result in what is described as a magnetic hysteresis, i.e. energy is required to switch the magnetisation direction.

If a magnetic field is applied to a ferromagnet there will be a sharp increase in magnetisation (reorientation of magnetic domains) followed by saturation, where the moment has reached a maximum value, M_{sat}. Once magnetised, the (global) magnetic moment that remains after the magnetic field is removed is likely to be less than M_{sat}, and is denoted, M_r, the remanent magnetisation. The field required to

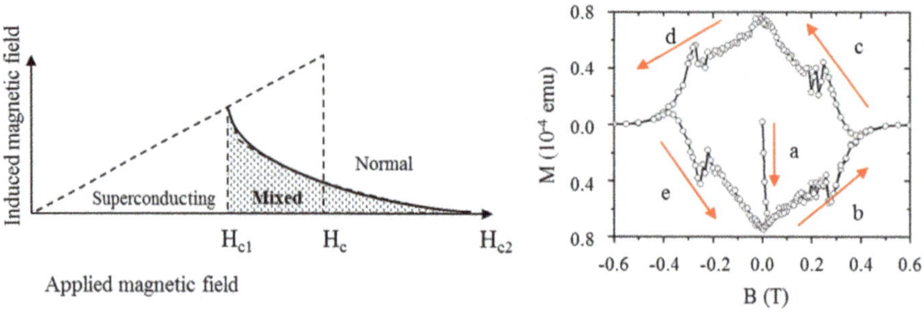

Figure 8.13. Left: Type I and II superconductors. Right: a five quadrant magnetisation measurement for a Type II superconductor (SC). (a) Initial application of magnetic field is screened by the superconductor (virgin state). (b) Above H_{c1}, as magnetic vortices start to permeate the SC, the magnetic moment starts to increase (diamagnetic screening current competes with magnetic vortices). (c) As the field is reduced the magnitude of the diamagnetic moment decreases, but the magnetic vortices remain. (d) As the field is decreased further (negative field), the direction of the magnetic vortices is shifted. Jumps in data indicate flux pinning. (e) Return of magnetisation as field is decreased, to follow (b) in the next quadrant.

switch from one magnetisation direction to another (i.e. switch polarity) is described as the coercive field, H_c.

Superconductors, on the other hand, start as perfect diamagnets: as a magnetic field is applied they will generate opposing magnetic fields, in other words the relative magnetic permeability, $\mu_r = 0$. There are two main types: Type I, which mostly comprises single element materials such as Nb or Al; and Type II superconductors, which typically comprise complex materials such as YBCO (YBaCuO ceramic) or MgB_2. With regards to their behaviour under applied magnetic field, Type I superconductors will be stable up to a given critical field, H_c, which is dependent on temperature and current applied. Type II superconductors, on the other hand, will exhibit what is described as a mixed-state above an initial critical field, H_{c1}. If $H > H_{c1}$ then magnetic vortices (with flux quanta $h/2e \approx 2.1 \times 10^{-15}$ Wb) will permeate the superconductor. The higher the applied field, the higher the density of magnetic vortices until a second critical field, H_{c2}, is reached and the superconductor collapses into the normal state, as shown in figure 8.13. The mix of superconducting and normal state within the Type II superconductor results in the non-obvious '5 quadrant' magnetisation loop shown in figure 8.13.

8.3.1 Vibrating sample magnetometer (VSM)

The VSM uses induction (equation 8.24) in order to measure the magnetic moment of a given sample. In its most basic sense, the sample is mounted on a stick that is held between the poles of a magnet and vibrated about a given point. The moving sample will in turn generate a change in local magnetic field that is then measured up by a series of pick up coils. This can be calibrated against samples of known moment (and similar geometry if possible, in order to reduce uncertainty). A schematic of a basic VSM is given in figure 8.14. Note that this includes the option to control the temperature of the sample space (usually with cryogens such as liquid helium or

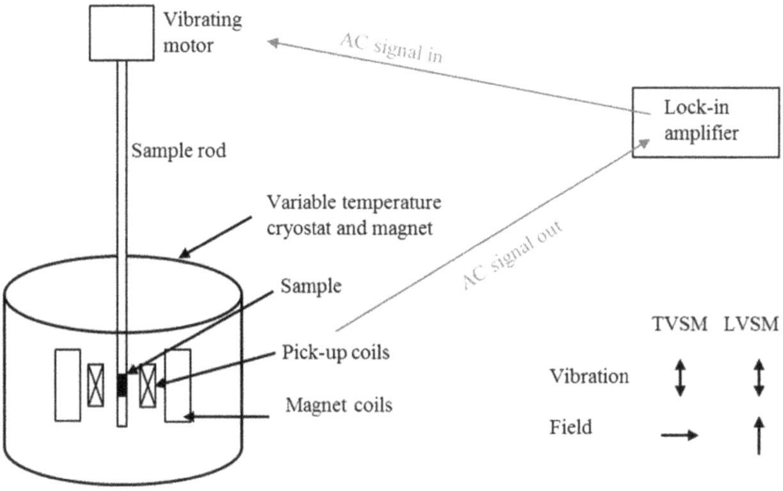

Figure 8.14. Schematic of a VSM capable of (magnetic) field and temperature variation. Field can be applied parallel to (longitudinal mode) or at right angles to (transverse mode) the sample vibration. A lock-in amplifier is typically used to reduce noise by locking the measurement to the frequency of sample vibration.

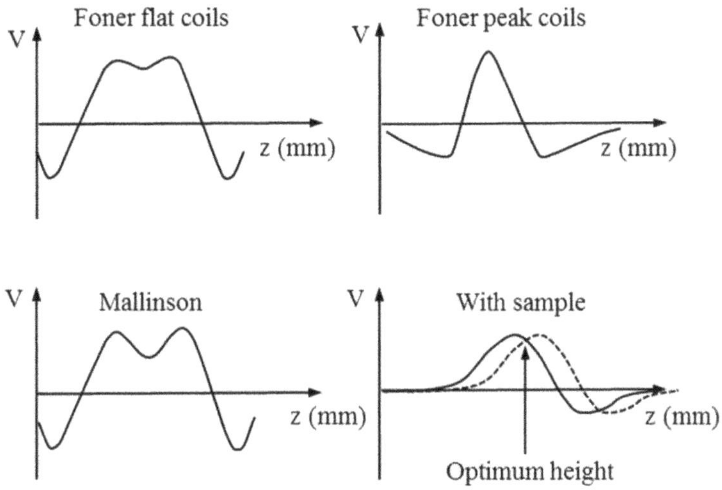

Figure 8.15. Sketch of '*z*-height' scans for different pick-up coils.

nitrogen), and that the field and vibration orientations can vary dependent on application.

There are several different types of magnetic pick-up coils—the Foner and Mallinson coils are just two examples. These will have a response that depends on the position of the sample with respect to the centre of the coils, as shown in figure 8.15. As such, the first step in any VSM measurement is to determine the '*z*-height position' or position of maximum pick-up, by performing what is often referred to as a 'height scan'. In this case, a small field is applied to magnetise the sample and it is moved through a small distance (normally <2 cm), whilst the voltage output of the coils is monitored. To minimise any offset due to response time of the electronics (i.e. experimental artefact), this scan is repeated in the forward and reverse directions and the overlap between the two curves taken as the 'actual' *z*-height. Once this has been set the measurement can start.

8.3.2 SQuID magnetometry

SQuID magnetometry utilises the sensitivity of a superconducting quantum interference device (SQuID) in order to measure the moment of a given magnetic sample. Similar to the VSM, a set of pick-up coils are used, however, in this instance they are comprised of superconducting wires wound as a second-order gradiometer that is inductively coupled to the SQuID device via connecting superconducting wires. Again, a current is induced by moving the sample through the pick-up coils. The SQuID is housed away from the sample and applied magnetic field so that it will not pick up small changes in magnetic field (relaxation of the magnet or fluctuations); only the induced current from sample movement within detector coils.

The pick-up coils for a second-order gradiometer are typically comprised of four windings, in clockwise and anti-clockwise directions, as shown in figure 8.16. As the sample is scanned through these coils a current is induced, which is passed through the SQuID sensor and converted to a voltage output (remember that superconductors have zero resistance, so voltage generation is not as direct as it was for the VSM). This voltage output will be proportional to the magnetic moment of the sample.

As a rule of thumb, SQuID magnetometers are of the order of 100 times more sensitive than VSM (sensitivity of $\approx 1 \times 10^{-7}$ emu and 1×10^{-5} emu, respectively). Improvement in technology over the last 20 years has decreased the difference in sensitivity between the two magnetometers, and hybrid versions (SQuID VSM) are now available from Quantum Design. As the sample needs to be scanned through

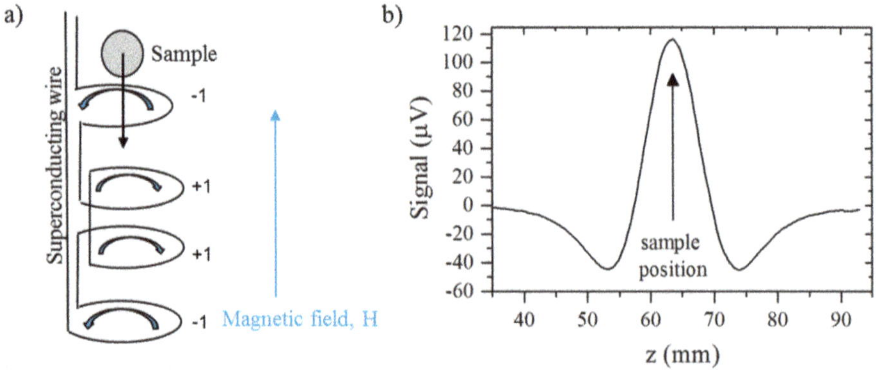

Figure 8.16. Basic components of a SQuID magnetometer. (a) Sketch of the SQuID second-derivative coilset commonly employed. Adapted from *Fundamentals of Magnetism and Magnetic Materials* [16]. (b) Example of a single '*z*-height' scan, which is fit in order to determine the magnetic moment. At maximum signal the sample position is located in the centre of the coils.

the pick-up coils for each measurement, however, the disadvantage with standard SQuID measurements are that they slower, and can often miss dynamic responses to magnetic field.

8.3.3 Extraction method

Similar to the VSM and SQuID techniques, the extraction method will use the voltage generated in a coil through which the sample to be measured is pulled. The integration of the resultant voltage curve (as a function of the extracted length, z) is then used in conjunction with a known calibration factor (dependent on the number of coils and their resistance) to determine the total magnetisation.

8.4 Magnetic imaging

There are several types of magnetic imaging techniques that are primarily employed in order to explore the impact of shape (geometry) and defects on the evolution of magnetic structure in response to an applied magnetic field. Brillouin light scattering (BLS) and XMCD were discussed in chapter 5; extensions of these techniques include x-ray induced photoelectron microscopy (XPEEM), which combines aspects of PES and XMCD to image magnetic domains.

For complex phase transitions, such as observed in metamagnetic materials, where there might be competition between antiferromagnetic (AFM) and ferromagnetic (FM) states, or magnetostructural coupling, magnetic imaging can be used to identify the controlling features of the magnetic phase transition. For example, phase impurities in the $Gd_5Ge_2Si_2$ system seed the onset of FM as a magnetic field is applied, resulting in asymmetric magnetic hysteresis loops, as shown in the Hall probe imaging example of figure 8.17. It is also often employed to study the dynamics of domain motion in transformer materials, where reduction of hysteresis is key to decreasing energy losses in the magnetic core.

In this section, several of the key magnetic imaging techniques will be described, with focus on the advantages and limitations of each technique, as summarised in table 8.2.

8.4.1 Magneto-optic Kerr effect (MOKE)

The magneto-optic Kerr effect (MOKE) utilises the interaction of light (i.e. an electromagnetic wave) with a magnetic surface. It is commonly employed as a qualitative magnetometer for thin films where rapid characterisation of the hysteresis loop is required. With regards to imaging applications, MOKE imaging is useful where information over a large area is preferred over high resolution (as the system will be diffraction limited to the wavelength of the laser used), or where high speeds are necessary (such as with time resolved Kerr microscopy).

The Kerr effect itself, is defined as the rotation of polarisation of light reflected off of a sample surface. This is similar to the Faraday effect: observation of rotation on transmission through a magnetised sample.

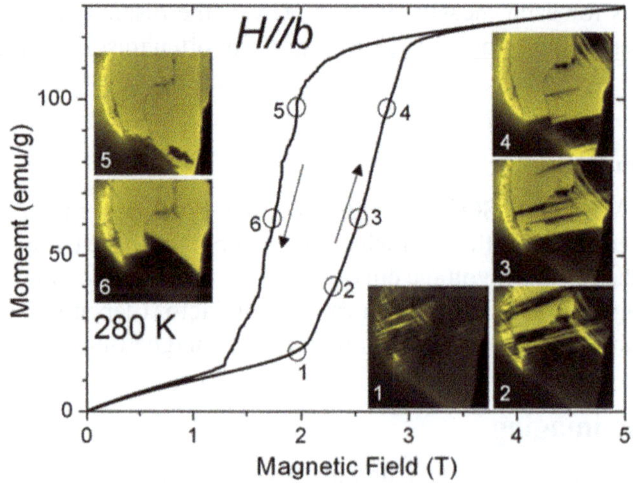

Figure 8.17. Magnetic imaging of a $Gd_5Ge_2Si_2$ single crystal highlighted the differences in the switching mechanism on field increase and decrease. Yellow indicates high moment, *M*, black indicates low moment. Data was obtained with a scanning Hall probe imager [17]. John Wiley & Sons. Copyright 2009 WILEY-VCH Verlag GmbH & Co. KGaA, Weinheim.

8.4.1.1 Fundamentals of the MOKE measurement

The Kerr effect, $\varnothing_{\mathrm{k}}$, is typically defined as the sum of the Kerr rotation, θ_{k}, and the Kerr ellipticity, $i\varepsilon_{\mathrm{k}}$:

$$\varnothing_{\mathrm{k}} = \theta_{\mathrm{k}} + i\varepsilon_{\mathrm{k}} \tag{8.26}$$

where the Kerr rotation describes a shift in phase due to different refractive indices of polarised light incident on the magnetic material, and the Kerr ellipticity describes different absorption rates. The measurement geometry, as shown in figure 8.18, will control which magnetisation states contribute most to the observed Kerr effect. For example, polar MOKE will primarily show a Kerr rotation due to out of plane moment, whereas the longitudinal and transverse Kerr effects will show a Kerr rotation for in-plane moment that is parallel or perpendicular to the incident wave, respectively.

A simple MOKE imaging system will comprise of a polaroid to polarise the incident beam of light (white light source would be preferable), with a lens to collimate or focus the beam. The reflected light is then passed through another polaroid (or half-wave plate) that is rotated to be 'just off extinction', i.e. it is rotated to minimise the light which passes through, then shifted by 2°–5° so that any change in contrast due to different magnetisations will be more obvious. The collected image will then be passed through an eyepiece or camera for analysis. Some examples of MOKE imaging are given in figure 8.19. Extensions of this technique include: scanning Kerr microscopy, where the incident light is focussed to a spot on the sample surface and scanned across the area of interest; and time resolved Kerr microscopy, which can be used to explore magnetisation dynamics [20].

Table 8.2. Key information for the different magnetic imaging techniques available.

Technique	Scanning area	Resolution	In-plane (IP) or out-of-plane (OP), magnetisation (*M*) or *B* field (*B*)	Information depth
MFM	<100 × 100 μm	~0.1 μm	OP, B	1 μm
SEMPA	<100 × 100 μm	~1 nm	IP, M	Surface(~0.5 nm)
Scanning Hall probe imaging	<5 × 5 mm	~2–20 μm	OP, B	Average
MOKE	<10 × 10 mm	~0.2 μm	IP, M	Surface(~10 nm)
Bitter technique	>2 × 2 mm	~1 μm	OP, B	1 μm
Lorentz TEM	<100 × 100 μm	<1 nm [18]	IP, B	Average through sample
STEM DPC (differential phase contrast)	<100 × 100 μm	~2 nm	IP, B	Average
Electron holography	<100 × 100 μm	~2 nm	IP, B	Average
SP-STM	<100 × 100 μm	Atomic	IP/OP*, M	Surface(~0.1 nm)
BLS	<10 × 10 mm	μm—250 nm	IP, M	Surface(~10 nm)
PEEM	<100 × 100 μm	<100 nm	IP, OP, M	Surface(~2 nm)
XPEEM	<100 × 100 μm	~5 nm [19]	IP, OP, M	Surface(~2 nm)

* Depends on the chosen tip

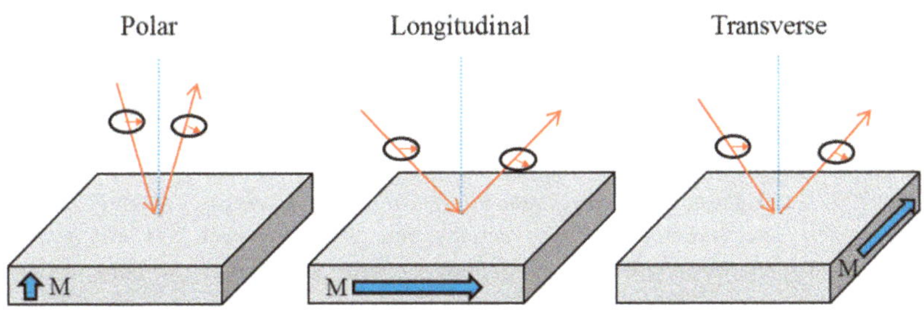

Figure 8.18. Kerr effect in different geometries.

8.4.2 Scanning Hall probe

The scanning Hall probe (SHP) imager uses a Hall bar device, as shown in figure 8.20, to measure the stray field at the surface of a material being studied. Much like MOKE imaging, it has the advantage of large scan areas, but without the sensitivity to vibrations or laser stability that can plague MOKE. The sensitivity of

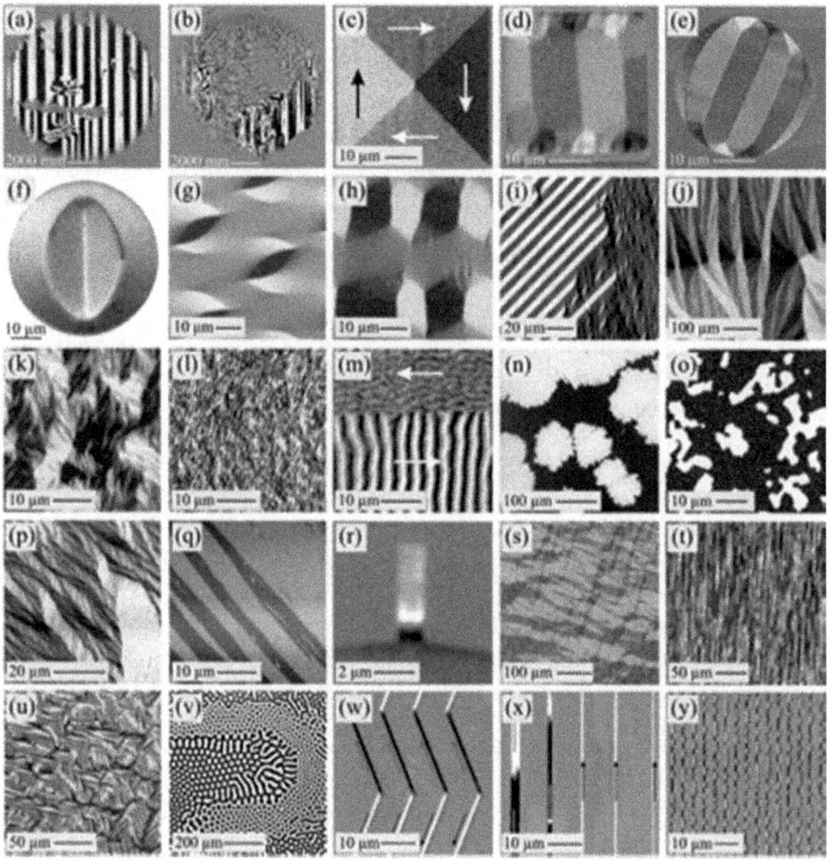

Figure 8.19. Examples of MOKE imaging applications. Reproduced from [21]. © IOP Publishing Ltd. CC BY 3.0. Original caption read: '*Magnetic domain images from electrical steel with (a) nearly perfect and (b) misaligned orientation of grains (samples: R. Schäfer, IFW Dresden), from patterned soft magnetic elements (c)* $Ni_{81}Fe_{19}$ *(after [295, 296]), (d), (e) amorphous FeCoSiB elements [297], and (f)* $Ni_{45}Fe_{55}$ *films [298]. Magnetic domain patterning by ion irradiation in (g), (h) extended FeCoSiB layers and (i) exchange-biased samples [299]. Complicated domain patterns in (j) CoFe single layers and (k)* $CoFe/SiO_2$ *multilayer structures ([234, 300]), soft magnetic (l) CoFe (reprinted from [234], copyright 2004, with permission from Elsevier) and (m)* $Ni_{81}Fe_{19}$ *films with weak out-of-plane magnetic anisotropy [234, 301] (*© 2001 *IEEE; reprinted, with permission, from [301]), and (n) Pt/Co/Pt multilayers with strong out-of-plane anisotropy films [302]. (o) Patch domains in synthetic spin valve sensors (reprinted with permission from [303]; copyright* 2001, *AIP Publishing LLC), (p) domains in exchange-biased CoFe/IrMn layers (reprinted figure with permission from [304], copyright* 2003, *AIP Publishing LLC), and (q) in a* $Tb_{45}Fe_{55}/Gd_{40}Fe_{60}$ *ferrimagnetic exchange spring system (reprinted figure with permission from [194], copyright* 2008 *by the American Physical Society). (r) MO contrast at the air bearing surface of a longitudinal recording head (after [140] with kind permission from Springer Science and Business Media [140], see also [139]). Domain formation in (s) LSMO (sample: A Steffen, FZ Jülich), (t) GaMnAs (sample: H Ohno, Tohoku University), (u)* $CoFe_2O_4$ *(sample: M Abes, B. Murphy, Kiel University, imaging: N O Urs, Kiel University), and (v)* $BiLu_2Fe_4GaO_{12}$ *(sample: M Gusev, Research Institute of Materials Science and Technology Zelenograd; imaging: M Kustov, Kiel University). Domain images from soft magnetic (w) Ni81Fe19 (sample: M Kläui, University Konstanz), (x)* $Co_{50}Fe_{50}$ *(sample: B Hausmanns, University Duisburg), and (y) Fe thin film nanowires [305] (© 2003 IEEE; reprinted, with permission, from [306]).*'.

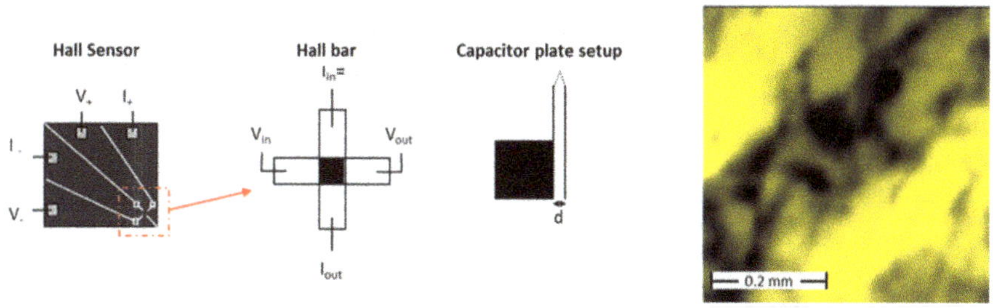

Figure 8.20. Basics of a Hall probe magnetic imaging system. Left: sketch of the Hall sensor typically used, with focus on the active area where the Hall bar is patterned using focussed ion beam etching. Right: example image obtained using this system [17], where yellow indicates high moment, black indicates low moment.

this technique is also higher, depending only on the type of material used to construct the Hall bar. Typically a voltage, V_H, will be measured, as defined by:

$$V_H = C_H I B \quad (8.27)$$

where C_H is a calibration constant specific to the Hall device used, I is the applied current, and B is the magnetic flux density.

There are two common versions of the SHP: one where the Hall bar is controlled by stepper motors (and thus the resolution is limited to 2 μm) [17], the other, where it is controlled by piezoelectrics (where the resolution can be higher, but the scan range will decrease) [22]. It is common to map the sample surface so that a constant height can be maintained. This can be achieved with use of capacitive plates that detect when the Hall bar touches the sample surface, and by ensuring that the sample is polished to fractions of a micron (typically using diamond paste).

Whilst this method may only measure the stray field, careful application can yield more detailed results. For example, when imaging superconducting samples, the current flowing (which in turns leads to the detected magnetic field) can be reconstructed using a series of images obtained at different heights above the surface in combination with the deconvolution procedure outlined in [23].

8.4.3 Magnetic force microscopy (MFM)

Magnetic force microscopy (MFM) is the magnetic analogue of atomic force microscopy where rather than the cantilever being sensitive to atomic forces it is magnetised so that it will now also be sensitive to magnetic forces (i.e. stray fields) as well. The advantage of MFM is that resolution can be particularly high; limited only by the size of the tip. However, for particularly soft materials, the magnetisation of the tip itself can cause a change in the magnetisation state of a sample, so interpretation of images requires some care. For an in-depth review of MFM, see [24].

8.4.4 Spin polarised scanning tunneling microscopy (SP-STM)

Spin polarised scanning tunnelling microscopy (SP-STM) is similar to STM, as described in chapter 4, but commonly uses a magnetic tip, which acts as a source or sink of spin polarised electrons. It was first used to measure the spatial distribution of spins on a Cr(001) single crystal surface using a ferromagnetic CrO_2 tip [25], but has since been developed to work with a multitude of different tips such as ferromagnetic Fe or antiferromagnetic Cr [26].

To understand the basic principles of SP-STM it is useful to remember the properties of magnetic tunnel barriers, or other tunnel junctions (as described in chapter 7). In the simple case of zero bias voltage, the differential conductance, which is measured by STM can be described by:

$$\frac{\mathrm{d}I}{\mathrm{d}U} \propto \rho_{\mathrm{T}}^{\uparrow}(E_{\mathrm{F}})\rho_{\mathrm{S}}^{\uparrow}(E_{\mathrm{F}}) + \rho_{\mathrm{T}}^{\downarrow}(E_{\mathrm{F}})\rho_{\mathrm{S}}^{\downarrow}(E_{\mathrm{F}}) \tag{8.28}$$

where differences in the resistivity, ρ, of the majority and minority states (spin $\uparrow$ or $\downarrow$) for the sample, S, or the tip, T, will change according to magnetisation of the sample. For low bias voltages, U, tunnelling will be elastic, which means that the spin will be conserved. These processes are loosely demonstrated in figure 8.21. At high voltages, loss of spin polarisation occurs due to events such as spin scattering (a non-elastic process).

One way to characterise the magnetic information available from $\mathrm{d}I/\mathrm{d}U$ is to calculate the asymmetry:

$$A(U) = \frac{\mathrm{d}I/\mathrm{d}U_{\uparrow}(U) - \mathrm{d}I/\mathrm{d}U_{\downarrow}(U)}{\mathrm{d}I/\mathrm{d}U_{\uparrow}(U) + \mathrm{d}I/\mathrm{d}U_{\downarrow}(U)} \tag{8.29}$$

where $\mathrm{d}I/\mathrm{d}U_{\uparrow\downarrow}$ is the differential conductance measured for different magnetisations (i.e. $\pm B$). The information obtained in this way can be used to build an image of the magnetic domains, and can produce similar data to that obtained using spin polarised inverse PES [26].

Other methods of magnetic imaging with SP-STM include constant-current mode, which has the highest resolution and is the typical measurement mode for STM; and local magnetoresistance mode, which requires the tip magnetisation to switch back and forth. Ideally for local MR the tip will be both soft and have a low magnetic moment so that it can be easily switched and the stray field is minimised. The magnetoresistance can then be measured using a lock-in technique [26]. Ideally this technique requires:

1. Clean tips as adsorbants may reduce the spin polarisation.
2. Minimised dipolar interaction between the tip and the sample.
3. 'Soft' tips so that the magnetisation direction can be easily switched with field.

The first requirement necessitates preparing the tip under ultra high vacuum (UHV) conditions ($<1 \times 10^{-8}$ mBar) and may ultimately limit the lifetime (of the tip). For

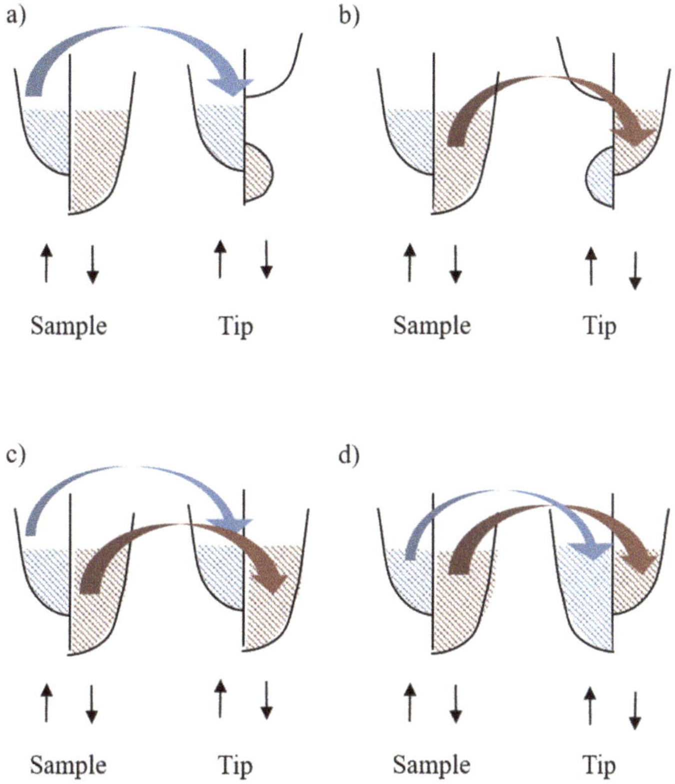

Figure 8.21. Schematic of the possible tunnelling current between: (a) and (b) a magnetised sample and a half metallic (ferromagnetic) tip when aligned or anti-aligned, with respect to each other; and (c) and (d) a magnetised sample and a ferromagnetic tip when aligned or anti-aligned, with respect to each other. The arrows indicated tunnelling current of spin up (blue) and spin down (red) states. If the bias current is low, the spin of the electron will be preserved and can therefore only move from occupied (sample) states to unoccupied (tip) states.

mechanically prepared tips, a narrow constriction is created in a wire (Fe for a ferromagnetic tip, Cr or MnNi for an antiferromagnetic tip) using electrochemical etching. This wire is then torn apart in UHV into approximately 10 nm wide tips that are ready for imaging. The problem with 'bulk' magnetic tips is that there can potentially be large stray fields which could distort the sample magnetisation (i.e. breaks requirement 2). The sensitivity of the tips is also limited by the shape anisotropy (i.e. they will be magnetised along the tip axis). The advantage of these tips is that soft magnets such as Fe could simply be surrounded by a coil to switch magnetisation direction using an alternating current (requirement 3).

The solution to large stray fields is to use thin film magnetised tips. In this case, a tungsten wire is electrochemically etched in a saturated solution of sodium hydroxide resulting in tips with width of approximately 20–50 nm. This is then heated in UHV to ensure good adhesion of the magnetic film, resulting in an increase of the tip width. It is then magnetically coated with several monolayers (ML) of a chosen magnet such as: 3–10 ML of Fe for in-plane sensitivity; 7–9 ML of Gd, or 10–15 ML of $Gd_{90}Fe_{10}$, for out-of-plane sensitivity; or 25–45 ML of Cr for antiferromagnetically aligned out-of-plane sensitivity. Whilst an antiferromagnetic tip may be difficult to switch, the stray field produced will be due to a single atom only, and therefore incredibly small. This makes it especially important for studying sensitive samples such as soft ferromagnets or superparamagnets.

8.4.5 Scanning electron microscopy with polarisation analysis (SEMPA)

The scanning electron microscope with polarisation analysis (SEMPA) is a magnetic imaging technique with particularly high resolution, but limited scan area. There is a wide variety of information that is accessible through interaction of electrons with a material, as was discussed in chapter 4. This is the focus of most e-beam based microscopy. In 1976 it was discovered by Chobrok and Hofmann that when an electron beam is focussed onto a ferromagnetic sample, the secondary electrons that are produced will be spin-polarised anti-parallel to the magnetisation vector at the surface [27]. This polarisation will be independent of the electron energy and the number of spin polarised electrons will depend only on the secondary electron yield.

This effect utilised by Koike and Hayakawa to develop the first SEMPA [28, 29], the advantages of which are long working distances, depth of field and an easily variable magnification (from mm to nm). It is, however, a surface sensitive technique that requires UHV to operate.

To detect the spin polarisation of the secondary electrons (as a result of their interaction with the material surface magnetisation) they are collected and focussed onto a (100) tungsten target. The spin–orbit interaction of the electron–atom scattering will result in an asymmetry of the diffraction pattern. The asymmetry of the diffraction pattern is detected for each of four reflections (four-fold symmetry of the given Bragg peak) by a series of micro-channel plates optimised to pick up single electron events. A schematic of this is given in figure 8.22.

The asymmetry of the diffracted secondary electrons is measured by the polarisation function:

$$P = \frac{N_{\text{up}} - N_{\text{down}}}{N_{\text{up}} + N_{\text{down}}} \tag{8.30}$$

where N_{up} and N_{down} correspond to two detector signals and the magnetisation at the surface can be approximated by:

$$M \propto \mu_{\text{B}}(N_{\text{up}} - N_{\text{down}}) \tag{8.31}$$

This type of imaging technique will yield mostly in-plane data, however, out-of-plane components of the magnetisation can be inferred from the observed domain images (where asymmetry goes to zero). An example of images obtained in this way is given in figure 8.23.

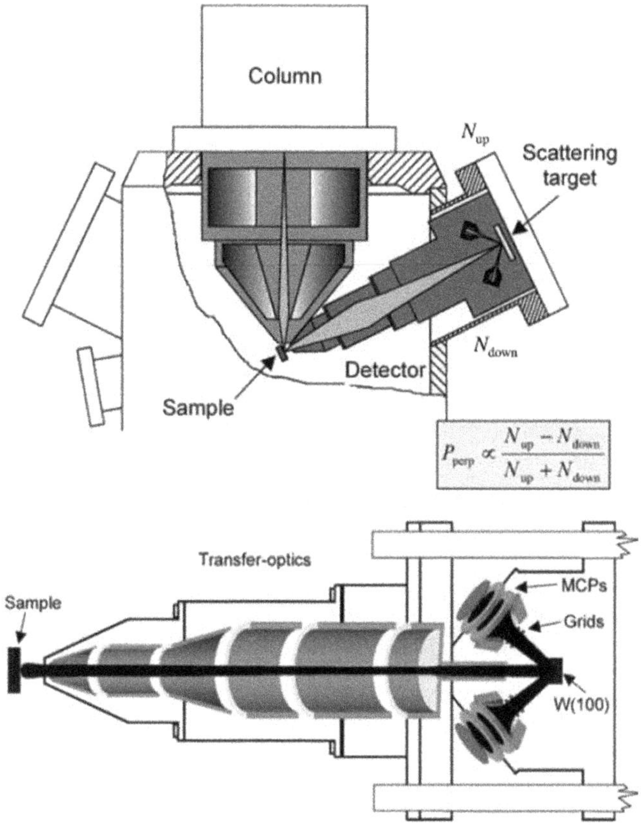

Figure 8.22. SEMPA schematic. Top: overview of the SEM. Bottom: the spin polarisation analyser. Reprinted by permission from Springer (GmbH): [Springer] [Applied Physics A] [A miniaturized detector for high-resolution SEMPA, R. Frömter, H.P. Oepen, J. Kirschner], [COPYRIGHT] (2003) [30].

8.4.6 Transmission electron microscopy (TEM) methods

There are several TEM methods available for magnetic imaging of thin specimens which primarily rely on the Lorentz force: deflection experienced by an electron moving through a magnetic field (equation (8.25)). These can broadly be classed as Lorentz TEM and electron holography methods. In the former, the objective lens of a standard scanning electron microscope is turned off so that the magnetic field at the sample is minimised. This results in deflection of the electron beam, which can be measured by a quadrant detector. Given that the electron beam is focussed perpendicular to the specimen, this technique will be sensitive to in-plane magnetic (or electrostatic) fields. Electron holography methods, on the other hand, will analyse the diffraction pattern produced from the interference of a reference beam that passes through vacuum and another that passes through the sample.

The advantage of Lorentz TEM methods includes high spatial resolution (depending on the specific mode) and a larger scattering cross-section for electrons compared to x-rays, which can yield higher contrast for thin films. A limitation is

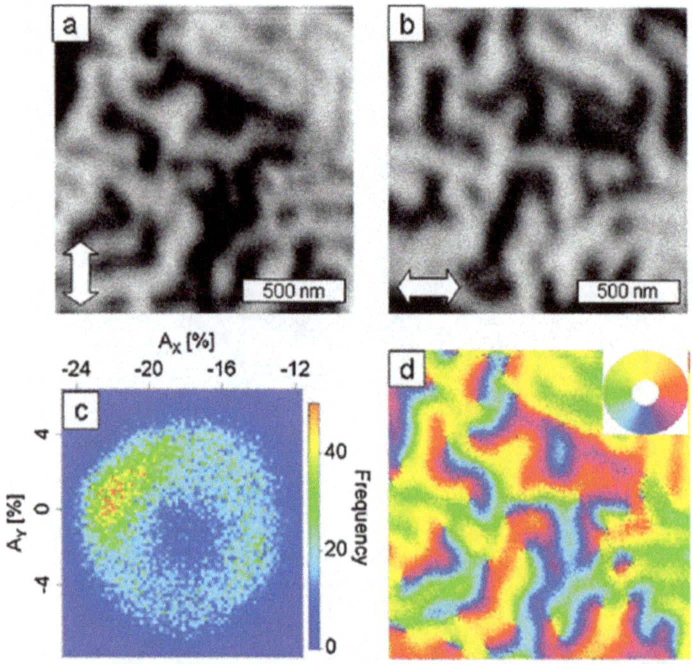

Figure 8.23. Maze domains observed with SEMPA. (a) & (b) *y*- and *x*-components of magnetisation, respectively. (c) Histogram of data from (a) and (b). (d) Combined colour map representation of the in plane magnetisation. Reprinted (figure) with permission from [31]. Copyright (2008) by the American Physical Society.

that the sample must be transparent to electrons (so that they can pass through), with thicker samples proving difficult to analyse. One of the challenges of Lorentz-TEM is the increase in spherical aberration of the objective lens used. Advances in this technique include spherical-aberration correction to improve resolution and chromatic aberration correction to enable analysis of thicker samples.

8.4.6.1 Lorentz transmission electron microscopy (TEM)

Traditional Lorentz TEM will use the change in contrast due to deflection of the electron beam in response to a magnetic gradient in the sample (the Lorentz force). This deflection is usually of the order of tens of microradians, which is miniscule compared to Bragg diffraction. There are two main modes that this technique can be operated in: Fresnel (out of focus) mode, and Foucault (objective in focus) mode. In the latter, one of the magnetic diffraction spots is blocked by a 'blocking aperture', which results in bright and dark areas: the magnetic contrast. These are both qualitative technique that will be sensitive to in-plane magnetisation only. However, if a stable out of plane feature is present—such as at a magnetic domain wall—then the sample can be tilted to provide a perpendicular component with respect to the electron beam. Whilst the information is qualitative, it is relatively quick, meaning that it is well suited to studying time dependent processes such as magnetisation reversal. Examples of Lorentz TEM include the observation of Néel skyrmions [32] and domain wall injection in permalloy ($Ni_{0.8}Fe_{0.2}$) nanowires [33].

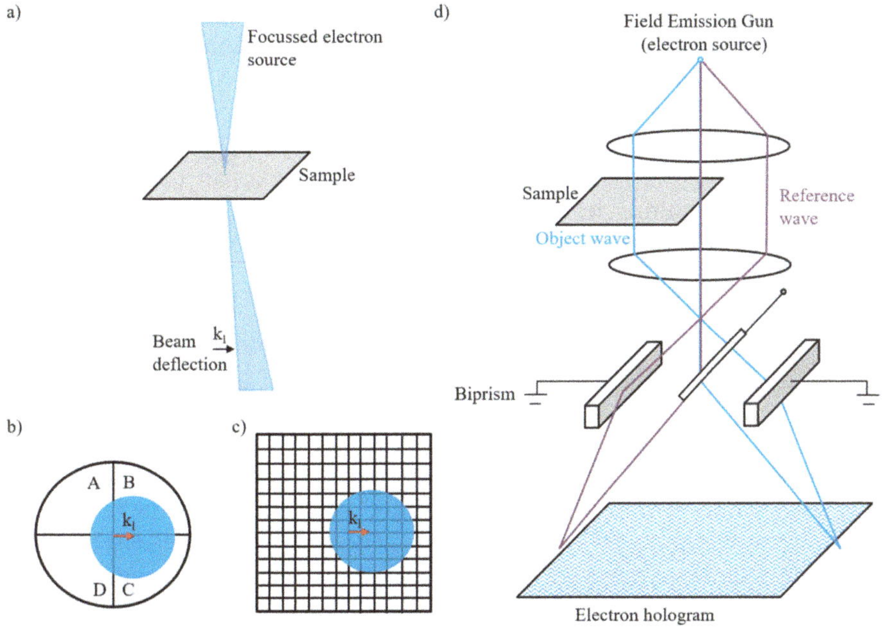

Figure 8.24. Schematic of a STEM DPC and a typical electron holography measurement. (a) STEM-DPC imaging mode, where the beam is deflected by k_l due to magnetic induction in the sample. (b) Projection of the deflected beam onto the four-quadrant detector. (c) Projection of the beam onto the pixelated detector. (d) Schematic of the electron holography mode.

8.4.6.2 Scanning transmission electron microscopy with differential phase contrast (STEM DPC)

In order to obtain quantitative information on the magnetic induction in a sample studied by TEM, information about the change in phase of the electron wave passing through the specimen is required. This is the basis of STEM DPC, where the farfield diffraction from an electron beam focussed onto a sample is projected onto a four-quadrant detector, as shown in figure 8.24. During the measurement, the objective lens is switched off so that the sample sits is an almost field-free environment. The difference in intensity measured at these separate segments (of the detector) can then be used to reconstruct a quantitative measurement of induction in the sample. For example, for the four quadrants of the detector—A, B, C, D—the following relationships can be used to determine magnetic induction in the x- and y-planes (B_x, B_y) as well as reconstruct a standard bright field image (BF):

$$B_y \propto (\mathrm{A} + \mathrm{D}) - (\mathrm{B} + \mathrm{C}) \tag{8.32}$$

$$B_x \propto (\mathrm{A} + \mathrm{B}) - (\mathrm{C} + \mathrm{D}) \tag{8.33}$$

$$\mathrm{BF} = (\mathrm{A} + \mathrm{B} + \mathrm{C} + \mathrm{D}) \tag{8.34}$$

The increase in measurement time due to scanning the sample makes this technique an unreliable observation of time dependent effects. This increase in acquisition time means that there is a possibility of a build up of contamination during a measurement. It can also sometimes be difficult to resolve reliable magnetic information due to diffraction contrast and defects or debris on the sample. A solution to this is to employ a pixelated detector, such as the Medipix detector used at the University of Glasgow [34] and also demonstrated in figure 8.24. By applying a suitable mask (such as determining only the shift, k_l at each point), you can remove diffraction information, thus resulting in cleaned up magnetic images [34]. In this case, the deflection can be converted to a magnetic induction by the following:

$$\beta_l = -\frac{e\lambda}{h} B_s t \tag{8.35}$$

$$\beta_l = -\frac{e\lambda}{h} \int \boldsymbol{B} \times \mathrm{d}\boldsymbol{l} \tag{8.36}$$

where β_l is the angular deflection of the beam, e is the charge on an electron, λ is the electron wavelength, h is Planck's constant, and $\mathrm{d}\boldsymbol{l}$ describes the integral element along the electron's path [34]. Examples of STEM DPC include observations of skyrmions in a cubic nanowedge of FeGe [35] and electric field imaging of single atoms [36].

8.4.6.3 Electron holography

Electron holography requires the interference of a reference beam (in vacuum) with electrons deflected by the sample. A bisprism, placed close to the image plane, as seen in figure 8.24, can selectively increase the area of overlap of the reference and sample beams by adjusting the voltage between the central wire of the biprism, and its ground. The interference between these two beams can then be used to determine the change in both the phase and intensity on passing through the sample. The phase shift, φ, observed by this technique will consist of a phase shift due to the mean inner potential, φ_e, and magnetic induction, φ_m [37]:

$$\varphi(x, y) = \varphi_\mathrm{e}(x, y) + \varphi_\mathrm{m}(x, y) \tag{8.37}$$

where φ_e is proportional to the mean inner potential, V_0, and a constant, C_E determined by the accelerating voltage of the electron beam (as well as its rest mass and wavelength), the specimen thickness, t. To separate this from the magnetic contribution to the phase shift either a spatially flipped set of holograms, or holograms under magnetic reversal are obtained. In the former, the sample is turned over after acquiring the first hologram, and a second hologram is obtained from the same region. The difference in symmetry of the mean inner potential and magnetic induction means that if carefully aligned the sum will provide $2\varphi_\mathrm{e}$ and the difference will provide $2\varphi_\mathrm{m}$. This measurement can then be used as a reference to subtract φ_e from all subsequent images. In practice, it is often easier to obtain holograms under magnetic reversal—where opposite magnetic fields are applied to the sample to change the magnetic state. This is only possible if the sample magnetisation process is reversable. From these images, a magnetic induction map can be generated by taking the Fourier transform of the resulting interference pattern. It should be noted that due to the alignment of the sample with respect to the reference 'path' these measurements can be time consuming.

Examples of electron holography include: imaging of magnetic induction in Fe_3O_4 nanoparticles as indicators of historical paleomagnetic fields [38]; generating magnetic flux maps of a skyrmion lattices in a $Fe_{0.5}Co_{0.5}Si$ single crystal [39]; and determining the magnetic microstructure in magnetotactic bacteria [40];

8.4.7 Photoexcited electron microscopy (PEEM) and scanning transmission x-ray microscopy (STXM)

As highlighted earlier, PEEM can also be used to image the magnetic structure of a surface. In this technique, the detector is not relying on electrons directly generated by photoemission, but on either (a) generation of Auger electrons (see section 5.3.1), or (b) secondary electrons induced by Auger emission (referred to as a secondary electron cascade). Magnetic contrast can be achieved if there is enough energy resolution in the hemispherical detector to distinguish Zeeman splitting.

An alternative approach is x-ray induced photoelectron microscopy (XPEEM), which combines aspects of XMCD and PES, where the sample is irradiated with a high intensity soft x-ray source, close to the absorption edge of a specific element. With a pulsed source, such as at the ESRF is Grenoble, France, time resolved magnetic imaging can also be achieved with XPEEM [41].

The primary contrast mechanism in this imaging will be chemically selective due to the unique nature of energy levels for threshold PEEM, and the absorption spectrum for XPEEM. It will also yield topographical information due to shadowing effects and local changes in electromagnetic fields as a result of topography. Secondary contrast mechanisms will include charging of the surface as it will distort the path of escaping electrons [42]. As such, the sample is ideally kept well grounded.

It has also been shown that x-ray magnetic linear dichroism (XMLD) can be used to image aniferromagnetic samples. In this configuration, the difference in two x-ray absorption spectra is determined, where the incident x-ray is linearly polarised and either the applied magnetic field or the polarisation vector are rotated by 90°. As the analysis is on the x-ray spectra and not photoemission (of electrons), this technique is referred to as scanning transmission x-ray microscopy (STXM) or transmission x-ray microscopy (TXM). Examples include: the analysis of domain formation in epitaxial layers of α-Fe_2O_3 [43]; concurrent observation of ferromagnetic domains either side of an antiferromagnet:ferromagnet interface using XMCD and XMLD [44]; and angular dependence of XMLD based PEEM to determine the spin axes of antiferromagnetic domains [45].

As a single XPEEM measurement will include magnetic, topographical and element specific information, comparison with an asymmetric image (i.e. where the polarisation has been rotated) is required to extract magnetic information. Similar to several of the magnetic imaging techniques described in this chapter, the asymmetry, A, can be determined to enhance the magnetic contrast:

$$A = \frac{I^+ - I^-}{I^+ + I^-} \tag{8.38}$$

where $I^{+,-}$ is the intensity of the transmitted x-ray source for right and left circularly polarised x-rays for XMCD (or linearly polarised x-rays at 90° to each other for XMLD).

To summarise, x-ray based imaging techniques can provide element specific magnetic imaging, with moderate resolution. PEEM relies on the analysis of electrons emitted by photoexcitation and can either measure the energy of these secondary electrons, or use the asymmetric contrast available with polarised sources. When operating in STXM mode, either circularly or linearly polarised x-rays can be used to extract information about ferromagnetic or antiferromagnetic domains.

8.5 Examples of magnetisation techniques in the literature

In the following sections, several examples of research articles are highlighted where magnetisation techniques were used to characterise or identify a given material. For each article, starter questions are provided, which are aimed at focussing the reader's attention on the key aspects of the article (with respect to this chapter).

8.5.1 In 'Depth profile of spin and orbital magnetic moments in a subnanometer Pt film on Co'

Suzuki *et al* used x-ray magnetic circular dichroism spectroscopy (XMCD) to measure the magnetic properties of ultrathin Pt layers deposited onto Co [46].

Reading through the article, try to answer the following:

1. What are the basic principles of XMCD? i.e. how does it work? and what does it measure?
2. What other experimental techniques were used?
3. Why were the XMCD experiments carried out at a synchrotron facility?
4. What value do the authors report for magnetic moment of the 0.15 nm Pt? and how do they justify any differences between this and published values?
5. How does the magnetisation vary as a function of depth in the Pt layer?

The aim of this work was to determine the magnetisation of Pt thin layers, typically used as capping layers in magnetic thin film devices, as a function of thickness. It was found that for these samples, a magnetic moment of 0.61 μ_B/atom was induced in Pt at the Pt:Co interface and that this decayed exponentially with thickness. The authors also found that the magnetisation depth profile measured here was consistent with resonant x-ray magnetic reflectometry

8.5.2 In 'Room-temperature helimagnetism in FeGe thin films'

Zhang *et al* showed that the magnetic transition temperature of FeGe, a material that exhibits helimagnetism due to the Dzyaloshinskii–Moriya interaction, can be increased in thin films [47].

Reading through the article, try to answer the following:

1. What experimental methods have been used in this paper?
2. Define the Dzyaloshinskii–Moriya interaction (DMI).
3. What feature of the samples studied here leads to a DMI?
4. What are magnetic skyrmions? And why are they of interest to the magnetic community?
5. How do REXS measurements differ from XRR or XRD?
6. What material properties does broadband microwave spectroscopy investigate? What is the more common name for this technique?

The aim of this work was to investigate the magnetic phase diagram for a well known helimagnet, when prepared as thin films. The authors found that the transition temperature could be systematically engineered in sputtered thin films through careful choice of deposition parameters (temperature, rate, Ar pressure) and substrate. They also demonstrate use of REXS, FMR and transverse field muon-spin rotation as complementary techniques for study of complex magnetic structures.

8.5.3 In 'Direct observation of magnetic monopole defects in an artificial spin-ice system'

Ladak *et al* showed that direct evidence of monopole defects in a honeycomb lattice using magnetic force microscopy [48].

Reading through the article, try to answer the following:

1. What do the authors mean by 'magnetic spin-ice'?
2. What do the authors mean when they refer to 'ice rules'?
3. What is meant by 'magnetic monopole' in this context?
4. What material(s)/device(s)/structure(s) have been investigated?
5. How were the magnetic states imaged?
6. How does this measurement work?
7. What leads to magnetic frustration in this (these) system(s)?

The aim of this work was to directly observe mobile excitations labelled 'monopole defects' that arise in frustrated magnetic systems. The authors observed monopole defects as sites of higher magnetic moment (magnetic charge of 3q rather than q), and demonstrate creation and movement of these defects within the honeycomb lattice.

8.5.4 In 'Quantum spin chains with frustration due to Dzyaloshinskii–Moriya interactions'

Hälg *et al* investigate a family of materials known as 'spin chains', which demonstrate geometric magnetic frustration due to differing strength of Dzyaloshinskii–Moriya (DM) interactions on inter and intrachain exchange [49].

Reading through the article, try to answer the following:

1. What experimental methods have been discussed in this paper?
2. What specific compounds have been investigated?
3. Define the Dzyaloshinskii–Moriya interaction.

4. What is the crystal structure and space group of $K_2CuSO_4Cl_2$?
5. What impact does Br substitution for Cl have on the crystal structure and why?
6. Why might C_p/T versus T help to identify key features of the heat capacity measurement?
7. What do the authors mean by a spin wave dispersion?

The aim of this work was to investigate a family of materials exhibiting DM interactions using a combination of techniques that would enable detailed characterisation of the magnetic structure within the crystal lattice. The authors concluded that the varying magnitude of inter and intra-chain DM interactions leads to the magnetic frustration exhibited in these material systems.

8.5.5 In 'Neutron diffraction study of the magnetic properties of the series of Perovskite-type compounds [(1 − x)La,xCa]MnO$_3$'

Wollan and Koehler distinguished between ferromagnetic and antiferromagnetic order using neutron diffraction. Note that figure 18 is a particularly useful breakdown of how the magnetic structure was calculated and led to different magnetic peaks [50].

Reading through the article, try to answer the following:

1. Sketch out the perovskite structure.
2. What types of magnetic ordering are present in these compounds?
3. How did the authors identify AFM Bragg peaks?
4. What impact does temperature have on the diffraction pattern?
5. How was μ_{hkl} determined?
6. How does the Mn^{3+}:Mn^{4+} ratio affect the magnetic ordering in this compound?
7. What crystal structures were observed?
8. How was M_{sat} measured?

The aim of this work was to classify the magnetic structures in the rich material system $[La_{1-x}Ca_x]MnO_3$. *Neutron diffraction can be used to comprehensively determine whether magnetic order is antiferromagnetic or not. The authors thus classified the magnetic structures observed in this family of compounds.*

8.6 Questions

8.6.1 Magnetic units, a quagmire

Unfortunately working with magnetic units is not always straightforward. In this set of questions we will be concentrating on switching between the more commonly used unit systems—CGS and SI. The following will be useful:

$$N_A = 6.022 \times 10^{23}\ \mathrm{mol}^{-1}$$
$$\mu_0 = 4\pi \times 10^{-7}\ \mathrm{H\,m}^{-1}$$
$$\mu_B = 9.274 \times 10^{-24}\ \mathrm{J\,T}^{-1}$$
$$B = \mu_0(H + M)\ \text{(SI)}$$
$$B = H + 4\pi M\ \text{(CGS)}$$
$$\chi = M/H$$
$$H_d = -NM\ \text{(SI)}$$

Fe atomic mass = 55.845
Al atomic mass = 26.98

Also, see appendices A and B.

1) *Magnetic flux density*
   - A. Convert magnetic flux of B = 100 Gauss to Tesla.
   - B. Convert magnetic flux of B = 1 mT to Gauss.
2) *Magnetic field strength*
   - A. Convert magnetic field of $H = 1$ Oe to A m^{-1}.
   - B. Convert magnetic field of $H = 2000$ A m^{-1} to Oe.
   - C. In free space, what would be the value of magnetic flux, B (in Gauss and Tesla), if a field of 1000 A m^{-1} is present?
   - D. In free space, what would be the value of magnetic flux, B (in Gauss and Tesla), if a field of 20 Oe was present?
3) *Magnetisation*
   - A. Convert magnetisation, $M = 20$ emu g^{-1} to Am^2 g^{-1}.
   - B. Convert magnetisation, $M = 50$ emu cm^{-3} to A m^{-1}.
   - C. $M = 3$ emu, $m = 0.05$ g, density = 5 g cm^{-3}.
     - i) Calculate magnetisation per unit volume in SI and CGS.
     - ii) Calculate the magnetisation per unit mass in SI and CGS.
     - iii) If the sample described above was Fe_3Al, what would the average magnetic moment per Fe atom be in SI units? In terms of μ_B?
   - D. The magnetisation of a sample is 100 emu, with a volume of 0.2 cm^2 and a mass of 1 g. Calculate the volume and mass magnetisation in emu g^{-1}, A m^{-2} g^{-1}, emu cm^{-2} and A m^{-1}.
4) *Permeability*

   If the magnetic permeability of a material, μ, is 1.2×10^{-6} H m^{-1}, calculate its relative permeability, μ_r and magnetic susceptibility, χ. Is it likely to be paramagnetic, ferromagnetic, or diamagnetic?

8.6.2 Magnetic interactions

1) *The magnetic dipolar interaction*

   The magnetic dipolar interaction is defined as:

$$E = \frac{\mu_0}{4\pi r^3}\left[\boldsymbol{\mu}_1 \cdot \boldsymbol{\mu}_2 - \frac{3}{r^2}(\boldsymbol{\mu}_1 \cdot \boldsymbol{r})(\boldsymbol{\mu}_2 \cdot \boldsymbol{r})\right]$$

Calculate E, in terms of the separation r, and moments μ_1, μ_2, for the following dipoles:

a) ↑ ↑
b) ↑ ↓
c) ↑ ←
d) ↑→
e) ↑↗
f) →→
g) ←→
h) ←↗

8.6.3 Extracting information from magnetisation measurements

1. *Magnetometry*
   A. For the data given in figure Q8.1, determine the saturation moment (σ_M) at 295 K in CGS and SI units if the mass of material measured was 60 μg.
   B. Estimate the coercive field, H_c, in CGS and SI at 295 K, 120 K, 110 K, 100 K, and 50 K.
   C. What are the corresponding coercive fields, B_c, in CGS and SI units?

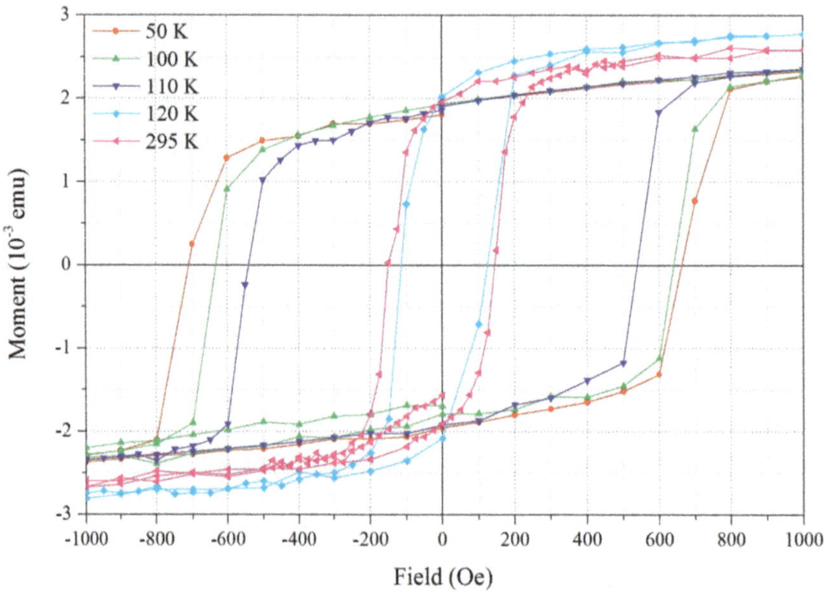

Figure Q8.1. Hysteresis loop of a FeO_x thin film.

2. *Demagnetisation factor*
   A. Estimate the demagnetisation field for the following magnetic measurements:
      i) A sphere of nickel with applied magnetic field of 1 T, volume of 1 cm^3, and measured magnetic moment of 1.45×10^3 emu.
      ii) A sphere of nickel in a field with magnetic strength of 100 Oersted, volume of 1 cm^3 and measured magnetic moment of 1.2×10^3 emu.
      iii) A sphere of material with applied magnetic field of 1000 Gauss, density of 8.9 g cm^{-3}, mass of 10 g and measured moment of 1.3×10^3 emu.
   B. How would the demagnetisation field differ for Q 2A if it were a plate of material with field applied perpendicular and parallel to the long direction?
3. *Magnetic moment*
   A. Determine the magnetic moment per Fe atom, in units of μ_B, for La ($Fe_{11}Si$) if the saturation moment at 5 K = 200 emu g^{-1}.
   B. If the chemical formula of the data given in figure 1 was Fe_3O_4 was what is the magnetic moment in units of μ_B per Fe atom?
   C. If the chemical formula of the data given in figure 1 was Fe_2O_3 what is the magnetic moment in units of μ_B per Fe atom?

References

[1] Mills A A 2004 The lodestone: history, physics and formation *Ann. Sci.* **61** 273–319

[2] Gilbert W *De Magnete (London, 1600)* trans. by Fleury Mottelay P 1893
Gilbert W *De Magnete (London, 1600)* (New York: Dover) 1958

[3] Needham J 1962 *Science and Civilisation in China* IV (Cambridge: Cambridge University Press)
Ronan C A and Needham J 1986 *The Shorter Science and Civilisation in China* III (Cambridge: Cambridge University Press) ch 1

[4] Portmann O, Vaterlaus A and Pescia D 2003 An inverse transition of magnetic domain patterns in ultrathin films *Nature* **422** 701–4

[5] Hubert A and Shäfer R 1998 *Magnetic Domains: The Analysis of Magnetic Microstrucures* (Berlin: Springer)

[6] Gupta S and Suresh K G 2015 *J. Alloys Compd.* **618** 562–606

[7] Nisoli C, Moessner R and Schiffer P 2013 *Rev. Mod. Phys.* **85** 1473–90

[8] Warner M and Cheung A C H 2013 *A Cavendish Quantum Mechanics Primer* (Cambridge: Periphyseos Press)

[9] Coey J M D 2010 *Magnetism and Magnetic Materials* (Cambridge: Cambridge University Press)

[10] Ruderman M A and Kittel C 1954 *Phys. Rev.* **96** 99–102

[11] Gercsi Z and Sandeman K G 2010 *Phys. Rev. B* **81** 224426

[12] Fert A R 1990 *Mater. Sci. Forum* **5960** 439–80

[13] Dmitrienko V E *et al* 2013 *Nat. Phys.* **10** 202

[14] Fert A *et al* 2017 *Nat. Rev. Mater.* **2** 17031

[15] Quinsat M *et al* 2017 *AIP Adv.* **7** 056318

[16] McElfresh M 1994 *Fundamentals of Magnetism and Magnetic Materials Quantum Design* (Purdue University)

[17] Moore J D *et al* 2009 *Adv. Mater.* **21** 3780–3

[18] McVitie S *et al* 2015 *J. Phys. Conf. Ser.* **644** 012026

[19] Schmidt Th *et al* 2013 *Ultramicroscopy* **126** 23–32

[20] Hiebert W K *et al* 1997 *Phys. Rev. Lett.* **79** 1134–7

[21] McCord J 2015 *J. Phys. D: Appl. Phys.* **48** 333001

[22] Sandhu A 2001 *Japan. J. Appl. Phys.* **40** part 1 6B

[23] Perkins G K *et al* 2002 *Supercond. Sci. Technol.* **15** 1140–6

[24] Kazakova O *et al* 2019 *J. Appl. Phys.* **125** 060901

[25] Wiesendanger R *et al* 1990 *Z. Phys. B—Condensed Matter* **80** 5–6

[26] Bode M 2003 *Rep. Prog. Phys.* **66** 523

[27] Chobrok G and Hofmann M 1976 *Phys. Lett. A* **57A** 257

[28] Koike K and Hayakawa K 1984 *Appl. Phys. Lett.* **45** 585–6

[29] Koike K and Hayakawa K 1984 *Jpn. J. Appl. Phys.* **23** L187–8

[30] Frömter R *et al* 1991 *Appl. Phys. A* **76** 869–71

[31] Frömter R *et al* 2008 *Phys. Rev. Lett.* **100** 207202

[32] Pollard S *et al* 2017 *Nat. Commun.* **8** 1–8

[33] Arita M *et al* 2014 *Mater. Trans.* **55** 403–9

[34] Krajnak M *et al* 2016 *Ultramicroscopy* **165** 42–50

[35] McGrouther D *et al* 2016 *New J. Phys.* **18** 095004

[36] Shibata N *et al* 2017 *Nat. Commun.* **8** 15631

[37] Shindo D 2005 *Mater. Trans.* **44** 2025–34

[38] Almeida T P *et al* 2016 *Sci. Adv.* **2** e1501801

[39] Park H *et al* 2014 *Nat. Nanotechnol.* **9** 337–42

[40] Dunin-Borkowski R E *et al* 2004 *Eur. J. Mineral.* **13** 685–9

[41] Kuksov A *et al* 2004 *J. Appl. Phys.* **95** 6530

[42] Schneidner C and Schonhense G 2002 *Rep. Prog. Phys.* **65** 1785–839

[43] Bezencenet O *et al* 2011 *Phys. Rev. Lett.* **106** 107201

[44] Nolting F *et al* 2000 *Nature* **405** 767–9

[45] van der Laan G 2011 *Phys. Rev. B* **83** 064409

[46] Suzuki M *et al* 2005 *Phys. Rev. B* **72** 054430

[47] Zhang S L *et al* 2017 *Scientific Reports* **7** 123

[48] Ladak S *et al* 2010 *Nature Physics* **6** 359

[49] Hälg M *et al* 2014 *Phys. Rev. B* **90** 174413

[50] Wollan E O and Koehler W C 1955 *Phys. Rev.* **100** 545

IOP Publishing

Characterisation Methods in Solid State and Materials Science

Kelly Morrison

Appendix A

Useful units and constants

Quantity	Symbol	Value
Speed of light in a vacuum	c	$2.997\,925 \times 10^{8}$ m s^{-1}
Planck's constant	h	$6.626\,069 \times 10^{-34}$ J s
Reduced Planck's constant	$\hbar$	$1.054\,562 \times 10^{-34}$ J s
Boltzmann's constant	k_B	$1.380\,651 \times 10^{-23}$ J K^{-1}
Avogadro's constant	N_A	$6.022\,142 \times 10^{23}$ mol^{-1}
Gas constant	R	8.314 47 J K^{-1} mol^{-1}
Gravitational constant	G	6.6742×10^{-11} N m^{2} kg^{-2}
Permitivity of free space	ε_0	$8.854\,189 \times 10^{-12}$ F m^{-1}
Permeability of free space	μ_0	$1.256\,637 \times 10^{-6}$ H m^{-1}
Fine structure constant	α	$7.297\,353 \times 10^{-3}$
Charge of an electron	e	$1.602\,177 \times 10^{-19}$ C
Mass of an electron	m_e	$9.109\,383 \times 10^{-31}$ kg
Mass of a proton	m_p	$1.672\,622 \times 10^{-27}$ kg
Mass of a neutron	m_n	$1.674\,927 \times 10^{-27}$ kg
Classical radius of an electron	r_e	$2.817\,938 \times 10^{-15}$ m
Bohr radius of the atom	a_o	$5.291\,771 \times 10^{-11}$ m
Bohr magneton	μ_B	$9.274\,009 \times 10^{-24}$ J T^{-1}
Nuclear magneton	μ_N	$5.050\,783 \times 10^{-27}$ J T^{-1}

doi:10.1088/2053-2563/ab2df5ch9

IOP Publishing

Kelly Morrison

Appendix B

Magnetic units

Quantity	Symbol	Conversion
Magnetic flux density	B	1 Gauss (G) = 10^{-4} Tesla (T, Wb m^{-2})
Magnetic flux density	Φ	1 Maxwell (Mx, Gcm^2) = 10^{-8} Weber (Wb, Vs)
Magnetic field strength	H	1 Oersted (Oe, Gb cm^{-1}) = $10^3/4\pi$ (A m^{-1})
Volume magnetisation	M	1 emu cm^{-3} = 10^3 A m^{-1}
	$4\pi M$	1 Gauss (G) = $10^3/4\pi$ (A m^{-1})
Intensity of magnetisation	J, I	1 emu cm^{-3} = $4\pi \times 10^{-4}$ (T, Wb m^{-2})
Magnetisation per unit mass	σ, M	1 emu g^{-1} = 1 Am^2 kg^{-1}
		1 emu g^{-1} = $4\pi \times 10^{-7}$ Wb m kg^{-1}
Magnetic moment	m	1 emu (or erg G^{-1}) = 10^{-3} Joules per Tesla (Am^2, J T^{-1})
Magnetic dipole moment	J	1 emu (or erg G^{-1}) = $4\pi \times 10^{-10}$ Wb m
Volume susceptibility	χ	1 emu cm^{-3} = 4π (dimensionless)
		1 emu cm^{-3} = $(4\pi)^2 \times 10^{-7}$ Henry per meter (H m^{-1}, Wb)
Mass susceptibility	χ_p	1 cm^3 g^{-1} (or emu g^{-1}) = $4\pi \times 10^{-3}$ m^3 kg^{-1}
		1 cm^3 g^{-1} (or emu g^{-1}) = $(4\pi)^2 \times 10^{-10}$ H m^2 kg^{-1}
Permeability	μ	1 (dimensionless) = $4\pi \times 10^{-7}$ H m^{-1} (or Wb)
Relative permeability	μ_r	—
Energy density per unit volume	W	1 erg cm^{-3} = 10^{-1} J m^{-3}
Demagnetisation factor	D	
Magnetic potential difference	U, F	1 Gilbert (Gb) = $10/4\pi$ Ampere (A)

Gaussian and CGS emu	$B = H + 4\pi M$
SI	$B = \mu_0(H + M)$
μ_0	$4\pi \times 10^{-7}$ H m^{-1}

doi:10.1088/2053-2563/ab2df5ch10

IOP Publishing

Kelly Morrison

Appendix C

Space and symmetry groups

Triclinic		48	Pban	97	I422	146	R3	**Cubic**	
1	P1	49	Pccn	98	$I4_122$	147	P-3	195	P23
2	P-1	50	Pban	99	P4mm	148	R-3	196	F23
Monoclinic		51	Pmma	100	P4bm	149	P312	197	I23
3	P2	52	Pnna	101	$P4_2cm$	150	P321	198	$P2_13$
4	$P2_1$	53	Pmna	102	$P4_2nm$	151	$P3_112$	199	$I2_13$
5	C2	54	Pcca	103	P4cc	152	$P3_121$	200	Pm-3
6	Pm	55	Pbam	104	P4nc	153	$P3_212$	201	Pn-3
7	Pc	56	Pccn	105	$P4_2mc$	154	$P3_221$	202	Fm-3
8	Cm	57	Pbcm	106	$P4_2bc$	155	R32	203	Fd-3
9	Cc	58	Pnnm	107	I4mm	156	P3m1	204	Im-3
10	P2/m	59	Pmmn	108	I4cm	157	P31m	205	Pa-3
11	$P2_1/m$	60	Pbcn	109	$I4_1md$	158	P3c1	206	Ia-3
12	C2/m	61	Pbca	110	$I4_1cd$	159	P31c	207	P432
13	P2/c	62	Pnma[a]	111	P-42m	160	R3m	208	$P4_232$
14	$P2_1/c$	63	Cmcm	112	P-42c	161	R3c	209	F432
15	C2/c	64	Cmce	113	$P\text{-}42_1m$	162	P-31m	210	$F4_132$
Orthorhombic		65	Cmmm	114	$P\text{-}42_1c$	163	P-31c	211	I432
16	P222	66	Cccm	115	P-4m2	164	P-3m1	212	$P4_332$
17	$P222_1$	67	Cmme	116	P-4c2	165	P-3c1	213	$P4_132$
18	$P2_12_12$	68	Ccce	117	P-4b2	166	R-3m	214	$I4_132$
19	$P2_12_12_1$	69	Fmmm	118	P-4n2	167	R-3c	215	P-43m
20	$C222_1$	70	Fddd	119	I-4m2	**Hexagonal**		216	F-43m
21	C222	71	Immm	120	I-4c2	168	P6	217	I-43m
22	F222	72	Ibam	121	I-42m	169	$P6_1$	218	P-43n
23	I222	73	Ibca	122	I-42d	170	$P6_5$	219	F-43c
24	$I2_12_12_1$	74	Imma	123	P4/mmm	171	$P6_2$	220	I-43d
25	Pmm2	**Tetragonal**		124	P4/mcc	172	$P6_4$	221	Pm-3m
26	$Pmc2_1$	75	P4	125	P4/nbm	173	$P6_3$	222	Pn-3n
27	Pcc2	76	$P4_1$	126	P4/nnc	174	P-6	223	Pm-3n

doi:10.1088/2053-2563/ab2df5ch11

28	Pma2	77	$P4_2$	127	P4/mbm	175	P6/m	224	Pn-3m
29	$Pca2_1$	78	$P4_3$	128	P4/mnc	176	$P6_3/m$	225	Fm-3m
30	Pnc2	79	I4	129	P4/nmm	177	P622	226	Fm-3c
31	$Pmn2_1$	80	$I4_1$	130	P4/ncc	178	$P6_122$	227	Fd-3m
32	Pba2	81	P-4	131	$P4_2/mmc$	179	$P6_522$	228	Fd-3c
33	$Pna2_1$	82	I-4	132	$P4_2/mcm$	180	$P6_222$	229	Im-3m
34	Pnn2	83	P4/m	133	$P4_2/nbc$	181	$P6_422$	230	Ia-3d
35	Cmm2	84	$P4_2/m$	134	$P4_2/nnm$	182	$P6_322$		
36	$Cmc2_1$	85	P4/n	135	$P4_2/mbc$	183	P6mm		
37	Ccc2	86	$P4_2/n$	136	$P4_2/mnm$	184	P6cc		
38	Amm2	87	I4/m	137	$P4_2/nmc$:	185	$P6_3cm$		
39	Aem2	88	$I4_1/a$	138	$P4_2/ncm$	186	$P6_3mc$		
40	Fmm2	89	P422	139	I4/mmm	187	P-6m2		
41	Fdd2	90	$P42_12$	140	I4/mcm	188	P-6c2		
42	Imm2	91	$P4_122$	141	$I4_1/amd$	189	P-62m		
43	Iba2	92	$P4_12_12$	142	$I4_1/acd$	190	P-62c		
44	Ima2	93	$P4_222$	**Trigonal**		191	P6/mmm		
45	Pmmm	94	$P4_22_12$	143	P3	192	P6/mcc		
46	Pnnn	95	$P4_322$	144	$P3_1$	193	$P6_3/mcm$		
47	Pccm	96	$P4_32_12$	145	$P3_2$	194	$P6_3/mmc$		

[a] Pbnm is the non-standard space group for #62 that is commonly used for manganites.

IOP Publishing

Characterisation Methods in Solid State and Materials Science

Kelly Morrison

Appendix D

Error propagation

The keystone of an experimental method is finding a way to communicate your level of confidence in a result. This is the purpose of the figure of merit discussed in chapter 3 when describing the Reitveld refinement method—i.e. how good is the simulated fit to your experimental data? When making simple measurements of, for example, length, time, or mass, it can be the level of confidence in the value you quote (i.e. the ruler was 30 ± 0.005 cm). For detection of x-rays or neutrons, it can be the respective certainty in the count rate at each measurement point (i.e. statistical error). In such a case, if you have a set of N values with some variance about a mean value, the sample standard deviation can be a good way of quantifying the uncertainty. The standard deviation, σ, is defined as:

$$\sigma = \sqrt{\frac{\sum_{i=1}^{N}(x_i - \bar{x})^2}{N}}$$

where $\bar{x}$ is the average (mean), and N is the total number of datapoints in your sample. Based on this we can define a confidence interval—how far outside of the standard deviation might we expect a datapoint to lie, as demonstrated by figure D1.

For example, 68% of datapoints would be expected to fall within the first standard deviation, σ, 95% of datapoints would be expected to lie within 2σ, 99.7% within 3σ and so on.

Notice that the standard deviation, σ, or error, decreases as N increases. This is not a linear relationship, however, rather proportional to the square root of N. Referred to as the 'law of diminishing returns' this is often the limitation of 'counting out' a statistical measurement: for a dataset collected in 1 h, to half the error will require counting for a further 2 h, to half it again will require counting for another 4 h (and so on). This often leads to the scenario where a compromise has to be made between the achievable certainty, and measurement time.

doi:10.1088/2053-2563/ab2df5ch12

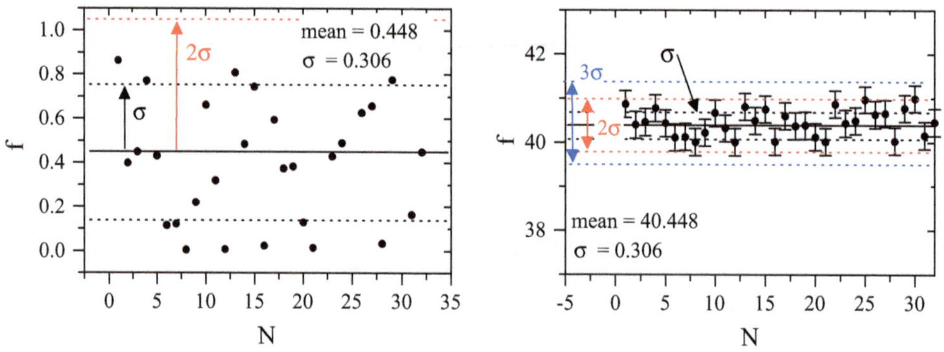

Figure D1. Example datasets with same standard deviation about a different mean value. Confidence intervals are indicated by dotted lines, where 99% of data is expected to lie within 2σ.

Another useful trick to remember is how to combine errors for often complex combinations of variables. For example, if you were to measure the time, t, it took for an object to travel a distance, d, in order to determine its speed:

$$v = \frac{d}{t}$$

there would be two variables (t, d) that could have significant uncertainties (σ_d, σ_t) associated with them. The error on the calculated speed, ($v \pm \sigma_v$) would be:

$$\sigma_v = \sqrt{\sigma_d{}^2/t^2 + d^2\sigma_t{}^2/t^4}$$

Sometimes also written as:

$$\frac{\sigma_v}{v} = \sqrt{\sigma_d{}^2/d^2 + \sigma_t{}^2/t^2}$$

This can be determined by using the general error propagation formula:

$$\sigma_f = \sqrt{\left(\frac{\partial f}{\partial x}\right)^2 \sigma_x{}^2 + \left(\frac{\partial f}{\partial y}\right)^2 \sigma_y{}^2 + \left(\frac{\partial f}{\partial z}\right)^2 \sigma_z{}^2 + \ldots}$$

where f is the function that you are trying to determine the total error for, and a, b, c… are the variables that define that function. Some common results of this are given below:

1) Addition or subtraction of two variables.

$$f = a + b$$
$$f = a - b$$
$$\sigma_f = \sqrt{\sigma_a{}^2 + \sigma_b{}^2}$$

2) Multiplication of two variables:

$$f = ab$$
$$\sigma_f = \sqrt{b^2\sigma_a^2 + a^2\sigma_b^2}$$
$$\frac{\sigma_f}{f} = \sqrt{\sigma_a^2/a^2 + \sigma_b^2/b^2}$$

3) Division of two variables:

$$f = a/b$$
$$\sigma_f = \sqrt{\sigma_a^2/b^2 + a^2\sigma_b^2/b^4}$$
$$\frac{\sigma_f}{f} = \sqrt{\sigma_a^2/a^2 + \sigma_b^2/b^2}$$

Notice that the relative error, σ_f/f is the same for case (2) and (3).

IOP Publishing

Characterisation Methods in Solid State and Materials Science

Kelly Morrison

Appendix E

Cartesian and spherical polar coordinate systems

A common mathematical trick when trying to calculate the area or volume of an object by integration is to choose a coordinate system that mirrors the 'symmetry' of the object. An example of this in chapter 3 was the use of the spherical polar coordinate system. This is particularly useful if you have a point source that has an isotropic distribution (same amplitude in all directions). A summary of the variables, ranges, volume (dV) and area (dA) integrals is given in the table below.

Coordinate system	Variable forms	Ranges	dV dA
Cartesian	x	$-\infty \leqslant x < \infty$	$dxdydz$
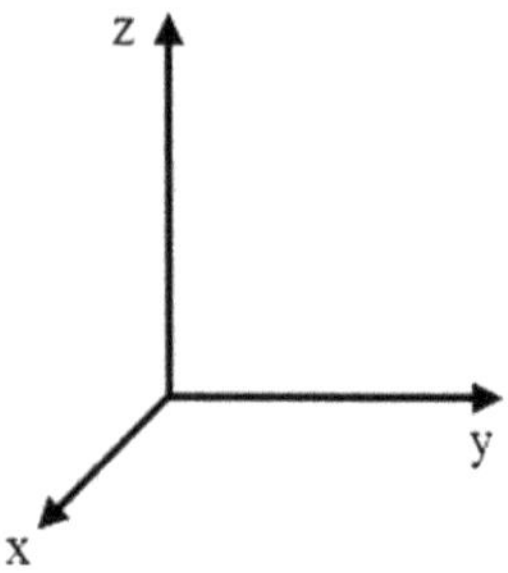	y	$-\infty \leqslant y < \infty$	
	z	$-\infty \leqslant z < \infty$	

(*Continued*)

Cylindrical	$x = r \cos\varphi$	$0 \leqslant r < \infty$	$r\, \mathrm{d}r\mathrm{d}\varphi\mathrm{d}z$
	$y = r \sin\varphi$	$0 \leqslant \varphi < 2\pi$	
	$z = z$	$-\infty \leqslant r < \infty$	

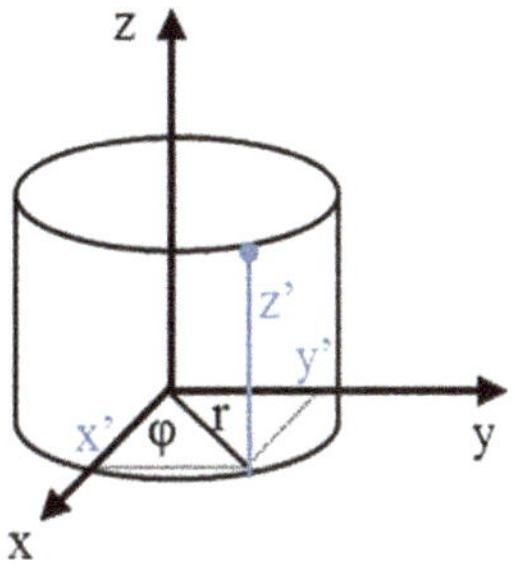

Spherical polar	$x = r \sin\theta \cos\varphi$	$0 \leqslant r < \infty$	$r^2 \sin\theta\, \mathrm{d}r\mathrm{d}\theta\mathrm{d}\varphi$
	$y = r \sin\theta \sin\varphi$	$0 \leqslant \varphi < 2\pi$	
	$z = r \cos\theta$	$0 \leqslant \theta < \pi$	

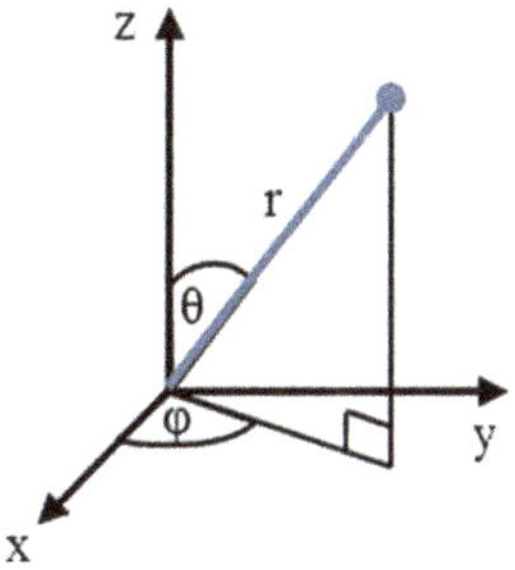

IOP Publishing

Characterisation Methods in Solid State and Materials Science

Kelly Morrison

Appendix F

Fullprof labscript

I will be walking you through the basics of using the Fullprof software to view, model and refine diffraction measurements, followed by some of the useful functions contained within. This is of course not the only software package available for Rietveld refinement, however, it is the most widely used (globally) open source version.

Further tutorials on the Fullprof (figure F1) studio can be found at https://www.ill.eu/sites/fullprof/php/tutorials.html.

1) Orientation
2) Load a datafile into **winplot**
    a) Change *x*-axis to show Q, or d-spacing

Figure F1. Screenshot of the fullprof toolbar.

    b) Normalise the dataset
    c) Edit title and axes labels.
3) Load a pcr file into **winplot**
    a) Display reflections list
4) Open the **pcr editor**
    a) Load one of the pcr files available at https://kmphysics.com/x-ray-diffraction-files/ or in the tutorial subfolder of the installation
    b) Make a note of the spacegroup of phases present
    c) Make a note of the lattice parameters
    d) Make a note of the atom positions

doi:10.1088/2053-2563/ab2df5ch14

e) Run the simulation/refinement
f) Make a note of which output files have now been generated.

5) Open **FpStudio**
a) Load the fst file that has now been generated.

Examples

The Heusler—Co_2MnSi—exhibits significantly different XRD patterns dependant on the position of individual Co, Mn and Si atoms. In the fully ordered state (L21) the basis can be described by Co at (¼, ¼, ¼); Mn at (0, 0, 0), Si at (½, ½, ½). The generic unit cell for this is shown in figure F2.

In this lab, you will be adapting the pcr file for the L21 Co_2MnSi structure to produce the three possible disordered states:

- B2—Mn and Si sites are equally occupied (by Mn and Si).
- DO3—Co and Mn sites are equally occupied (by 2Co and Mn).
- A2—Equal probability of 2Co, Mn, Si at each site.

The unit cells for these are shown in figure F3.

This will allow you to:

1. Simulate the x-ray diffraction pattern (pcr editor)
2. Observe it (winplot)—including the indexing of the peaks, and
3. Check that the crystal structure you have created agrees with figures F2 and F3 (Fpstudio).

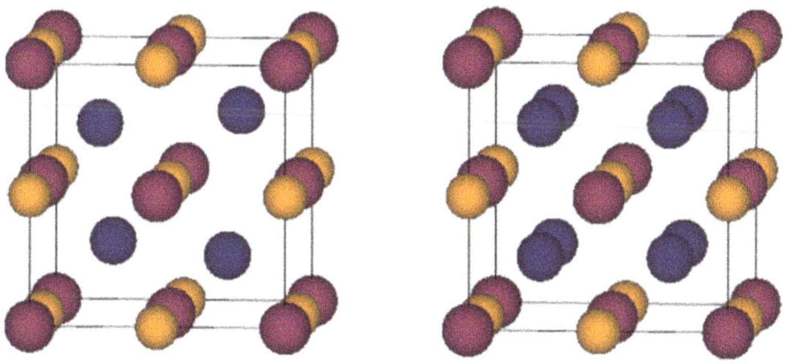

Figure F2. Unit cell for the (a) half-Heusler (XYZ) and (b) Heusler (X_2YZ) structures. X atoms (blue) Y atoms (purple) Z atoms (orange).

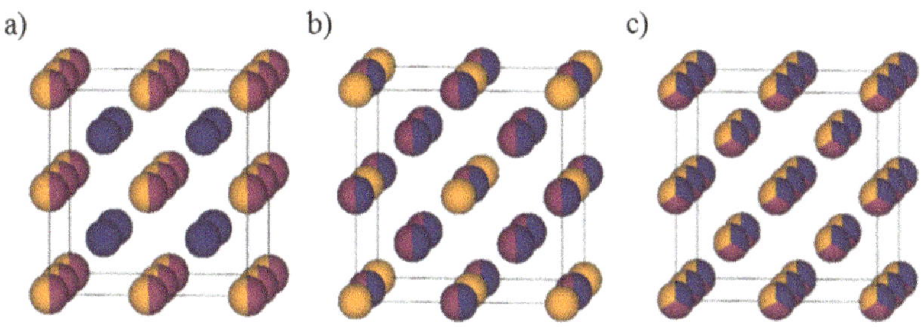

Figure F3. Unit cell for the (b) Heusler (Co_2MnSi) structure with (a) B2 disorder (Mn-Si site swap), (b) DO3 disorder (Co-Mn swap), and (c) A2 disorder (Co-Mn-Si swap). Co atoms (blue) Mn atoms (purple) Si atoms (orange).

1) Load the L21 file into the pcr editor:
   a) Open the pcr editor by clicking on the icon indicated in figure F1
   b) The screen shown in figure F4 should appear

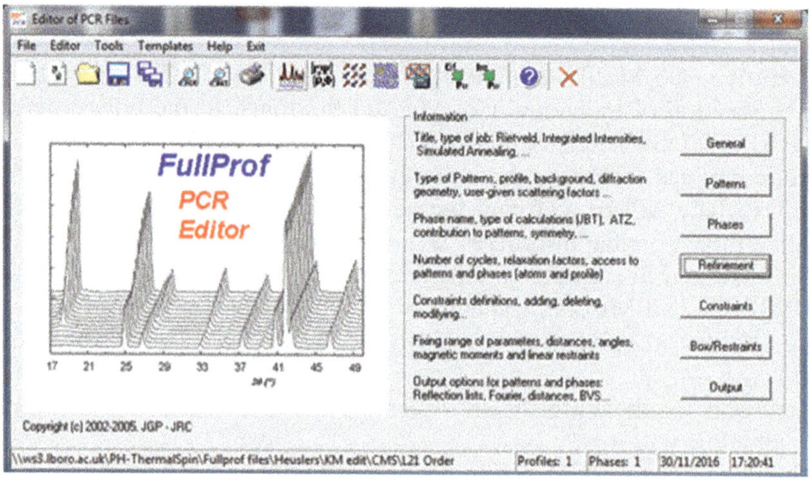

Figure F4. Screenshot of Fullprof PCR Editor main menu.

   c) To open the pcr file go to 'File' ➔ 'Open' and choose the pcr file 'L21.pcr'.
2) Check the spacegroup:
   a) From the ED PCR window click 'Phases'
   b) The following screen should appear (figure F5)

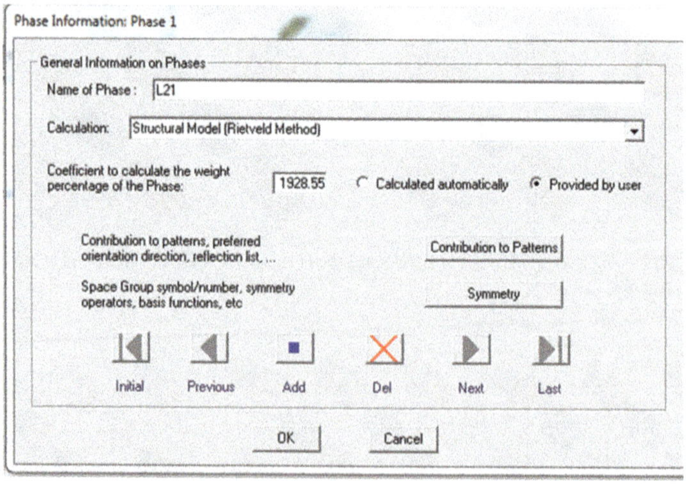

Figure F5. Screenshot of Fullprof PCR Editor Phase Information sub menu.

   c) Click on 'symmetry'
   d) The following screen should appear (figure F6).

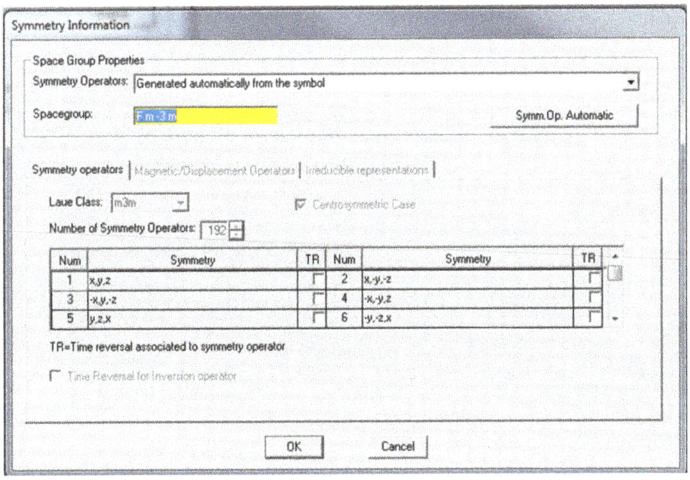

Figure F6. Screenshot of Fullprof PCR Editor Symmetry Information sub menu.

You should see here that the space group is Fm-3m. Note that you can choose the space group number (1–230) or the symbols that represent it (Fm-3m, or R3c or…). If you change this, then you need to click 'Symm. Op. Automatic' so that the symmetry operations that are relevant for that spacegroup are updated.

3) Check the atom positions:
   a) From the ED PCR window click 'Refinement'
   b) The following screen should appear (figure F7)

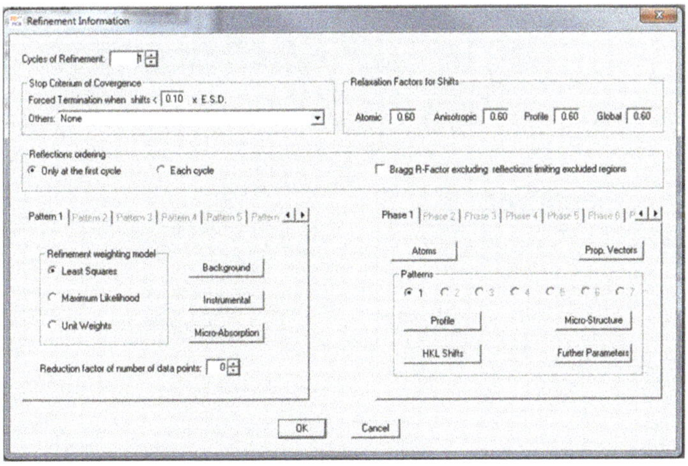

Figure F7. Screenshot of Fullprof PCR Editor Refinement sub menu.

   c) Click on 'Atoms'
   d) The following screen should appear (figure F8).

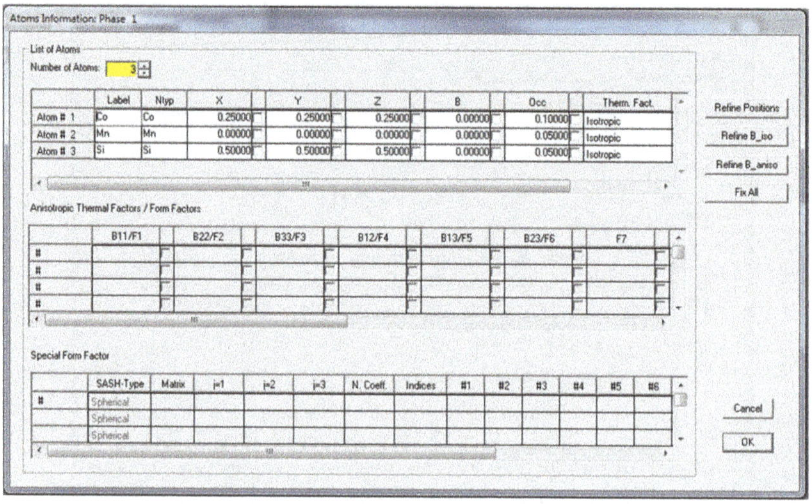

Figure F8. Screenshot of Fullprof PCR Editor Atoms Information sub menu.

In the top box of this screen you can see the positions of the atoms [Co at (¼, ¼, ¼); Mn at (0, 0, 0), Si at (½, ½, ½)], as described earlier.

There are several columns here: (x, y, z) correspond to the position of the atoms; Occ. corresponds to the relative occupancy of that site (useful when we start to introduce disorder); Ntyp corresponds to the element at that site. All other columns are not important for this tutorial.

4) Run the pcr file—this will generate additional files that can be read by winplot and Fpstudio. You can do this by exiting all the submenus and clicking the refinement button from the main ED PCR window (figure F9):

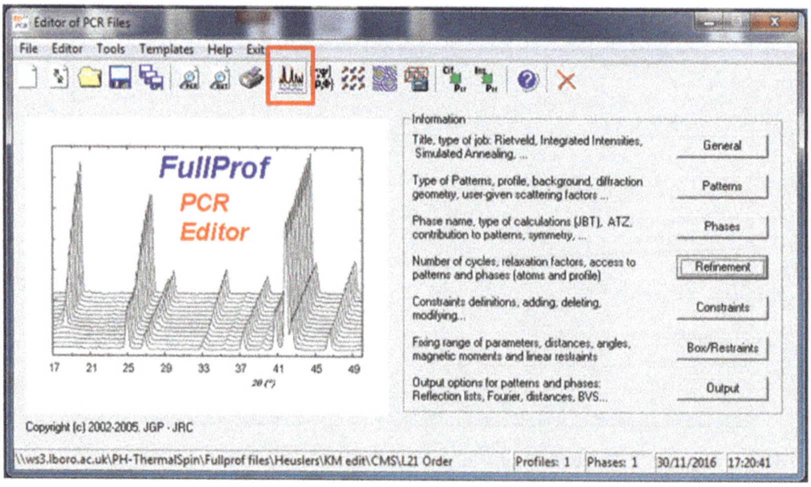

Figure F9. Screenshot of Fullprof PCR Editor Refinement sub menu, refinement button.

This will open the simulation window (figure F10):

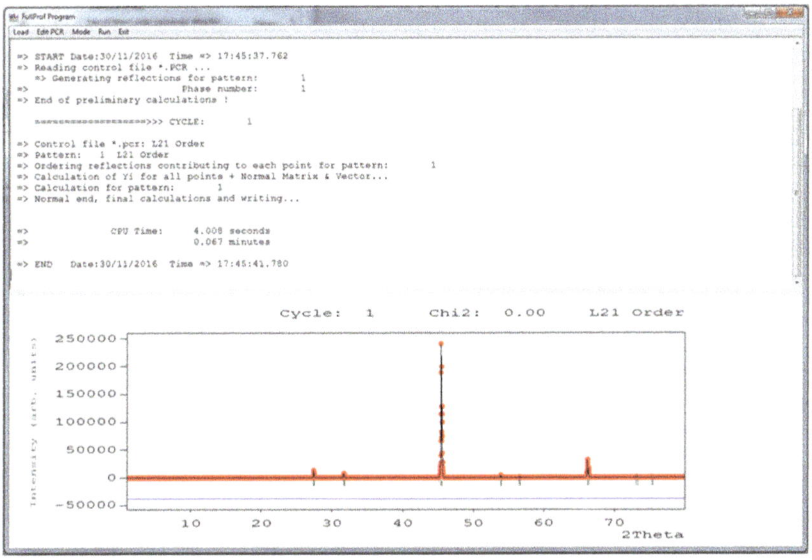

Figure F10. Screenshot of the Fullprof Simulation window.

Once the simulation has finished, close it and start checking the pattern produced.

5) From the Fullprof toolbar click on the Winplot icon indicated in figure F1 to open the window shown in figure F11.

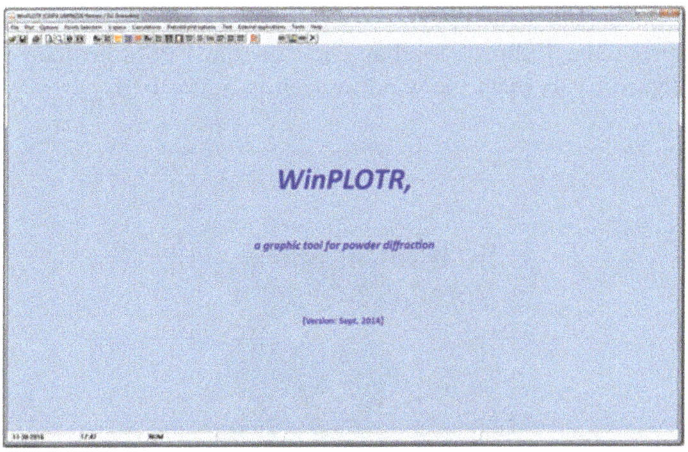

Figure F11. Screenshot of the WinPLOTR window.

6) Load the prf file—this is a file that was produced when you ran the simulation in ED PCR.
   a) Click 'File' ➔ 'Open Rietveld/profile file'
   b) Make sure '101: Fullprof PRF file' is selected
   c) Choose the L21 prf file—it should be in the folder you saved your pcr file in.
7) Load the Miller indices for each peak:
   a) Click 'Text' ➔ 'Write reflexions text' ➔ 'Write indices'
   b) You should now see the corresponding Miller indices for each Bragg peak (figure F12).

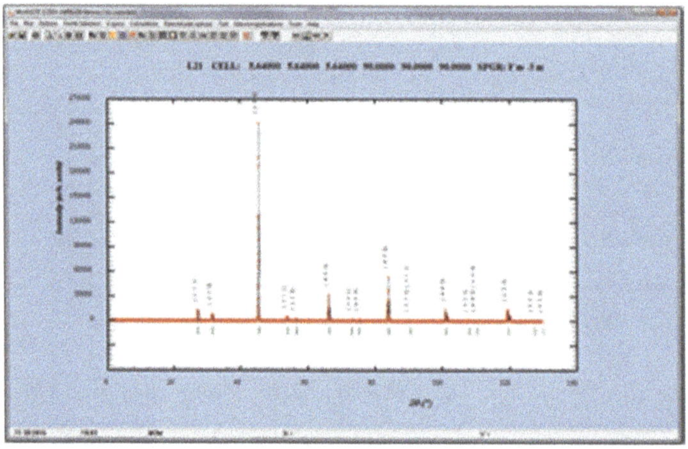

Figure F12. Screenshot of the WinPLOTR window with refinement data loaded.

Notice that there is a clear peak at (111)—this will disappear as you introduce disorder into this system.

8) Check the crystal structure.
   a) From the Fullprof toolbar click on the FPStudio icon indicated in figure F1 to open the window seen in figure F13

Figure F13. Screenshot of FPStudio.

b) Load the crystal structure by clicking 'File' → 'open' and finding the L21.fst file (it should be in the folder you saved your pcr file in).

The structure you've loaded should resemble figure F14.

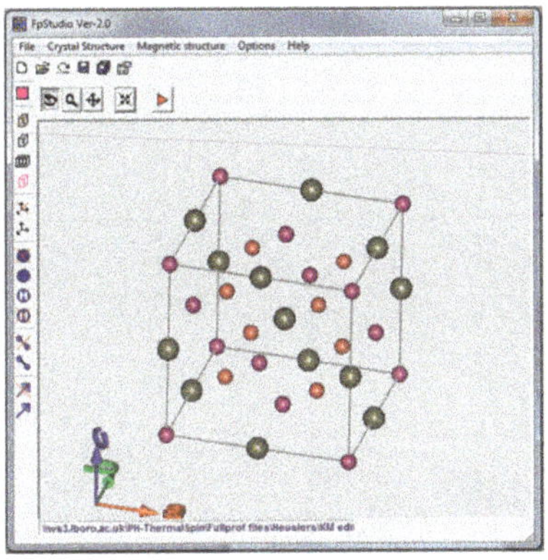

Figure F14. Screenshot of FPStudio with a crystal file loaded.

9) Now start constructing the disordered states.
   a) By referring to figure F3, convince yourself of the change in Basis. For example, for B2 disorder the Mn and Si sites are equally occupied (by Mn and Si). This would mean that you would need to define Si and Mn atoms at (0, 0, 0) and (½, ½, ½)
   b) From ED PCR, with L21.pcr loaded, open the refinement window (step 3) and add/remove atoms as needed
   c) Save the pcr with a new filename—perhaps 'B2.pcr'
   d) Run the simulation (Step 4)
   e) Check the XRD pattern (steps 5–7). For B2 you should see the pattern shown in figure F15

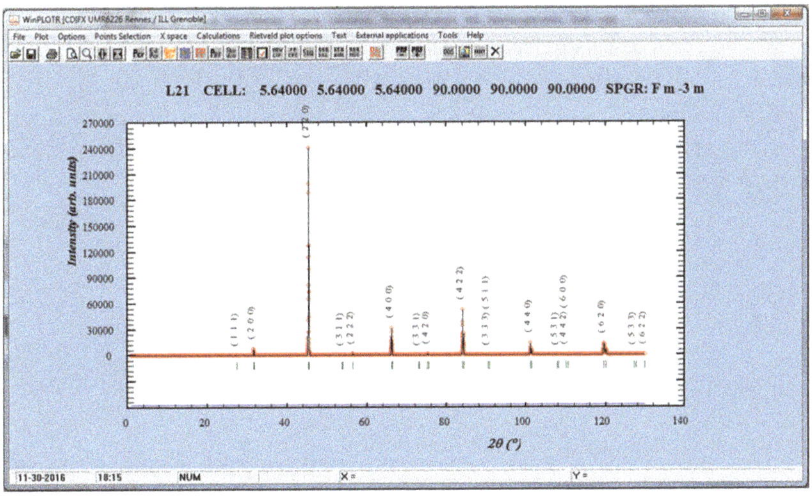

Figure F15. Screenshot of WinPLOTR with a refinement of a Co_2MnSi structure with B2 order loaded.

f) Check the crystal structure—this is especially important if something goes wrong (step 8).

Note that you may need to consider the occupancy…

For reference, the XRD patterns for the disordered Co_2MnSi states (figure F16):

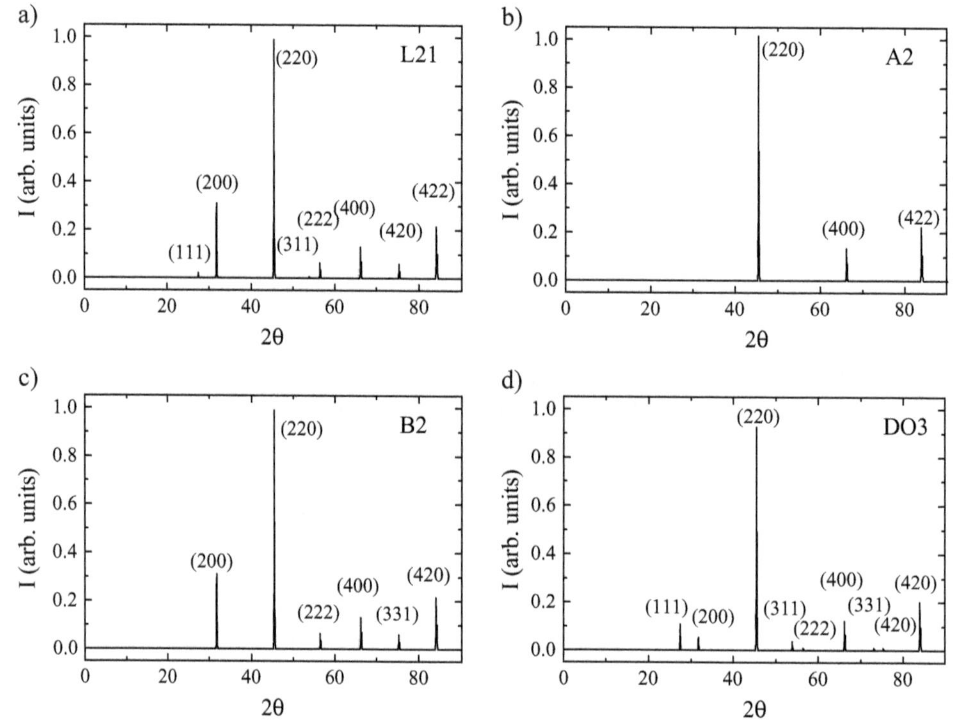

Figure F16. Summary of expected XRD patterns for (a) L21 order, (b) A2 disorder, (c) B2 disorder, and (c) DO3 disorder in the Co_2MnSi structure.

Note that the indexing is slightly different here compared to published data due to how the unit cell was defined.

Appendix G

GenX labscript

The aim of this computer lab is not just to familiarise you with the use of GenX to fit x-ray reflectivity data, but also to gain a deeper understanding of the various contributing factors to the reflectivity profile that you may observe.

GenX is an opensource fitting package for reflectivity data available at http://genx.sourceforge.net/ [1]. There are plenty of tutorials available in its online manual page http://genx.sourceforge.net/doc/.

Background

In chapter 3 we derived the following reflectivity profiles for a substrate and single layer film, respectively:

$$R(Q) \propto \frac{16\pi^2}{Q^4}\beta_s^2$$

$$R(Q) \propto \frac{16\pi^2}{Q^4}\left[\beta_1^2 + (\beta_1 - \beta_s)^2 - 2\beta_1(\beta_1 - \beta_s)\cos(LQ)\right]$$

where: β_1 is the scattering length density (SLD) of the layer; β_s is the scattering length density of the substrate; L is the thickness of the single layer; and Q is the momentum transfer vector ($\frac{4\pi\sin\theta}{\lambda}$).

As more layers are added this (Fourier based) treatment of the reflectivity profile becomes more complicated and it is easier/more accurate to use the dynamical scattering approach (see [2] p 104 onwards for more). In short, however, it is useful to remember that as with a diffraction grating, the fringe separation will be related to the inverse of the film thickness (grating spacing), and that a combination of multiple layers will result in 'beat' patterns in the reflectivity data. Hopefully you will see examples of this in this lab.

Part A. Getting started

1) To start, it is easier to work from a framework file for GenX, you can find this at https://kmphysics.com/x-ray-reflectivity-basic-models/: **#glass fit.hgx**.

What you should notice after loading this file is that data has appeared in the data and output boxes (as defined in figure G1*). Several parameters have also appeared in the grid tab (found in the simulation box). Note that the basic commands are indicated in* figure G2.

2) Run a single simulation.

   At this point the simulated data will appear on top of the provided data (symbols), with a scaled difference between simulated and experimental data below the main plot (FOM difference plot).

3) Change the figure of merit.

   *To do this click **settings**, **optimizer**, and choose a different figure of merit (FOM) from the drop-down box. Commonly used FOM for this type of data are **chi2bars** and **logR1**.*

4) Run a single simulation and observe how the FOM scan on the lower half has changed.

5) Select some parameters to fit in the simulation box (grid tab shown in figure G3).

   You should be able to judge this from the FOM difference plot, but if in doubt, intensity is probably a good starting point (just as with Rietveld refinement).

6) Run a continuous simulation.

   If you select the FOM tab in the outputs box (figure G4) you will be able to observe how this (hopefully) decreases with time. You could also monitor the

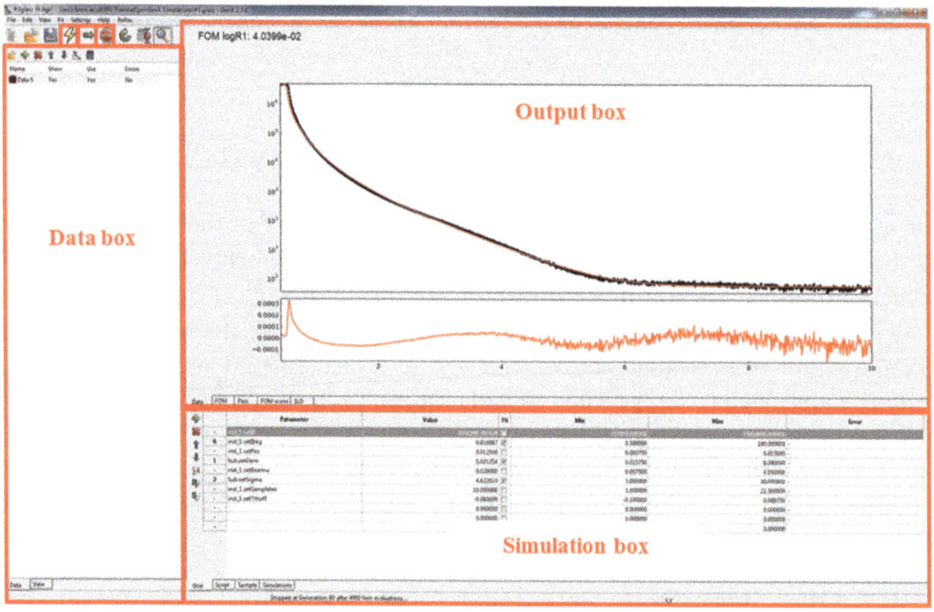

Figure G1. Overview of the GenX window.

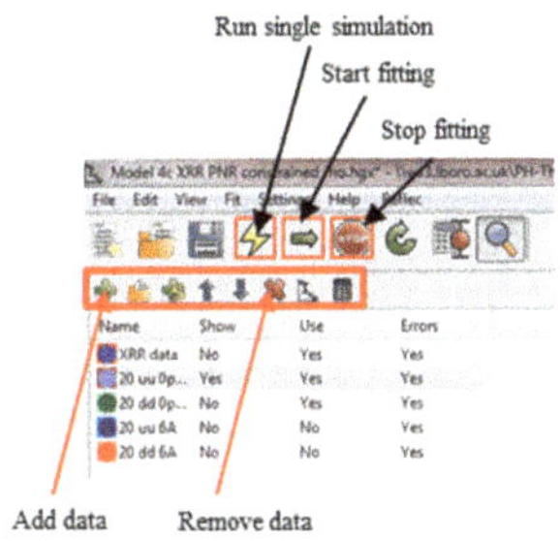

Figure G2. Key command window for GenX (I have assumed that the open/save file icons are obvious).

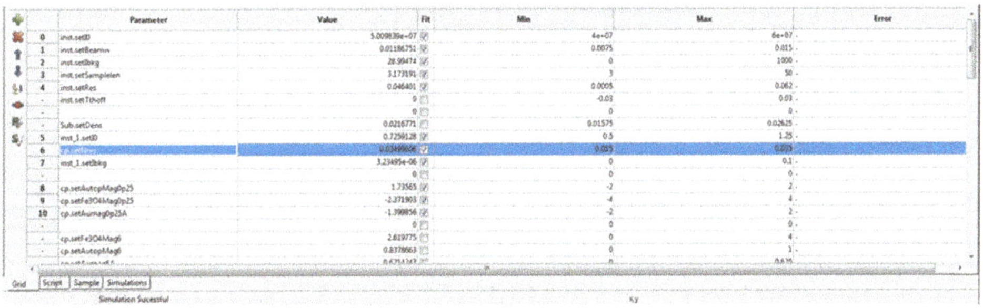

Figure G3. Illustration of the different tabs available in the simulation box: *Grid, Script, Sample, Simulations.*

parameters (Pars tab) to see how the range on the parameter(s) being fitted changes with time (and to observe whether the parameter is 'hitting a wall').

7) Try to obtain a good fit for the glass substrate by adding more parameters to the fit. You will need to stop the continuous simulation, before you make any changes.

   Key parameters include sample length (inst_1.setSamplelen), substrate roughness (Sub.setSigma), substrate density (Sub.setDens), Instrument max intensity I_0 (inst_1.setI_0), resolution (inst_1.set Res), beam width (intr_1.setBeamw) and background (inst_1.setbkg).

   Also, the chi2bars FOM will help to optimise for intensity and sample length, however the logR1 FOM will help to optimise for higher Q (substrate roughness).

8) Before you finish, check the SLD profile, found in the SLD tab of the outputs box. For a substrate it should be fairly simple (see figure G5).

Part B. Adding layers

In the sample tab of the simulation box, you should see some information about the sample you have just simulated/fit, in this case the information seen in figure G6.

You can add layers to this by first inserting a stack (give it a name), then inserting a layer (refer to figure G6).

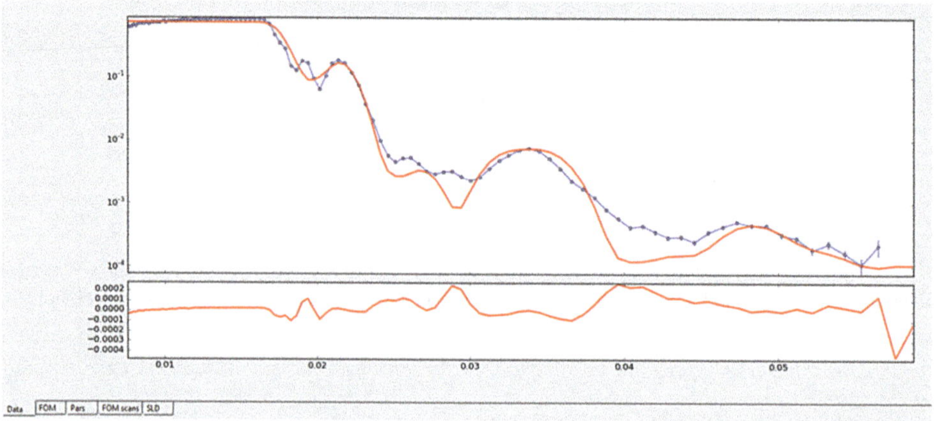

Figure G4. Illustration of the different tabs available in the output box: *Data, FOM (Figure of Merit), Pars (parameters), FOM scans, SLD (scattering length density).*

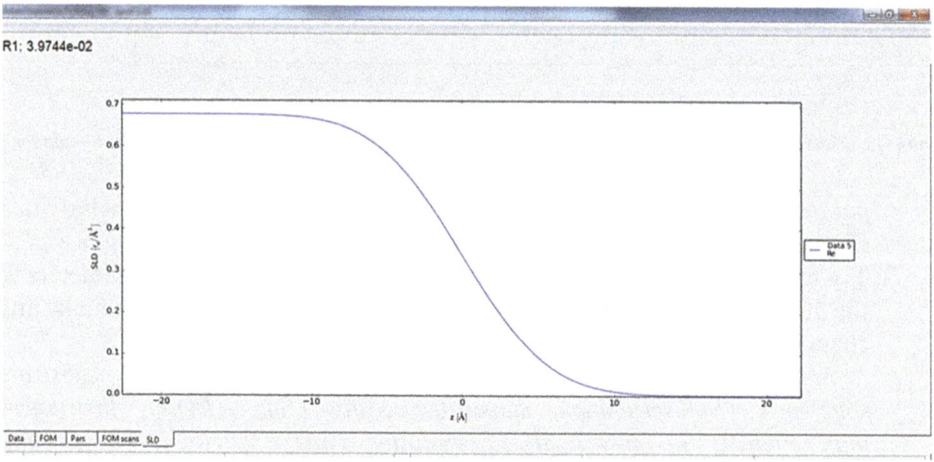

Figure G5. SLD profile for a glass substrate.

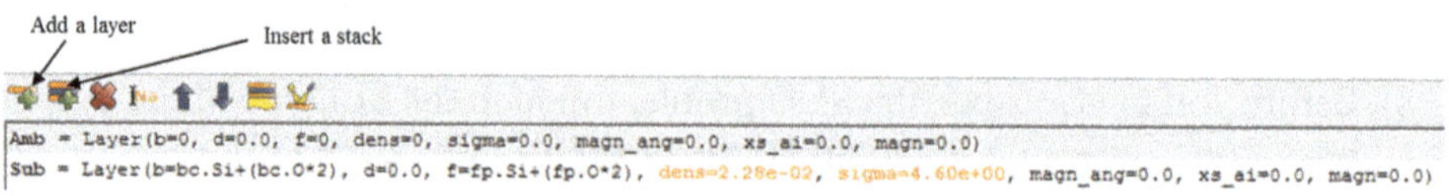

Figure G6. Sample information.

1) Insert a layer of Fe_3O_4. Double-click on the line of text that has just appeared for that layer to open the layer editor (see figure G7).
2) You will need to define the scattering lengths for this layer, this can be done by calling on a database hidden in the program. For x-rays the scattering length is contained in the object 'fp', for neutrons it is contained in the object 'bc'; each element is represented by its chemical symbol.

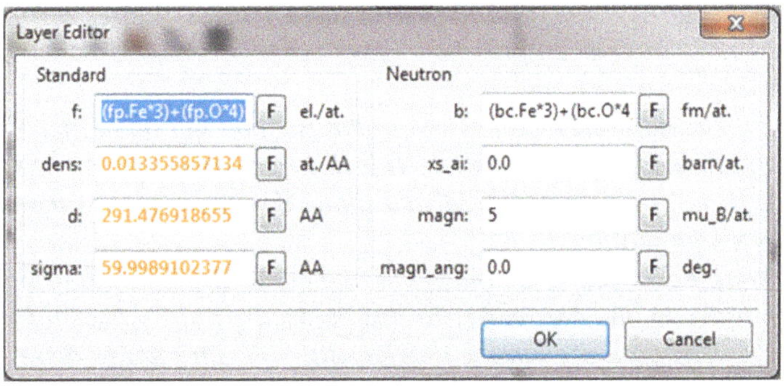

Figure G7. Layer editor accessed from the sample tab in the simulations box by inserting a stack then inserting a layer.

For example, for Fe you would type ***fp.Fe*** *for the standard/x-ray column and* ***bc.Fe*** *for the neutron column* (*see* figure G7).

3) You will also need to define the density (atoms per Å^3), thickness (Å), and roughness (Å). Some good values to start with are 0.015, 800, and 10, respectively.
4) As we are not simulating/fitting neutron data in any great detail here, you can leave the absorption (xs_ai), and magnetisation (magn/magn_ang) boxes empty.
5) Run a single simulation.

   You should see that multiple fringes have appeared and that the critical edge has shifted.
6) Change the roughness of the Fe_3O_4 layer to see how it affects the fringe amplitude.

   You can do this either from the sample tab in the simulation box, by double clicking on the Fe_3O_4 layer, or by adding a parameter in the grid tab of the simulation box (more later).

Notes:

Correlated roughness: roughness of substrate and layer is the same.
How does the SLD profile change as you vary the roughness?

Part C. Adding new data

Before loading new data to be fit you will need to make sure that the import settings are correctly set. In the data box you should see a dataset listed (e.g. Data 5).

1) Right click the dataset and select 'import settings'. Make sure that y error is set equal to 2 (this specifies the column number that is designated as the errors in your measurement).
2) Import a new dataset: select your dataset, choose File -> import -> import data, and select the example dataset: 'FeOx in glass.dat'.

You should now see a dataset with more fringes (closer to the simulation you just ran) in your outputs box (data tab).

3) Run a single simulation.

   Again, the difference between your dataset and the simulation will be shown on the bottom half of the data tab. It is likely that the intensity is very different so choose an appropriate FOM and allow the intensity to vary as you run a continuous simulation.

   You will now need to add new parameters to the grid tab of the simulation box, in order to find a fit closer to this dataset. All the values for the substrate will have been optimised in the previous fit; you can remove these from the list as they should not need to be fit.

4) Add some new parameters. Good values to fit/vary are the density, the roughness, and the thickness of the Fe_3O_4 layer; the instrumental background may also have changed.
5) See if you can obtain a reasonable fit for the Fe_3O_4 film.

   You may need to consider the effect of oxidation on the film, i.e. perhaps a thin layer of fully oxidised iron (Fe_2O_3) forms on the surface. Also, how might you simulate the effect of strain relaxation (as observed with the hut clusters weekly paper)?

6) **Extension** Load the FeOx:Pt data (an additional Pt layer on top of the Fe_3O_4 layer) and try to fit.

Part D. Simulating data

Now that you've navigated most of the basic features of GenX you can try simulating some data. First you will need to change the dataset from actual data to a simulation dataset.

1) Add a new dataset, then select this dataset and click on the calculator icon. This will open the data calculations box.
2) In the calculations box, under 'import from' select 'simulation' and click 'apply'.
3) You will need to define the instrument for the new dataset. In the simulations tab of the simulations box, double-click on the Data X: Specular(d.x, inst) and select 'inst_1' from the drop-down instrument menu. (This has been predefined as an x-ray based instrument).
4) Delete the old dataset by selecting it and clicking on.
5) Run a single simulation. The only data you'll see in the data tab of the outputs box is the simulation.
6) Simulate the following layers and estimate the thickness from the fringe separation ($t \sim 2\pi/\Delta Q$; $Q = 4\pi \sin\theta/\lambda$; λ has been set as 1.54 Å here):
   a. 5 nm Pt on glass
   b. 10 nm Fe_3O_4 on glass
   c. 5 nm Pt on $SrTiO_3$
   d. 80 nm Fe_3O_4:5 nm Au:10 nm Pt.

Part E. Changing the instrument parameters

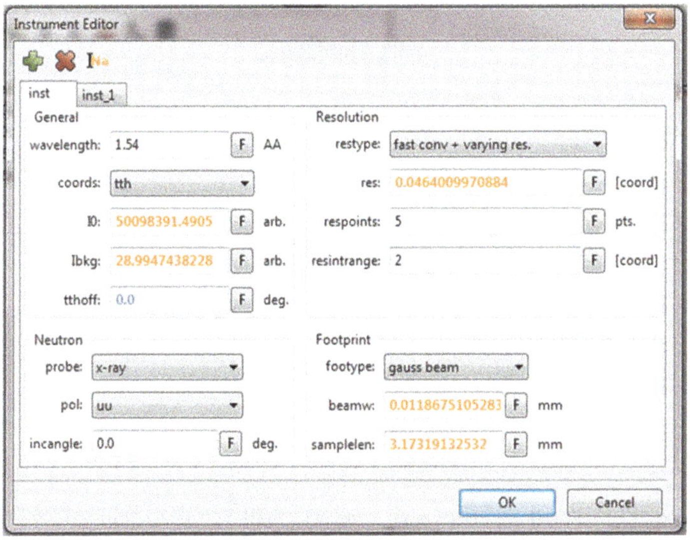

Figure G8. Instrument editor, showing the instrument parameters for polarised neutron reflectivity.

You can start to explore the fundamental difference between x-ray and neutron diffraction by changing the instrument defined in the simulation.

1) In the simulations tab of the simulations box, double-click on the Data X: Specular(d.x, inst_1) and change the instrument to 'inst'.
2) Run a single simulation and note how the profile changes.
3) To edit the specific properties of each 'instrument' you can select the instrument editor from the sample tab of the simulations box (see figure G8). Here you can specify whether your x-axis is in 2θ, or Q, or whether you are using x-rays, neutrons, polarised neutrons etc.

   Note: if you were to apply for beamtime to do neutron reflectivity (polarised or unpolarised), a good argument for why you need neutrons would be to simulate the pattern for both x-rays and neutrons where the contrast may be different.

References

[1] Björck M and Andersson G 2007 *J. Appl. Cryst.* **40** 1174–8

[2] Sivia D S 2011 *Elementary Scattering Theory: For X-ray and Neutron Users* (Oxford: Oxford University Press)

IOP Publishing

Characterisation Methods in Solid State and Materials Science

Kelly Morrison

Appendix H

Fourier optics labscript

Safety: Follow the normal laboratory safety rules. The use of a laser in this experiment requires care that eyes are not exposed to a direct un-attenuated (or focussed) beam. A beam expander attached to the laser should facilitate this.

Equipment:
- Laser;
- Beam expander;
- Eyepiece;
- Two convex lenses;
- Diffraction screen;
- Aperture;
- Slide holder.

Optional:
- Camera

Introduction: The Fourier transform (FT) is an extremely useful tool in the evaluation of diffraction phenomena. For any plane wave (optical, acoustic, x-ray etc) incident on an object, the angular distribution of the diffracted intensity is obtained by taking the FT of the transmission function of the object and then squaring the modulus of this transform. This diffraction pattern can then be treated as an object, the transform of which gives rise to an image of the original object. This transform can readily be realised by the use of simple lenses *as long as the Fraunhofer condition is met* (i.e. far field). The focal plane of the lens is the FT plane, i.e. the distribution of intensity at the focal plane of the lens is an accurate representation of the FT of the diffracting object. Since the image is formed from the diffraction pattern, alteration of the pattern leads to changes in the image and the concept of

doi:10.1088/2053-2563/ab2df5ch16

spatial filtering. Before the development of computers optical FT were widely used to solve problems in crystallography.

AIMS:

1. To show how diffraction can be explained and understood using FTs.
2. To show how properties of FT theory can be demonstrated using diffraction.
3. Using Fourier series to analyse a diffraction grating.
4. To apply Fourier theory to spatial filtering.

Some background

Fourier transforms

Physically an FT gives us a quantitative picture of frequency content of a function. For units $u = 1/x$ if x is in s m^{-1}, u is in frequency (s^{-1})/(m^{-1}).

$$f(x) = \int_{-\infty}^{\infty} F(u)e^{i2\pi ux}\mathrm{d}u \tag{H.1}$$

$$F(u) = \int_{-\infty}^{\infty} f(x)e^{-i2\pi ux}\mathrm{d}x \tag{H.2}$$

Refer to chapter 2 if you need a reminder on the general results of FT in optics and diffraction experiments.

Experimental set-up

Ensure that you have a collimated beam. You can do this simply by testing the radius of the beam on a piece of white paper as you move it away from the source. There are two commonly used methods for producing a collimated beam, as shown in figure H1. The first is to couple a lens placed one focal length away from a pinhole source (figure H1(a)). The second method is to use a beam expander on a laser, which comprises a lens combination that magnifies and collimates the beam from a smaller (often LED) source. You will also need two lenses, a slit holder, an eyepiece (for magnification), various objects (most likely a selection of slides) and a screen (to reconvolve the image onto).

Next align components at the same height in a straight line. Ensure that lenses are the correct distance from one another, as specified in figure H1, where f is the focal length of the nearest lens.

H.1 Classical diffraction examples

Set-up the experiment as shown in figure H1(c) and make a note of the diffraction pattern formed when the following objects are placed in the object plane (using the slit holder):

A) **Single slit**

- What happens if you increase the size of the slit?
- Does this agree with what you would expect from figure 2.2 (chapter 2)?

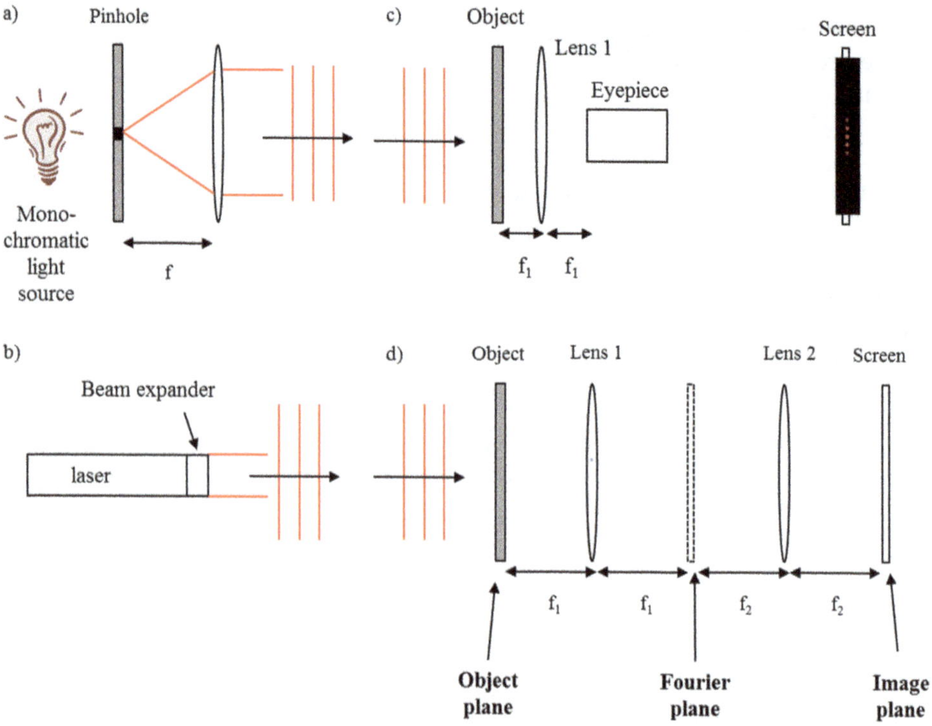

Figure H1. Basic experimental set-up to observe diffraction patterns seen in chapter 1 and produced during this exercise. (a) and (b) Methods of producing a collimated beam. (c) Use of an eyepiece to magnify the diffraction image. (d) The 4f arrangement.

- Confirm that you can reform the original image with the 4f arrangement shown in figure H1(d).

B) **Double/multiple slit**

- How does the diffraction pattern differ from the single slit?
- What happens if you increase the slit spacing?

If you do not have double slits of varying width/separation, you could calculate what you would expect with reference to chapter 2, section 2.6.1.

C) **BCC/FCC structure**

- How does the distance between diffraction ‘spots’ change for the image and the object?
- What would happen if the ‘atom’ spacing was decreased?

D) **○, □, ▽ and ▯ apertures.**

- Make a note of the diffraction pattern formed in each case.
- How would you relate the symmetry of the aperture with the diffraction pattern formed

H.2 Test of the *shift theorem*

Shift Theorem: $f(x) \leftrightarrow F(u); f(x-a) \leftrightarrow \exp^{-2\pi iua}; \exp^{2\pi iua} f(x) \leftrightarrow F(u-a)$

Place a single slit at the object plane and make a note of the diffraction pattern produced. What happens if you shift the slit along the horizontal axis (perpendicular to the beamline)? Does this agree with the above?

H.3 Test of the *addition theorem*

Addition theorem: $f(x) \leftrightarrow F(u)$, $g(x) \leftrightarrow G(u)$; $f(x) + g(x) \leftrightarrow F(u) + G(u)$

Exercise:

Using the definition of the FT and its inverse, show that:

$f(x) + g(x) \leftrightarrow F(u) + G(u)$

Now try this with Fourier optics:

- If $f(x)$ is the single slit aperture function, $f(x) = \mathrm{rect}(x/a)$, what will be the expected $F(u)$?
- If $g(x)$ is a rectangular aperture, $g(x) = \mathrm{rect}(x/a)\mathrm{rect}(y/b)$, what would be the expected $G(u)$?
- Now take a piece of black card and use a scalpel to cut a single slit and ▯ aperture, a small distance, x, away from each other.
- Sketch the diffraction pattern formed. Does this agree with what you would expect?

H.4 Test of the convolution theorem

Convolution theorem: $f(x) \otimes g(x) \leftrightarrow F(u)G(u)\; f(x)g(x) \leftrightarrow F(u) \otimes G(u)$

- Place the ▯ aperture and a diffraction grating at the object plane (convolution of the two functions).
- Make a note of the diffraction pattern formed.
- Does this agree with what you would expect?

1.5 Spatial filtering

The aim here is to filter out part of the image by using an appropriate filter in the Fourier plane.

- Set-up the experiment as shown in figure H1(d) with the ○ aperture as the object.
- Use a cross mask to block out part of the ○ diffraction pattern (at the Fourier plane). What image is formed by the second lens after this?
- What type of mask would you need to convert the ○ into a ▽?
- Test this.

Kelly Morrison

Appendix I

Selected solutions

2.9.2 Normalisation of the Fourier transform pair

2.2 Defining reciprocal space in terms of u (1/λ)
$N = 1$
2.3 Defining reciprocal space in terms of ω (2π/t)
$N = 1/\sqrt{2\pi}$

2.9.3 Further questions

3.1 A broadened aperture function

$$f(x) \otimes g(x) = \sqrt{2\pi}F(q)G(q) = \frac{\sqrt{2\pi}A \sin\left(\frac{qa}{2}\right)}{\pi q} e^{-\sigma^2 q^2/2}$$

3.7.1 Structure factor and reflection rules

1. *The structure factor of a simple cubic lattice of sodium atoms*
   a) Basis: (0,0,0)
   b) $\mathbf{r} = 0\mathbf{a_1} + 0\mathbf{a_2} + 0\mathbf{a_3}$
   c) $S_G = \sum f_j e^{-i.\boldsymbol{G}.\boldsymbol{r}_j}$

   $\boldsymbol{G}.\, \boldsymbol{r}_j = 2\pi(hx_j + ky_j + lz_j)$

   $S_G = f_{\mathrm{Na}} e^{-i2\pi(0h+0k+0l)} = f_{\mathrm{Na}}$

   Therefore a Bragg peak will be observed for any combination of {h,k,l} ($S_G = fNa$).
2. *The structure factor of a face centred cubic (FCC) lattice of sodium atoms*
   a) Basis: (0,0,0), (1/2,1/2,0), (1/2,0,1/2), (0,1/2,1/2)
   b) $\mathbf{r_1} = 0\mathbf{a_1} + 0\mathbf{a_2} + 0\mathbf{a_3}$

doi:10.1088/2053-2563/ab2df5ch17

$\mathbf{r_2} = 1/2\mathbf{a_1} + 1/2\mathbf{a_2} + 0\mathbf{a_3}$
$\mathbf{r_3} = 1/2\mathbf{a_1} + 0\mathbf{a_2} + 1/2\mathbf{a_3}$
$\mathbf{r_4} = 0\mathbf{a_1} + 1/2\mathbf{a_2} + 1/2\mathbf{a_3}$

c) $S_G = \sum_j f_j e^{-i.G.r_j}$

$$S_G = f_{\mathrm{Na}}(e^{-i2\pi(0h+0k+0l)} + e^{-i2\pi(1/2h+1/2k+0l)} + e^{-i2\pi(1/2h+0k+1/2l)} + e^{-i2\pi(0h+1/2k+1/2l)})$$
$$= f_{\mathrm{Na}}(1 + e^{-i\pi(h+k)} + e^{-i\pi(h+l)} + e^{-i\pi(k+l)})$$

Therefore a Bragg peak will be observed if {h,k,l} are all even or all odd ($S_G = 4fNa$).

3. *The structure factor of a body centred cubic (BCC) lattice of sodium atoms*
   a) Basis: (0,0,0), (1/2,1/2,1/2)
   b) $\mathbf{r_1} = 0\mathbf{a_1} + 0\mathbf{a_2} + 0\mathbf{a_3}$
   $\mathbf{r_2} = 1/2\mathbf{a_1} + 1/2\mathbf{a_2} + 1/2\mathbf{a_3}$
   c) $S_G = \sum_j f_j e^{-i.G.r_j}$

$$S_G = f_{\mathrm{Na}}(e^{-i2\pi(0h+0k+0l)} + e^{-i2\pi(1/2h+1/2k+1/2l)})$$
$$= f_{\mathrm{Na}}(1 + e^{-i\pi(h+k+l)})$$

Therefore a Bragg peak will be observed if (h+k+l) is even ($S_G = 2fNa$).

4. *The structure factor for NaCl*
   a) Basis Na: (0,0,0), (1/2,1/2,0), (1/2,0,1/2), (0,1/2,1/2)
   Basis Cl: (1/2,0,0), (0,1/2,0), (0,0,1/2), (1/2,1/2,1/2)
   b) Na:
   $\mathbf{r_1} = 0\mathbf{a_1} + 0\mathbf{a_2} + 0\mathbf{a_3}$
   $\mathbf{r_2} = 1/2\mathbf{a_1} + 1/2\mathbf{a_2} + 0\mathbf{a_3}$
   $\mathbf{r_3} = 1/2\mathbf{a_1} + 0\mathbf{a_2} + 1/2\mathbf{a_3}$
   $\mathbf{r_4} = 0\mathbf{a_1} + 1/2\mathbf{a_2} + 1/2\mathbf{a_3}$
   Cl:
   $\mathbf{r_5} = 1/2\mathbf{a_1} + 0\mathbf{a_2} + 0\mathbf{a_3}$
   $\mathbf{r_6} = 0\mathbf{a_1} + 1/2\mathbf{a_2} + 0\mathbf{a_3}$
   $\mathbf{r_7} = 0\mathbf{a_1} + 0\mathbf{a_2} + 1/2\mathbf{a_3}$
   $\mathbf{r_8} = 1/2\mathbf{a_1} + 1/2\mathbf{a_2} + 1/2\mathbf{a_3}$
   c) $S_G = \sum_j f_j e^{-i.G.r_j}$

$$S_G = f_{\mathrm{Na}}\left(e^{-i2\pi(0h+0k+0l)} + e^{-i2\pi(1/2h+1/2k+0l)} + e^{-i2\pi(1/2h+0k+1/2l)} + e^{-i2\pi(0h+1/2k+1/2l)}\right) + f_{\mathrm{Cl}}\left(e^{-i2\pi(1/2h+0k+0l)/} + e^{-i2\pi(0h+1/2k+0l)} + e^{-i2\pi(0h+0k+1/2l)} + e^{-i2\pi(1/2h+1/2k+1/2l)}\right)$$

$$S_G = f_{\mathrm{Na}}\left(1 + e^{-i\pi(h+k)} + e^{-i\pi(h+l)} + e^{-i\pi(k+l)}\right) + f_{\mathrm{Cl}}\left(e^{-i\pi h} + e^{-i\pi k} + e^{-i\pi l} + e^{-i\pi(h+k+l)}\right)$$

Extinction conditions only if $(1 + e^{-i\pi(h+k)} + e^{-i\pi(h+l)} + e^{-i\pi(k+l)}) = 0$ and $(e^{-i\pi h} + e^{-i\pi k} + e^{-i\pi l} + e^{-i\pi(h+k+l)}) = 0$
Therefore, Bragg peaks observed only if {h,k,l} are all odd ($S_G = 4f_{Na} - 4f_{Cl}$) *or all even* ($S_G = 4f_{Na} + 4f_{Cl}$).

d) $< S_{h,k,l} > = \left(\frac{f_{Na} + f_{Cl}}{2}\right)(1 + e^{-i\pi(h+k)} + e^{-i\pi(h+l)} + e^{-i\pi(k+l)}$

$+ e^{-i\pi h} + e^{-i\pi k} + e^{-i\pi l} + e^{-i\pi(h+k+l)})$

Extinction conditions only if $(1 + e^{-i\pi(h+k)} + e^{-i\pi(h+l)} + e^{-i\pi(k+l)} + e^{-i\pi h} + e^{-i\pi k} + e^{-i\pi l} + e^{-i\pi(h+k+l)} = 0$
Therefore, Bragg peaks observed only if {h,k,l} are all even ($S_G = 4(f_{Na} + f_{Cl})$).

5. *Reflection rules*
Lattice parameter $a = 4.0495$ Å
$\lambda = 1.54$ Å
$2d \sin\theta = n\lambda$

$$\frac{1}{d^2} = \frac{h^2 + k^2 + l^2}{a^2}$$

(100): $d = a$ $2\theta = 2\sin^{-1}(\lambda/2d) = 21.92$
(110): $d = a/\sqrt{2}$ $2\theta = 31.2$
(111): $d = a/\sqrt{3}$ $2\theta = 38.5$
(200): $d = a/2$ $2\theta = 44.7$
(220): $d = a/2\sqrt{2}$ $2\theta = 65$
Reflections:

	SC	FCC	BCC	Observed?
100	√	x	x	?
110	√	x	√	x
111	√	√	x	√
200	√	√	√	√
220	√	√	√	√

The sample is FCC.

3.7.2 Extracting key information from x-ray diffraction data

1. *Superstructure lines*

a) Basis Au: (0,0,0)
Basis Cu: (1/2,1/2,0), (1/2,0,1/2), (0,1/2,1/2)
Au:
$\mathbf{r}_1 = 0\mathbf{a}_1 + 0\mathbf{a}_2 + 0\mathbf{a}_3$
Cu:
$\mathbf{r}_2 = 1/2\mathbf{a}_1 + 1/2\mathbf{a}_2 + 0\mathbf{a}_3$

$\mathbf{r}_3 = 1/2\mathbf{a}_1 + 0\mathbf{a}_2 + 1/2\mathbf{a}_3$
$\mathbf{r}_4 = 0\mathbf{a}_1 + 1/2\mathbf{a}_2 + 1/2\mathbf{a}_3$

$$S_G = \sum f_j e^{-i.\boldsymbol{G}.\boldsymbol{r}_j}$$
$$S_G = f_{\mathrm{Au}} e^{-i2\pi(0h+0k+0l)} + f_{\mathrm{Cu}}(e^{-i2\pi(1/2h+1/2k+0l)} + e^{-i2\pi(1/2h+0k+1/2l)} + e^{-i2\pi(0h+1/2k+1/2l)})$$
$$S_G = f_{\mathrm{Au}} + f_{\mathrm{Cu}}(e^{-i\pi(h+k)} + e^{-i\pi(h+l)} + e^{-i\pi(k+l)})$$

b) *None. As $f_{\mathrm{Au}} \neq f_{\mathrm{Cu}}$ the terms above can never fully cancel one another for any combination of $\{h,k,l\}$*
c) $S_G = \frac{(f_{Au}+f_{Cu})}{2}(1+e^{-i\pi(h+k)} + e^{-i\pi(h+l)} + e^{-i\pi(k+l)})$
d) *Extinction conditions: if $\{h,k,l\}$ are not ALL odd or ALL even.*
e) *Superstructure lines are the additional 'forbidden' reflections not observed for (d), but observed for (a). For {100}, {110}, {211}, {221}, {201} there is a combination of odd and even $\{h,k,l\}$ hence, reflections would not be observed for the disordered alloy (but would for the ordered alloy).*

2. *Heusler alloys*
a) $a = b = c$, therefore likely cubic structure (α, β, γ not given)

$$\frac{1}{d^2} = \frac{h^2 + k^2 + l^2}{a^2}$$

$n\lambda = 2d\sin\theta$
$d = \lambda/2 \sin\theta$
$\lambda = 1.54$ A
(100): $d = a$ $2\theta = 2\sin^{-1}(\lambda/2d) = 15.48$
(110): $d = a/\sqrt{2}$ $2\theta = 21.97$
(111): $d = a/\sqrt{3}$ $2\theta = 27$
(200): $d = a/2$ $2\theta = 31.3$
(220): $d = a/2\sqrt{2}$ $2\theta = 44.8$
(211): $d = a/\sqrt{6}$ $2\theta = 38.5$
(100): $d = a$ $2\theta = 2\sin^{-1}(\lambda/2d) = 15.48$
etc.
Marked reflections in order are: (111), (200), (220), (311), (222), (400)

b) Basis Co: (1/4,1/4,1/4)
Basis Mn: (0,0,0)
Basis Si: (1/2,1/2,1/2)
Co:
$\mathbf{r}_1 = 1/4\mathbf{a}_1 + 1/4\mathbf{a}_2 + 1/4\mathbf{a}_3$
Mn:
$\mathbf{r}_2 = 0\mathbf{a}_1 + 0\mathbf{a}_2 + 0\mathbf{a}_3$

Si:
$\mathbf{r}_3 = 1/2\mathbf{a}_1 + 1/2\mathbf{a}_2 + 1/2\mathbf{a}_3$

$$S_G = \sum f_j e^{-i.\boldsymbol{G}.\boldsymbol{r}_j}$$
$$S_G = f_{Co} e^{-i2\pi(1/4h+1/4k+1/4l)} + f_{Mn} e^{-i2\pi(0h+0k+0l)} + f_{Si} e^{-i2\pi(1/2h+1/2k+1/2l)}$$
$$S_G = f_{Co} e^{-i\pi(h/2+k/2+l/2)} + f_{Mn} + f_{Si} e^{-i\pi(h+k+l)}$$

For (220)
$S_G = f_{Co} e^{-i2\pi} + f_{Mn} + f_{Si} e^{-i4\pi} = f_{Co} + f_{Mn} + f_{Si} \sim 25 + 14 + 27 \sim 66$
For (111)
$S_G = f_{Co} e^{-\frac{i3}{2}\pi} + f_{Mn} + f_{Si} e^{-i3\pi} = f_{Mn} - f_{Si} - f_{Co} \sim 25 - 14 - 27 \sim -16$
For (200)
$S_G = f_{Co} e^{-i\pi} + f_{Mn} + f_{Si} e^{-i2\pi} = f_{Mn} + f_{Si} - f_{Co} \sim 25 + 14 - 27 \sim 12$
i.e. total constructive interference for (220), partially destructive for others.

c) *From the above structure factor, there would appear to be none. (Note that the Basis is not correct for this compound but a guess). Based on the observed reflections, the extinction conditions appear to be {h,k,l} NOT all even or all odd.*

3. *Orthorhombic compounds*
   a) Orthorhombic
   $\frac{1}{d^2} = \frac{h^2}{a^2} + \frac{k^2}{b^2} + \frac{l^2}{c^2}$
   $n\lambda = 2d\sin\theta$
   $d = \lambda/2\ \sin\theta$
   (100): $d = a 2\theta = 2\ \sin^{-1}(\lambda/2a) = 16.15$
   (002): $d = c/2\ 2\theta = 2\ \sin^{-1}(\lambda/c) = 22.92$, etc.
   Marked reflections in order are: (002), {(200)/(112)/(020)}, {(004)/(220)}, {(204)/(312)/(024)/(132)}
   b) $a \neq b$
   The lattice spacings, d, are therefore no longer equivalent.
   The peak positions are therefore at different 2θ.

4. *Rhombohedral/hexagonal compounds*
   a) Hexagonal
   $\frac{1}{d^2} = 4\frac{h^2 + hk + k^2}{3a^2} + \frac{l^2}{c^2}$
   $d = \lambda/2sin\theta$
   (100): $d = a\sqrt{3}/2 2\theta = 2sin^{-1}(\lambda/a\sqrt{3}) = 18.45$
   (012): $d = \left(\frac{4}{3a^2} + \frac{4}{c^2}\right)^{-1/2} = 3.91 2\theta = 2sin^{-1}(\lambda/2 * 3.91) = 22.7$
   etc.

Marked reflections in order are: (012), {(110)/(104)}, {(113)/(006)}, (024), {(300)/(214)/(018)}

b) $h + k + l = 3n$

c) Basis La: (0,0,1/4)
Basis Ca: (0,0,1/4)
Basis Mn: (0,0,0)
Basis O: (1/2,0,1/4)
La/Ca at (0,0,1/4), (0,0,3/4)
Mn at (0,0,0), (0,0,1/2)
O at (1/2,0,1/4),(−1/2,−1/2,1/4), (0,1/2,1/4), (0,−1/2,3/4), (−1/2,0,3/4), (1/2,1/2,3/4)

$$S_G = \sum f_j e^{-i.\boldsymbol{G}.\boldsymbol{r}_j}$$

$$S_G = \left(0.7f_{\mathrm{La}} + 0.3f_{\mathrm{Ca}}\right)\left(e^{-i\pi l/2} + e^{-i3\pi l/2}\right) + f_{\mathrm{Mn}}\left(e^{-i\pi l} + 1\right) + f_O\left(e^{-i\pi(h+\frac{l}{2})} + e^{-i\pi(\frac{l}{2}-h-k)} + e^{-i\pi(k+\frac{l}{2})} + e^{-i\pi(-k+\frac{3l}{2})} + e^{-i\pi(-h+\frac{3l}{2})} + e^{-i\pi\left(h+k+\frac{3l}{2}\right)}\right)$$

$f_{\mathrm{La/Ca}}$ terms zero when l is odd or even
f_{Mn} terms zero when l is odd
f_{O} terms zero when $(-h + k + l) = 3n$*
**which can be true when l is odd, thus satisfying complete extinction described for (c) except, with the additional requirement that l is odd.*

3.7.3 Space groups

1. *Pbnm*
   a) Multiplicity is 8
   b) Multiplicity is 4
2. *Pnma*
   a) Multiplicity is 4
   b) Multiplicity is 4

3.7.4 Identifying key information in XRR data

2) *Extracting key information from XRR data*

i. $2\theta_1 \sim 0.84$
$2\theta_2 \sim 0.94$
$Q = (4\pi\sin\theta)/\lambda$
$\Delta Q = 4\pi/\lambda(|\sin\theta_1 - \sin\theta_2|) = 0.007\,117$ Å^{-1}
$t = 2\pi/\,\Delta Q = 88.2$ Å $\sim$ 88 nm for thinner layer

ii. $2\theta_1 \sim 2.5$
$2\theta_2 \sim 5.2$
$Q = (4\pi\sin\theta)/\lambda$
$\Delta Q = 4\pi/\lambda(|\sin\theta_1 - \sin\theta_2|) = 0.192$ Å^{-1}

$t = 2\pi/\Delta Q = 32.7$ Å ~ 3.3 nm for thinner layer
$\Delta(2\theta) \sim 0.5/5 \sim 0.1$
$2\theta_1 \sim 0.5$
$2\theta_2 \sim 0.6$
$\Delta Q = 4\pi/\lambda(|\sin\theta_1 - \sin\theta_2|) = 0.007\,117$ Å^{-1}
$t = 2\pi/\Delta Q = 882.3652$ Å ~ 88 nm for thinner layer
Therefore:
Layer 1 ~ 3.3 nm
Layer 2 ~ 84.7 nm

3) *Extracting key information from XRR data*
a) Nb: $\Delta(2\theta) \sim 0.6$, $t \sim 14.7$ nm
Pt: $\Delta(2\theta) \sim 0.7$, $t \sim 12.6$ nm
Ta: $\Delta(2\theta) \sim 0.6$, $t \sim 14.7$ nm
Cu: $\Delta(2\theta) \sim 1$, $t \sim 8.8$ nm
Fe_3O_4: $\Delta(2\theta) \sim 0.35$, $t \sim 25.2$ nm
b) *Additional layers, most likely the formation of an oxide on the surface.*

3.7.6 Calibrating detectors—some basic geometry

1) 1*D detector*
a) If a feature is just resolved, then it can be detected by a pixel of width, 0.5 mm.
The total height of reflected beam at θ (as seen in figure Q3.10) can be described by:
$D_1\tan(\theta) = h$
where h is the position of the specular reflection. As $\theta = 0.5$ and $D_1 = 3$ m, $h = 0.026\,18$ m $= 13.09$ mm.
For the next resolvable feature:
$D_1\tan(\theta + \delta\theta) = 0.02618 + 0.0005 = 0.02668$
$\tan(\theta + \delta\theta) = \frac{0.02668}{3} = 0.0088933$
$\theta + \delta\theta = \tan^{-1}(0.0088933) = 0.509537$
$\delta\theta = 0.509537 - 0.5 = 0.009537 \sim 0.01°$
b) $D_1\tan(\theta) = h$
where h is the position of the specular reflection. As $\theta = 0.25$ and $D_1 = 3$ m, $h = 0.013\,09$ m $= 13.09$ mm.
For the next resolvable feature:
$D_1\tan(\theta + \delta\theta) = 0.01309 + 0.0005 = 0.01359$
$\tan(\theta + \delta\theta) = \frac{0.01359}{3} = 0.00453$
$\theta + \delta\theta = \tan^{-1}(0.00453) = 0.2595481$
$\delta\theta = 0.2595481 - 0.25 = 0.0095481 \sim 0.01°$
c) Feature every four pixels roughly corresponds to $\Delta\theta \sim 0.04°$.
$t \sim \frac{2\pi}{\Delta Q} \sim 716.2\lambda$
If $\lambda = 10$ Å, $t \sim 7162$ Å ~ 716.2 nm

7.7.1 Extracting information from bulk measurements

1) *Resistivity*

a) Resistance = $V/I = 4.3 \times 10^{-3} / 0.1 \times 10^{-3} = 43\ \Omega$
Resistivity = $RA/l = R(L_3L_4)/L_1 = 43(2 \times 10^{-3} * 2 \times 10^{-3})/5 \times 10^{-3} = 34.4\ \text{m}\Omega\text{m}$

b) Resistance = $V/I = 52.7 \times 10^{-3} / 0.1 \times 10^{-6} = 527\ \text{k}\Omega$
Resistivity = $RA/l = R(L_3L_4)/L_1 = 527 \times 10^3(2 \times 10^{-3} * 2 \times 10^{-3})/7 \times 10^{-3} = 30.11\ \Omega\text{m}$

c) Resistance = $V/I = 67.8 \times 10^{-3}/0.1 = 6.78\ \text{m}\Omega$
Resistivity = $RA/l = R/(\pi d^2/4)/L_1 = 6.78 \times 10^{-3}(7.85 \times 10^{-9})/26 \times 10^{-3} = 2.05 \times 10^{-9}\ \Omega\text{m}$

2) *Thermal transport measurements*

a) $S = V/\Delta T = 52 \times 10^{-3}/(295\text{–}290) = 10.4\ \text{mV K}^{-1}$

b) $\kappa = Q\Delta x/A\Delta T = 0.02 * 5 \times 10^{-3}/(2 \times 10^{-3} * 2 \times 10^{-3} * 5) = 5\ \text{W K}^{-1}\ \text{m}^{-1}$

c) Resistivity = $RA/l = R(L_3L_4)/L_1 = [(170\text{–}52) * 1 \times 10^{-3}/0.1 \times 10^{-3}] * (2 \times 10^{-3} * 2 \times 10^{-3})/5 \times 10^{-3} = 0.944\ \Omega\text{m}$

d) *The voltage generated is with respect to another material. Often it is compared against Pt.*

e) $zT = (10.4 \times 10^{-3})^2 * 292.5/(0.944 * 5) = 0.0067$

7.7.2 Extracting information from thin film measurements

1) *Sheet resistivity*

a) $R_A = (0.756 + 0.756 + 0.758 + 0.757)/4 \times 10^{-1} = 7.57$
$R_B = (14.25 + 14.23 + 14.23 + 14.25)/4 \times 10^{-1} = 142.4$
$R_A/R_B = 0.053$ (but it is the ratio that is important, so use R_B/R_A)
$R_B/R_A = 8.8$
From figure, $f \sim 0.55$
$R_s = \pi/\ln 2((R_A + R_B)/2) f = 186.7\ \Omega$/square

b) $R_A \sim R_B$
$e^{-\pi RA/Rs} + e^{-\pi RA/Rs} = 1$
$2\, e^{-\pi RA/Rs} = 1$
$\pi R_A/R_s = \ln 2$
$R_s = \pi R_A/\ln 2 = 34.3\ \Omega$/square or $645.4\ \Omega$/square
i.e. differs by a factor of 4 compared to part (a)

2) *Hall measurements*

a) $R_A = (0.756 + 0.757 + 0.756 + 0.758)/4 \times 10^{-1} = 7.57$
$R_B = (0.756 + 0.762 + 0.756 + 0.762)/4 \times 10^{-1} = 7.59$
$R_A \sim R_B$
Therefore, $R_s = \pi(R_A + R_B)/2\ln 2 = 34.36\ \Omega$/square

$p_s = 8 \times 10^{-8} IB/q|\Sigma V_i| = 8 \times 10^{-8} \times 1 \times 10^{-3} \times 10\,000/(1.6 \times 10^{-19} * [0.23 + 0.25 + 0.22 + 0.24 + 0.24 + 0.26 + 0.23 + 0.25] \times 10^{-3}) = 5 \times 10^{15}/1.92 = 2.6 \times 10^{15}\ \text{cm}^{-2}$

$\mu = 1/(qp_sR_s) = 1/(1.6 \times 10^{-19} \times 2.6 \times 10^{15} \times 34.36) = 69.96 \times 10^{1}\ \text{cm}^2\ \text{V}^{-1}\text{s}^{-1}$

b) If we assumed that $R_A \sim R_B$ (as I did above) then given that $R_A/R_B = 1.0026$ this would suggest R_s varies no more than 1% (i.e. $f \sim 1$).

c) $\rho = R_s d = 34.36 * 54 \times 10^{-9} = 1.86 \times 10^{-6}\ \Omega\text{m}$

$p = p_s/d = 2.6 \times 10^{15}\ \text{cm}^{-2}/54 \times 10^{-7} = 4.815 \times 10^{21}\ \text{cm}^{-3}$

8.6.1 Magnetic units, a quagmire

1) *Magnetic flux density*

A. 10 000 G = 1 T.

Therefore $B = 10$ mT

B. $B = 1000$ G

2) *Magnetic field strength*

A. 1 Oe = $10^3/4\pi = 79.58\ \text{A m}^{-1}$

B. $2000\ \text{A m}^{-1} = 2000/10^3 * 4\pi = 25.13$ Oe

C. $B = \mu_0(H + M)$ (SI) $= 4\pi \times 10^{-7} * 1000 = 1.26$ mT

$B = H + 4\pi M$ (CGS) $= 1000/10^3 * 4\pi = 12.6$ G

D. $B = \mu_0(H + M)$ (SI) $= 4\pi \times 10^{-7} * 20 * 10^3/4\pi = 2$ mT

$B = H + 4\pi M$ (CGS) $= 20$ G

3) *Magnetisation*

A. 1:1 for these units. Therefore, $M = 20\ \text{Am}^2/\text{kg}$.

B. $1\ \text{emu cm}^{-3} = 10^3\ \text{A m}^{-1}$

$M = 50 \times 10^3\ \text{A m}^{-1}$

C. Mass magnetisation: $M = 100/1\ \text{emu g}^{-1} = 100\ \text{emu g}^{-1} = 100\ \text{A m}^{-2}\ \text{kg}^{-1}$

Volume magnetisation: $M = 100/0.2\ \text{emu cm}^{-2} = 500\ \text{emu cm}^{-3} = 5 \times 10^5\ \text{A m}^{-1}$

4) *Permeability*

$\mu_r = \mu/\mu_0 = 1.2 \times 10^{-6}/4\pi \times 10^{-7} = 0.9549$

$\chi = \mu_r - 1 = -0.0451$

$\chi < 1$, therefore diamagnetic

8.6.2 Magnetic interactions

1) *The magnetic dipolar interaction*

a) $\frac{\mu_0}{4\pi r^3}\mu_1\mu_2$

b) $-\frac{\mu_0}{4\pi r^3}\mu_1\mu_2$

c) 0

d) 0

e) $\frac{\sqrt{2}\mu_0}{8\pi r^3}\mu_1\mu_2$

f) $-\frac{\mu_0}{2\pi r^3}\mu_1\mu_2$

g) $\frac{\mu_0}{2\pi r^3}\mu_1\mu_2$

h) $\frac{\sqrt{2}\mu_0}{4\pi r^3}\mu_1\mu_2$

8.6.3 Extracting information from magnetisation measurements

1. *Magnetometry*

A. $M \sim 2.5 \times 10^{-3}$ emu.
$\sigma_M = 2.5 \times 10^{-3} / 60 \times 10^{-6}$ g = 42 emu g^{-1} = 42 A m^{-2} kg^{-1}

B.

T	H_c (CGS, Oe)		H_c (SI, A m^{-1})
	1 Oe	=	$10^3/4\pi$ A m^{-1}
295	150 Oe		11.9 kA m^{-1}
120	125 Oe		9.95 kA m^{-1}
110	540 Oe		42.99 kA m^{-1}
100	640 Oe		50.96 kA m^{-1}
50	660 Oe		52.55 kA m^{-1}

C.

T	B_c (CGS, G)		B_c (SI, T)
	1 G	=	10^{-4} T
295	149 G		0.0149 T
120	125 G		0.0125 T
110	540 G		0.054 T
100	640 G		0.064 T
50	660 G		0.066 T

2. *Demagnetisation factor*

A.

i) $N = 1/3$ in SI
Volume magnetisation = $1.45 \times 10^3/1$ emu cm^{-3} = 1.45×10^3 emu cm^{-3}
In CGS
$H_d = -NM = -4\pi/3 * (1.45 \times 10^3)$ Oe = -6.073 kOe = -0.6 T
In SI
$H_d = -NM = -1/3 * (1.45 \times 10^3 \times 10^3)$ A m^{-1} = -4.83×10^5 A m^{-1}
$B_d = \mu_0 H_d = 4\pi \times 10^{-7} \times -4.83 \times 10^5 = -0.6$ T

ii) $N = 1/3$ in SI
Volume magnetisation $= 1.2 \times 10^3/1\ \text{emu cm}^{-3} = 1.2 \times 10^3\ \text{emu cm}^{-3}$
In CGS
$H_d = -NM = -4\pi/3 * (1.2 \times 10^3)\ \text{Oe} = -5\ \text{kOe} = -0.5\ \text{T}$
In SI
$H_d = -NM = -1/3 * (1.2 \times 10^3 \times 10^3)\ \text{A m}^{-1} = -4.0 \times 10^5\ \text{A m}^{-1}$
$B_d = \mu_0 H_d = 4\pi \times 10^{-7} \times -4.0 \times 10^5 = -0.5\ \text{T}$

iii) $N = 1/3$ in SI
Volume magnetisation $= 1.3 \times 10^3/10 * 8.9\ \text{emu cm}^{-3} = 1157\ \text{emu cm}^{-3}$
In CGS
$H_d = -NM = -4\pi/3 * (1157)\ \text{Oe} = -4.8\ \text{kOe} = -0.48\ \text{T}$
In SI
$H_d = -NM = -1/3 * (1157 \times 10^3)\ \text{A m}^{-1} = -3.86 \times 10^5\ \text{A m}^{-1}$
$B_d = \mu_0 H_d = 4\pi \times 10^{-7} \times -3.86 \times 10^5 = -0.48\ \text{T}$

3) $N \sim 0.1$ parallel to long direction
$N \sim 0.9$ parallel to long direction
Perpendicular Parallel
i) $H_d = -1.62\ \text{T} = -0.18\ \text{T}$
ii) $H_d = -1.35\ \text{T} = -0.15\ \text{T}$
iii) $H_d = -1.3\ \text{T} = -0.14\ \text{T}$

4) *Magnetic moment*

A. Look up atomic mass for La, Fe and Si to determine the mass of 1 mole (and the number of Fe atoms per mole $= 11N_a$).
$m_{La} = 138.91$
$m_{Fe} = 55.85$
$m_{Si} = 28.065$
Mass of 1 mole = 781.345 g = 0.781 kg
Fe atoms per kg $= 11N_A/0.781$
$M_s = 200\ \text{emu g}^{-1} = 200\ \text{Am}^2\ \text{kg}^{-1}$
Therefore magnetic moment per Fe atom $= 200\ \text{Am}^2\ \text{kg}^{-1}/(11N_A/0.781) = 2.36 \times 10^{-23}\ \text{Am}^2/\text{Fe atom}$
Divide by μ_B $(=9.274 \times 10^{-24}\ \text{J T}^{-1})$
$m = 2.54\ \mu_B/\text{Fe atom}$.

B. Look up atomic mass for Fe and O to determine the mass of 1 mole (and the number of Fe atoms per mole $= 3N_a$).
$m_{Fe} = 55.85$
$m_O = 15.999$
$M_{sat} = 2.3 \times 10^{-3}/60 \times 10^{-6} = 38.3\ \text{emu g}^{-1} = 38.3\ \text{Am}^2\ \text{kg}^{-1}$
Mass of 1 mole = 0.2315 kg
Fe atoms per kg $= 3N_A/0.231$

Therefore magnetic moment per Fe atom = $38.3/(3N_A/0.231)$
Divide by μ_B (=9.274×10^{-24} J T^{-1})
$m = 0.53\ \mu_B$/Fe atom.

C. Look up atomic mass for Fe and O to determine the mass of 1 mole (and the number of Fe atoms per mole = $2N_a$).
$M_{sat} = 38.3$ Am^2 kg^{-1}
Mass of 1 mole = 0.1597 kg
Fe atoms per kg = $2N_A/0.1597$
Therefore magnetic moment per Fe atom = $38.3/(2N_A/0.1597)$
Divide by μ_B (=9.274×10^{-24} J T^{-1})
$m = 0.55\ \mu_B$/Fe atom.

Lightning Source UK Ltd.
Milton Keynes UK
UKHW020849241021
392712UK00003B/54